STUDENT ACTIVITIES MANUAL

Mathematics for Elementary Teachers

A CONTEMPORARY APPROACH

Seventh Edition

STUDENT ACTIVITIES MANUAL

Lyn Riverstone
Oregon State University

Karen A. Swenson

Mathematics
for Elementary Teachers

A CONTEMPORARY APPROACH

Seventh Edition

Gary L. Musser
Oregon State University

William F. Burger

Blake E. Peterson
Brigham Young University

JOHN WILEY & SONS, INC.

Cover Art: Michael Jung

To order books or for customer service, please call 1-800-CALL-WILEY (225-5945).

ISBN 0-471-70119-X

Printed in the United States of America

10 9 8 7 6 5 4 3 2 1

Printed and bound by Courier Kendallville, Inc.

PREFACE

This manual has been designed to illustrate and provide practice in the use of effective learning strategies for the mathematics classroom. These strategies are used to provide the pre-service teacher with opportunities to gain an understanding of the mathematical concepts and content of the elementary school program while modeling effective mathematics techniques. Students who use this manual as a companion to the textbook *Mathematics For Elementary Teachers, A Contemporary Approach* by Gary Musser, William Burger, and Blake Peterson will not only enhance their own learning, but will also begin to place that learning within the framework of the elementary classroom. You may be using this manual as a part of a course focusing on mathematical content or on the methods of teaching mathematics, or perhaps your course combines content and methods. In whatever situation, this manual has features to enrich your learning. Each section can be identified by its unique icon. These components are described in the following paragraphs.

Hands-on Activities. This is the developmental phase of each chapter. It is through hands-on activities that mental images of a concept are developed. The purpose of this section is not only to teach "how to" perform the mathematical tasks, but also to foster understanding of the underlying concepts.

These activities provide experience with a variety of materials and how they can be used in the elementary school classroom. And since research shows that prospective teachers use manipulatives in their teaching in a manner similar to how they, themselves, learn, this experience is vital. In these activities, it is beneficial to approach them as a student just learning the concept. Imagine how an elementary school student would respond.

An objective is stated for each activity. While there are many forms for writing objectives, as future teachers, it is important to be aware of this aspect of any activity used. Not all concepts in the textbook are developed in the Hands-on Activities. Also, some of the activities herein develop the concepts more quickly out of necessity than would be practical in an elementary classroom. Yet they serve as a foundation for an expanded series of activities for elementary students.

This manual has been created to be as self-contained as possible. Some activities refer to Materials Cards. These are found at the back of the manual and are numbered according to the activity. Some commercial materials are available and can be used if you have them; for example, the multibase pieces in Chapters 2 and 4 can be replaced with Dienes blocks and the centimeter strips in Chapters 3 and 6 can be replaced with Cuisenaire rods. Other items used include cubes, dice, fraction bars, a metric tape, graph paper, and a protractor. Some activities also refer to eManipulatives that are found on the website for your textbook (see www.wiley.com/college/musser) and Physical Manipulatives that are part of the packet you may have purchased along with this manual. These are optional materials that may be used in place of the Materials Cards when doing the Hands-on Activities.

Exercise. In this section, you move from the concrete objects used in the hands-on section to visual representations and numerical problems. The focus of this section is on application of the material presented in the textbook and in the hands-on section of the manual. The exercises are presented in creative formats that you will find more interesting than usual pencil-and-paper exercises.

Connections to the Classroom. The problems in this section will help you to review the concepts from the chapter, as well as connect these concepts to the elementary classroom. The formats of these problems are of various types, including (i) analyzing common student error patterns, (ii) formulating questions you could ask to lead a student to his/her own understanding of a certain problem or concept, and (iii) describing a student's thought process.

Mental Math. This section provides a chance to develop flexibility in thinking about mathematics. You are encouraged to extend your ability to think about mathematics and to reason about the answers to mathematical problems. No paper and pencils allowed!

Directions in Education. At then end of each chapter is a section that stands on its own, where we look at issues in the mathematics education community. In addition to a brief summary, possible questions and resources are included to guide you as you explore how these issues will impact your future classroom. Both explicitly and implicitly, the recommendations of the National Council of Teachers of Mathematics are woven throughout our look at learning and teaching mathematics.

Solutions. The solutions section, found at the end of the manual, provides a chance for students to verify results. Solutions and explanations (where appropriate) are included for all sections of the manual. You should use this section only after having worked through the manual activities so that the experience more closely matches the experiences of students in the elementary classroom.

7th Edition Notes:

The 7th Edition of the *Student Activities Manual* is a major revision of the 6th edition, entitled the *Student Resource Handbook*. Many portions of the Exercise and Mental Math sections herein were authored by Karen Swenson in the Student Resource Handbook, while the Directions in Education were authored by Marcia Swanson. Both Karen and Marica have retired as the primary authors and Lyn Riverstone has assumed that role. Portions of many of the original Hands-on Activities appear in this new edition, though most have been revised by Lyn. The Connections to the Classroom section is completely new to this edition.

Lyn wishes to thank Gary Musser for his patience, guidance, and superb editing, Blake Peterson for his continued support, Dianne Hart, Marie Franzosa, and Scott Peterson for their help in testing these materials in the classroom, and Heather Margason for her help in preparing the solutions for this edition and **dedicates** this book to Michael and Henry Riverstone.

CONTENTS

1 Introduction to Problem Solving

THEME:

1. Understand the Problem
2. Devise a Plan
3. Carry Out the Plan
4. Look Back

HANDS-ON ACTIVITIES

In today's classrooms, teachers emphasize problem-solving as the core of the mathematics curriculum. By solving many different math problems, students are not only gaining a deeper understanding of mathematical concepts, they are also building problem-solving skills they will need throughout their lives. These skills extend beyond the classroom to the problems people face as part of everyday life and apply even to non-mathematical problems. When your car breaks down or you have a communication problem with a colleague, you use problem-solving skills and strategies to try to find solutions.

George Pólya, author of *How to Solve It*, identified four distinct stages of successful problem solving: (1) understand the problem; (2) devise a plan; (3) carry out the plan; and (4) look back. These steps outline a practical method that can be used when solving any problem. Problem solving is a complex process and it takes practice to learn how to choose an appropriate strategy or "tool" to solve a problem. A problem may have many correct solution paths, and as a teacher it will be important for you to be able to recognize, use, demonstrate, and explain these.

The activities in this chapter give you practice using Pólya's four-step process as well as some of the common problem-solving strategies, such as Guess and Test and Look for a Pattern. Add these tools to your problem-solving "toolbox" for use on future problems and in your own classroom.

OBJECTIVE: **Solve problems using the Guess and Test strategy**

Guess and Test is often one of the first problem-solving strategies students learn. Not only is this strategy useful in solving many types of problems, but applying it can help students gain a better understanding of the problem-solving process and can lead to the development of other strategies. The first activity will guide you through the problem-solving process using Pólya's four steps.

Refer to the following problem for #1-4: A student opens a 125-page book and announces, "The sum of the two facing page numbers is a square number." What are the page numbers the student has opened to?

1. The first stage in Pólya's four steps is to **understand the problem**. It can be helpful at this stage to write down what you know and what you need to find. This may involve making sure you understand all the words in the problem, listing the given information, and, perhaps, restating the problem in your own words.

   Known information:

   The word "sum" means: ________________________________

   A square number is: ________________________________

   Some examples of square numbers are: ________________________

   List any other important information from the problem:

   If you think that it will be helpful, restate the problem in your own words:

2. The second of Pólya's four steps is to **devise a plan**. This step involves not only choosing a strategy, but also deciding how to apply it.

   What are some clues that the Guess and Test strategy may be appropriate for solving this problem?

   Next, describe how you will apply the Guess and Test strategy to solve this problem. Do not write out the solution yet.

   The type of number(s) I will make a guess for are: ____________________

   I will know my guess gives me an answer to the problem if:

   ____

3. **Carry out the plan** is the third of Pólya's steps. This is where you write out your solution to the problem and then answer the question. A table may help you to present your work in an organized way.

   Answer in a complete sentence: ______________________________

   Reflect on your guess and test process above. Did you make random guesses or did you list your guesses in a systematic way? ____________ If randomly, do you now see a way to guess systematically? If you did use a systematic method, describe it here.

4. The last step is to **look back** over your solution. Looking back may involve solving the problem with a different strategy to be sure your answer is correct, generalizing the results of the problem, or finding other answers to the problem.

   Have you found all the solutions to the problem? If so, how do you know there are no other solutions? If not, find all the solutions.

5. Solve the following problem using the Guess and Test strategy and write out your problem-solving process according to Pólya's four steps.

   Mariella is selling raffle tickets for a school fundraiser. Adult tickets cost $4 and student tickets cost $3. If she has collected $65 so far, how many adult tickets did she sell?

   Check your answer with a classmate. Did you agree? If not, perhaps the two of you can come up with a more complete solution

OBJECTIVE: **Use the Draw a Picture problem solving strategy**

Problems that involve physical situations, geometric figures, or measurements are often better understood when the Draw a Picture strategy is used. Drawing a picture to give a visual representation of a problem is most often part of the understanding the problem stage of problem solving. This, in turn, can be helpful when devising a plan.

1. Consider the following problem: If it takes 20 minutes to cut a log into 5 pieces, how long would it take to cut a log into 10 pieces?

   Convince yourself that the Draw a Picture strategy is helpful by first trying to solve this problem *without* a picture.

   When asked to solve this problem, a student wrote "10 ÷ 5 = 2 and 2 × 20 = 40. My answer is 40 minutes." Can you explain this student's thinking? Did you get the same answer as the student?

2. Next, draw a picture to represent the problem in #1.

   What information is shown in your drawing that was missing from the student's understanding of the problem?

   Show how to use the picture you drew to solve the problem in #1. Is your answer different than the one you got the first time you solved?

ACTIVITY 1.1.3

OBJECTIVE: **Use variables to solve problems involving even and odd whole numbers**

You Will Need: **Materials Card 1.1.3**

The Use a Variable strategy is useful when you want to prove a general result. In the next activity, you will see how the use of a hands-on model can help in the development of variable expressions to represent even and odd whole numbers and their sums.

1. Materials Card 1.1.3 contains shapes formed by rows of square tiles.
   a. Find all the shapes that are rectangles two units wide and record the number of tiles in each shape, from smallest to largest.

   ______ , ______ , ______ , ______ , ______ , ______
   1st 2nd 3rd 4th 5th 6th

   What do you notice about the numbers in this sequence?
   What is the width of each rectangle? ______

   b. Make sketches of the 7th, 8th, and 9th rectangles in the sequence. Label each sketch with the total number of tiles in the rectangle.

   c. Write a description of the 100th rectangle in this sequence. What is the 100th even number?

   d. Use a variable to write an expression for the nth even number. ______

   e. If the even numbers beginning with 2 are written in increasing order, in which position is the number 1084? ______

2. Next, find all the shapes on Materials Card 1.1.3 that are not rectangles two units wide.
   a. Record the number of tiles in each shape, from smallest to largest.

   ______ , ______ , ______ , ______ , ______ , ______
   1st 2nd 3rd 4th 5th 6th

   What do you notice about the numbers in this sequence?

   b. Make sketches of the 7th, 8th, and 9th shapes in the sequence. Label each sketch with the total number of tiles.

   c. Write a description of the 100th shape in this sequence. What is the 100th odd number?

   d. Use a variable to write an expression for the nth odd number. ______

   e. If the odd numbers beginning with 1 are written in increasing order, in which position is the number 307? ______

3. The shapes from #1 and #2 can be used to visualize sums of even and odd numbers. Two even-numbered shapes can be put together, as shown next.

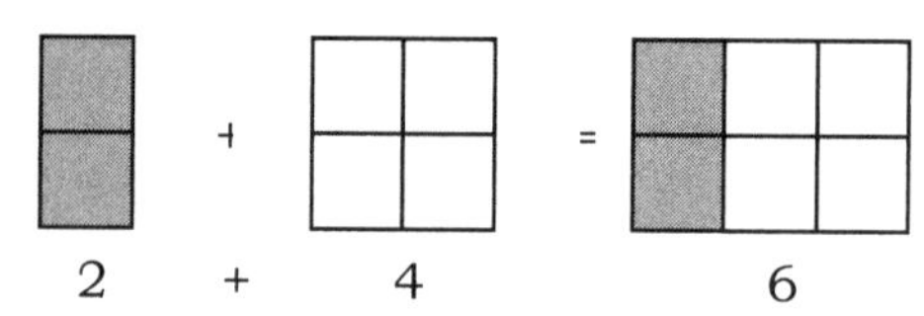

   a. Add two more even numbers using this visual technique. Record your answer and tell if the sum is odd or even. What do you notice about adding two even numbers?

   b. The examples in part (a) give you a way to visualize why the sum of any two even numbers may be even. However, you cannot prove that this is always the case simply by considering a few examples. To prove that the sum of *any* two even numbers is even, you need to use variables. Suppose that $2x$ and $2y$ are any even numbers. Show that the sum $2x + 2y$ is even.

4. Next, consider the sum of two odd numbers. Two odd-numbered shapes can be put together as shown next.

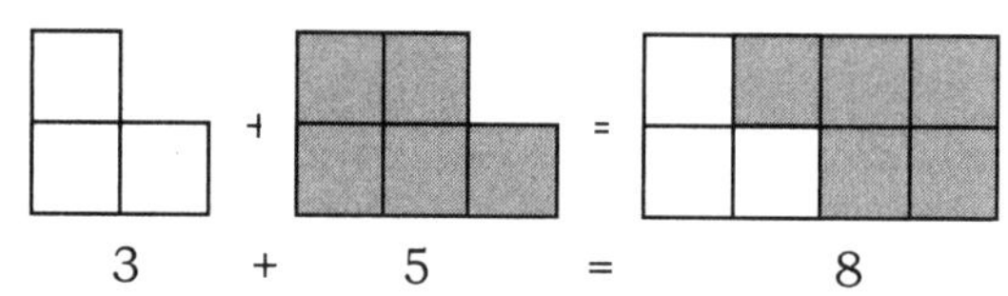

   a. Add two more odd numbers using this technique. Record your answer and tell if the sum is odd or even. What do you notice about adding two odd numbers?

   b. Use variables to show that the sum of two odd numbers is always even.

5. Now examine the sum of an even and an odd number.
   a. Choose an even-numbered shape and an odd-numbered shape and put them together. Record the new shape and the addition equation that is modeled. Try several examples. What do you notice about adding one even number and one odd number?

   b. Use variables to prove that your result in part (a) is always true.

OBJECTIVE: **Combine the Make a List and Look for a Pattern problem-solving strategies**

You Will Need: **Materials Card 1.2.1 (eManipulative option: *Towers of Hanoi*)**

Make a List and Look for a Pattern are two strategies that are often used together. Making an organized list of special cases of a problem can help to generate a pattern. In the next activity, you will apply these two problem-solving strategies as you investigate the mathematics of the legend of the Tower of Hanoi.

LEGEND: Monks of ancient Hanoi needed to move a tower of sacred stones. The stones were kept on poles at all times and were so heavy they could only be moved one at a time. Never could a larger stone be placed on a smaller one.

1. Place disks 3, 2, and 1 on peg A in that order. Your task is to move your stack to either peg B or C following these rules:
   - Move only one disk at a time.
   - Never put a larger disk on top of a smaller one.
   - Disks must stay on a peg, except when being moved to another peg.
   - Each time a disk is moved – count one move.

   What do you think is the minimum number of moves required to transfer the stack to peg B or peg C? _______ Compare your answer with a classmate's. Do you agree?

   Make a prediction for the minimum number of moves required to move 4 disks from one peg to another. _______

2. Repeat this activity using 4 disks and 5 disks. Record the minimum number of moves in this chart:

Number of Disks	**Minimum Number of Moves**
3	
4	
5	

3. Look for a pattern in the minimum number of moves you recorded in the table in #2. What do you notice?

4. To identify a pattern, it can be helpful to observe how successive terms in a sequence of numbers are related. In problem #3 you may have noticed that the minimum number of moves required for 4 disks is 8 more than the minimum number of moves for 3 disks. Also, the minimum number of moves for 5 disks is 16 more than the number of moves for 4 disks. Using this pattern, what do you think is the minimum number of moves required for 6 disks?

Although the previous pattern does tell you how to find the minimum number of moves for n disks using the minimum number of moves for $(n - 1)$ disks, it does not help you easily find the minimum number of moves for 100 disks, for example. You'd have to find all the terms in the sequence before the 100th. What you need is a **closed formula**; that is, a formula that will tell you the minimum number of moves for any given number of disks.

When trying to find a closed formula that represents each number in a sequence, it is not always obvious at first. Rewriting each term and looking at each one as a product or sum, for instance, will often reveal the pattern. Rewrite each number in the "Number of Moves" column by comparing it to a power of 2.

Record a formula here for the number of moves required for n disks: ______________

What role does the number of disks play in the expression you wrote for the minimum number of moves?

5. Use the formula you wrote in #4 to find the number of moves for six disks: ___________
   Check your answer using the six disks from Materials Card 1.2.1.

6. Suppose that we started our table with $n = 1$ and $n = 2$. Would the formula we found satisfy these two instances also? Explain.

ACTIVITY 1.2.2

OBJECTIVE: **Combine problem-solving strategies to discover a formula for triangular numbers**

The "Handshake Problem" is a famous problem often used to introduce students to the concepts involved in the sequence of triangular numbers. In the next activity, you will investigate these numbers by combining several of the problem-solving strategies introduced in this chapter.

The Handshake Problem:

There are 31 people at a party. If everyone shakes hands with each of the other party-goers exactly once, how many handshakes will there be?

1. To gain a better understanding of this problem, first consider a simpler problem. That is, what if there were 4 people at the party?

   a. Draw a picture to represent the handshakes between 4 people, if each person shakes hands with each of the other people exactly once.

   How many handshakes are there per person? _______ How many handshakes are there altogether for 4 people? _______ Explain why the total is not 4×3 (number of people $\times$ number of handshakes per person).

b. Now, make drawings to represent the handshakes between 2 people, 3 people and 5 people. Then, complete the following table with the information you discover in your drawings.

# of people	# of shakes per person	Total # of shakes
2		
3		
4		
5		

c. Examine the table in part (b) and describe any patterns you notice.

Can you predict the number of handshakes for 6 people? ________
Check your result by drawing a picture.

d. Based on your work in parts (a) – (c), are you able to write a closed formula that could be used to solve the handshake problem for 31 people, without the need to find the number of handshakes for 30 people? If so, state it here and then answer the question. In your answer, the leading digit should be 4 and your last digit should be 5. Does your formula work?

2. When counting the handshakes for 5 people, you may have drawn a picture such as the one at the right. You could first count the number of handshake for the first person; there are 4 handshakes. When you count the handshakes for the second person, how many new handshakes must be recorded? ________ Continue counting new handshakes for the third, fourth and fifth people and record them here: ______, ______, and ______.

   Write a *sum* that represents the total number of handshakes for 5 people.

3. The number 10 is called a triangular number since it can be written as 1 + 2 + 3 + 4. That is, a **triangular number** can be written as a sum of consecutive counting numbers, beginning with 1. The name "triangular number" comes from the fact that you can represent sums such as these in triangular arrays. Do you see how the sum 1 + 2 + 3 + 4 is represented in this diagram?

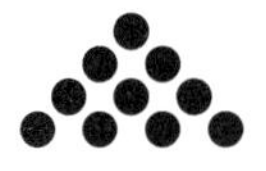

   a. Each of the numbers in the "Total # of shakes" column in the table you completed for problem #1b is a triangular number. Sketch the triangular arrays of dots for each number and label each sketch with its corresponding sum.

   b. In computing the number of handshakes for 31 people, what numbers must be added? Describe the sum, but don't add the numbers yet.

4. To find the number of handshakes for 31 people without having to add the consecutive counting numbers from 1 to 30, you need to develop a formula. One way to do this is to pair up numbers as shown in the following diagram.

$$1 + 2 + 3 + 4 + \cdots + 27 + 28 + 29 + 30$$

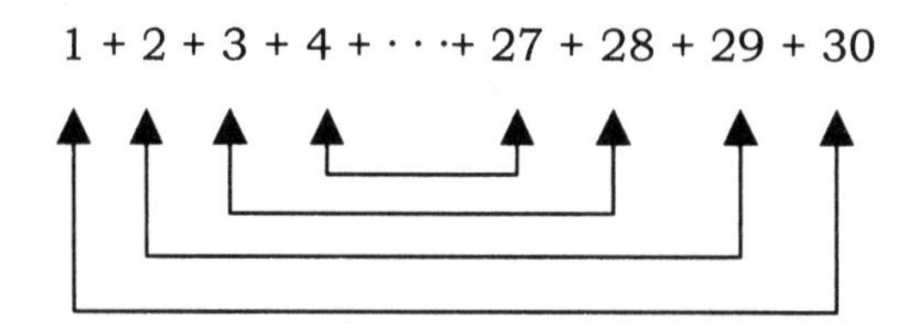

   a. What is the sum of each pair? _______ How many pairs are there? _______ What is the sum of the first 30 counting numbers? _______ How many handshakes are there for 31 people? _______

   b. Use the ideas from part (a) to explain how to find the sum of the consecutive counting numbers from 1 to n.

   c. Write a formula for the nth triangular number:

$$1 + 2 + \cdots + (n - 1) + n = \_\_\_\_\_\_\_\_\_\_$$

   d. How is the formula in part (c) related to counting handshakes? Would you use the formula for the nth triangular number to find the number of handshakes for n people? Explain.

5. Use your formula from #4c to find the following:

   a. the 100th triangular number

   b. the number of handshakes for 100 people

6. What problem-solving strategies did you find useful in solving the problems in this activity?

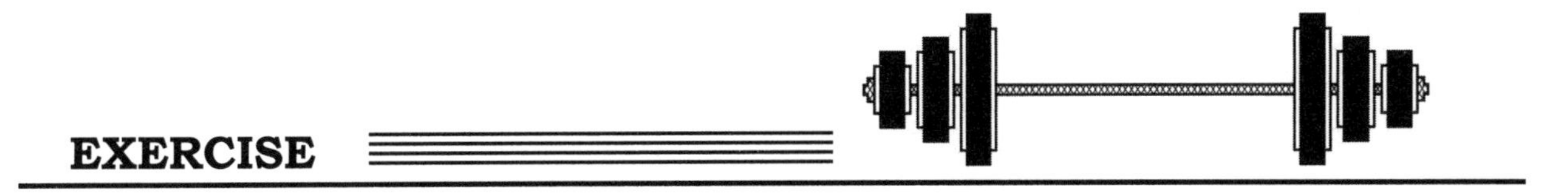

EXERCISE

A MAZE OF PATTERNS

Looking for patterns is one of the strategies for problem solving. Many patterns have been known and studied for centuries, but for those who are new to patterns, it takes some practice to develop the skill. Try these...

In the patterns below, letters have been placed where numbers should go. Find the number that would replace each letter in such a way as to continue the pattern. After completing the patterns, follow the directions given below for completing the maze.

PASCAL'S TRIANGLE:

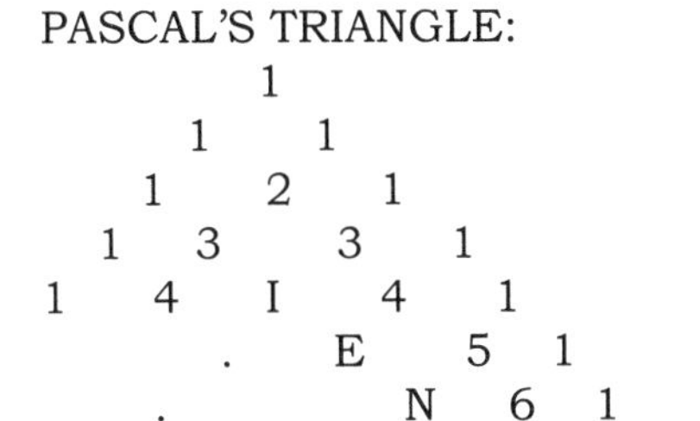

NUMERICAL SEQUENCES:

1, 4, 7, 10, O

1, 4, 9, 16, W

FIBONACCI SEQUENCE:

1, 1, 2, 3, 5, 8, K, W

TRIANGULAR NUMBERS:

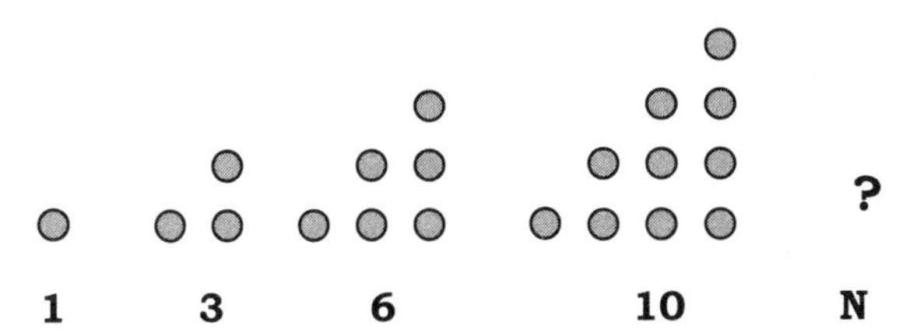

In the maze below, shade in the segment of the path that connects the diamond containing a letter from above with the diamond containing the replacement number. For example, from Pascal's Triangle, you would shade the segment between I and 6. Now see if you can find a path through the maze from start to finish which travels only on shaded segments.

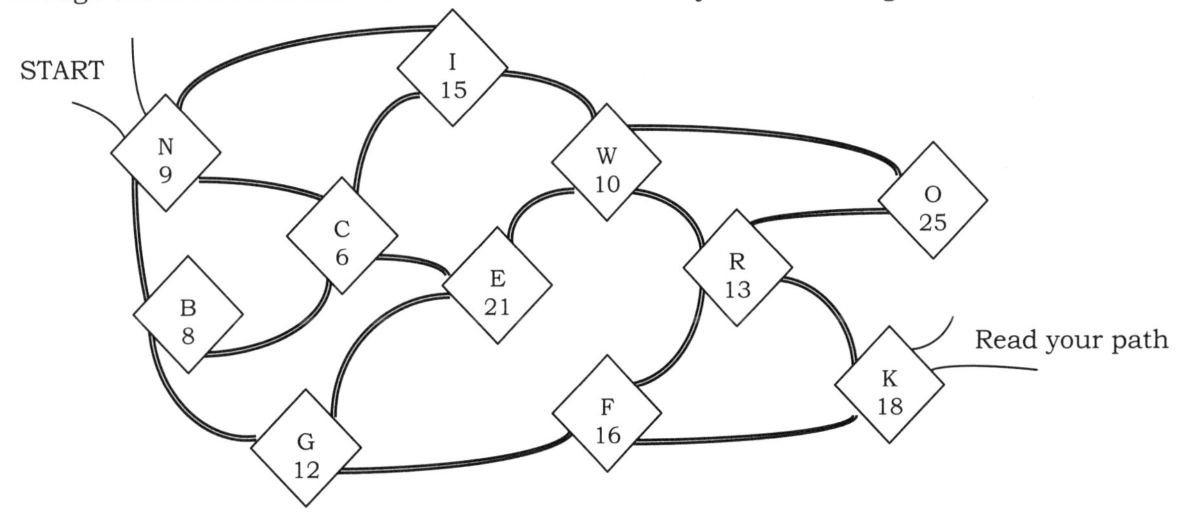

CONNECTIONS TO THE CLASSROOM

1. A teacher asked his class to solve the following problem: *A farmer wants to build a fence to enclose a triangular field that is 180 feet on each side. Fence posts cost $8.00 each. If he is going to put a post in each corner and a post every 6 feet along each side, how much will the farmer spend on fence posts?*

   A student likes to use the Guess and Test problem-solving strategy. When asked to solve this problem, she says "I don't know how to use Guess and Test for this one."

   a. Is Guess and Test an appropriate strategy for this problem? Why or why not? Is there a better strategy to use? If so, which one?

   b. Without simply telling the student how to solve the problem, explain how you could help her. Can you think of some questions you might ask the student to help her decide on an appropriate strategy to use?

2. A teacher asked her class to solve the following problem (from Activity 1.1.1): *Mariella is selling raffle tickets for a school fundraiser. Adult tickets cost $4 and student tickets cost $3. If she has collected $65 so far, how many adult tickets did she sell?*

   a. A student wants to employ the Use a Variable problem-solving strategy. In trying to devise his plan, the student says "I remember another problem like this one where we wrote two equations. I think I need to write two equations, where x represents the number of adult tickets Mariella sold and y represents the number of student tickets. I know one of the equations is $4x + 3y = 65$, but I can't figure out the other equation." This student is likely thinking of another problem he's solved that had more given information than this one. Could the Guess and Test strategy be helpful in this situation? If so, how?

   b. Add the following information to the problem: *Mariella has sold 20 tickets so far.* Does this new information lead to a unique solution? If yes, explain why.

   c. A student wants to use the Guess and Test strategy to solve the new problem. How would the guess and test process for this new problem be different from the process you used when you solved the original problem in Activity 1.1.1?

MENTAL MATH

How can you cut a cake into eight pieces with exactly three cuts?

How can you cut a doughnut into twelve pieces with exactly three cuts?

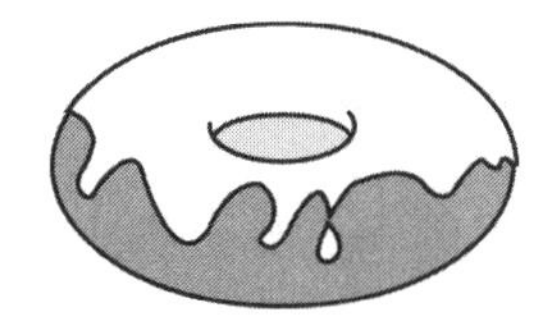

DIRECTIONS IN EDUCATION
Mathematics as Problem Solving
Hey, What's the Problem?

The National Council of Teachers of Mathematics, in *Curriculum and Evaluation Standards for School Mathematics,* recommended that problem solving should be the focus of school mathematics. The goal of having all students become mathematical problem solvers was again endorsed in the NCTM's *Principles and Standards for School Mathematics.* They emphasize that problem solving should permeate the entire mathematical curriculum. Rather that treating problem solving as a distinct topic to be covered only after students have developed computational skills, problem solving can provide the context for developing skills and conceptual understanding.

What are the goals for problem-solving instruction?

An instructional program for improving problem-solving abilities should include activities designed to help students develop

- **A vision for problem solving.** Students are surrounded by a variety of problems, both mathematical and nonmathematical, in their everyday world. Students need to be encouraged to recognize and pursue those problems with the knowledge they possess.

- **A framework for problem solving.** Problem solving is a complex process involving a variety of thinking skills. One framework, proposed by George Pólya, includes skills such as understanding the problem situation, formulating a plan to solve the problem, selecting and implementing appropriate strategies, and analyzing and evaluating the solution obtained.

- **Strategies for problem solving.** Students bring a variety of informal problem-solving strategies with them as they come to school. Nevertheless, they benefit from adding a variety of other strategies such as looking for a pattern, drawing a picture, or solving a simpler problem to their "toolbox" of strategies.

- **Positive attitudes and beliefs about problem solving.** Fundamental to the development of problem-solving ability is a positive attitude toward problem solving. Students need an open mind, an attitude of curiosity and exploration, and the willingness to be flexible and persistent.

- **Reflective patterns of thinking while solving problems.** In the process of solving problems, times of reflection and evaluation enhance the learning experience. Students need to be encouraged to monitor their progress, to think about the merits of one strategy in comparison to other alternatives, and to reflect on why one approach might have been more successful than others attempted.

- **Skills in solving problems in cooperative situations.** In the context of working with peers, students have the opportunity to develop skills such as clarifying their own ideas and evaluating the ideas of others. Discussions help students gain insight into problem-solving strategies and alternative approaches for solving problems.

How can I evaluate the problem-solving efforts of my students?

Because problem-solving instruction focuses on the process as well as the solution, evaluation of problem solving involves more than just looking for the "right" answer. The *Curriculum and Evaluation Standards* suggest that the assessment of students' problem-solving abilities should include evidence that they can "formulate problems; apply a variety of strategies to solve problems; solve problems; verify and interpret results; and generalize solutions." Techniques for evaluating progress in the problem solving include the following:

- **Observing and questioning students.** Either through informal observation and questioning or through a more structured interview, teachers can gather useful information about their students' performance, attitudes, and beliefs.

- **Using assessment data from students.** As students are asked to reflect on their own problem-solving efforts, they often can provide useful insights by completing student reports or attitude inventories.

- **Using holistic scoring techniques.** In evaluating the written work of students, various approaches assign points based on the phases of the process, on criteria related to specific thinking processes, or on the overall impression of the solution.

- **Using written assessment instruments.** Thoughtfully developed multiple-choice or short-answer completion items can be used to help teachers evaluate the ability of their students to carry out the various problem-solving strategies.

As you think about problem-solving instruction, ask yourself:

1. Do I have a sufficient collection of problem-solving strategies in my "toolbox" to teach students? If not, what can I do to improve my own abilities?
2. What can I do in my classroom to help my students develop a vision for problem solving?
3. What can I do to build students' positive attitudes and beliefs about problem-solving?

2 Sets, Whole Numbers, And Numeration

THEME: Sets and Place Value

HANDS-ON ACTIVITIES

Teachers often find that the most challenging topics to teach are the ones we know the best - the things we've known for so long we don't remember how we learned them. Sometimes, it's hard to imagine someone having difficulty understanding something we know so well. For example, how would you help a student who does not understand the meaning of the numeral "24"? For you, this is easy. For a child who is just learning about numbers and how our numeration system works, the concepts of place value and regrouping are new.

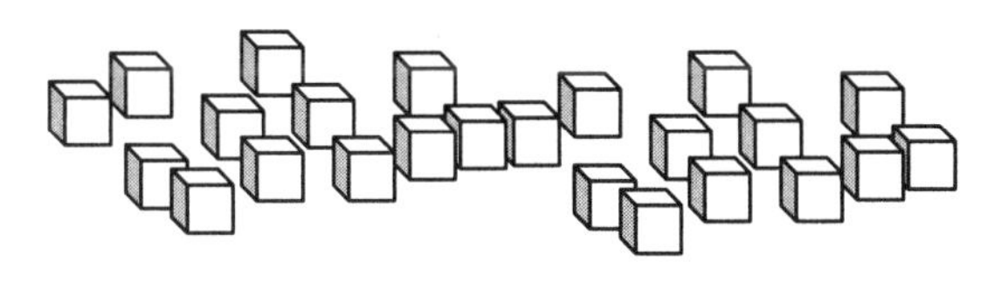

On one hand, the numeral 24 represents the number of objects in this set of blocks, *twenty-four.*

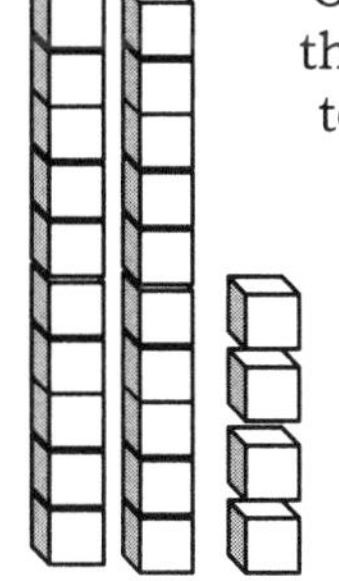

On the other hand, when we group the blocks by ten, we get *two* sets of ten and *four* single blocks. This is represented by the digits "2" and "4" in the numeral 24, using place value.

The first activities in this chapter introduce the ideas of sets, which are the basis for the concept of "number". Sets will be used again in Chapter 3 when you investigate whole-number operations, such as addition and subtraction. Next, a popular manipulative found in elementary classrooms, base pieces, will be introduced as a model for our numeration system. In order to help you discover the difficulties your future students may have when learning our base-ten numeration system, you will be working with base pieces in other bases, such as base three and base five. Lastly, functions, which are special types of relationships between sets, are introduced through the use of arrow diagrams.

OBJECTIVE: **Use Venn diagrams to picture relationships between sets**

You Will Need: **Materials Card 2.1.1 (eManipulative option: *Venn Diagrams*)**

Children develop the concept of "number" by working with collections of objects, called **sets**. A common classroom activity involves sorting or categorizing objects, such as buttons, into sets. In the next activity, you will learn the terminology of sets as well as describe and picture relationships between sets.

1. Relationships among sets are represented using **Venn diagrams** where the interiors of circles represent the sets. In a set, the objects are called **elements** or **members** of the set. In the simple Venn diagram #1 shown on Materials Card 2.1.1, the rectangle represents the **universal set**, or all the elements under consideration. For this activity, the universal set will be all buttons from Materials Card 2.1.1. Let *A* be the set of all round buttons. Place the buttons in the appropriate regions within the rectangle.

   a. Identify the buttons that are in set *A* and record them by number in the Venn diagram to the right.

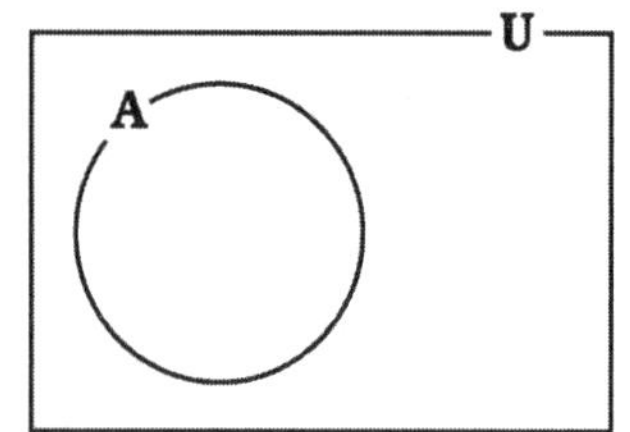

   b. What buttons are not in set *A*?

   ____

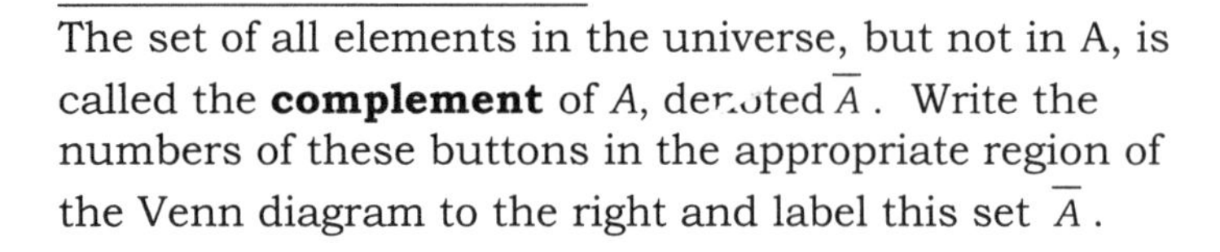

   The set of all elements in the universe, but not in A, is called the **complement** of *A*, denoted $\overline{A}$. Write the numbers of these buttons in the appropriate region of the Venn diagram to the right and label this set $\overline{A}$.

   c. For the set *A* given below, identify the buttons in the set by their numbers. Then write a brief description of the complement of set *A* and list the elements of $\overline{A}$.

   A = All round black buttons = ______________________

   $\overline{A}$ = ____ = ______________________

2. Venn diagram #2 on Materials Card 2.1.1 shows the general Venn diagram involving two sets. For the following sets, place your buttons in the appropriate region of the Venn diagram.

   a. *A* is the set of black buttons   *B* is the set of gray buttons

   What do you observe about the relationship between these sets?

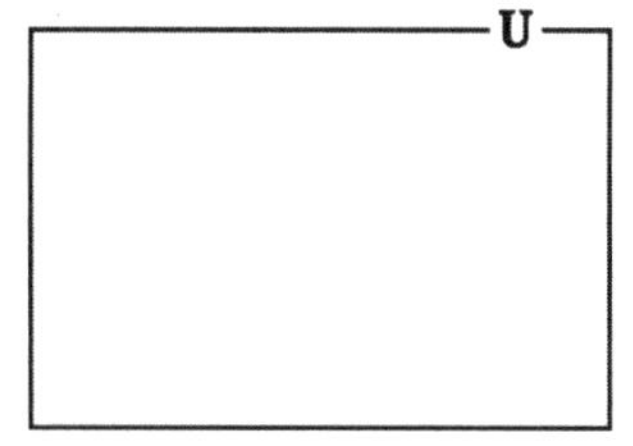

   A and *B* are called **disjoint sets** because they have no elements in common. Show another way to draw the Venn diagram for disjoint sets.

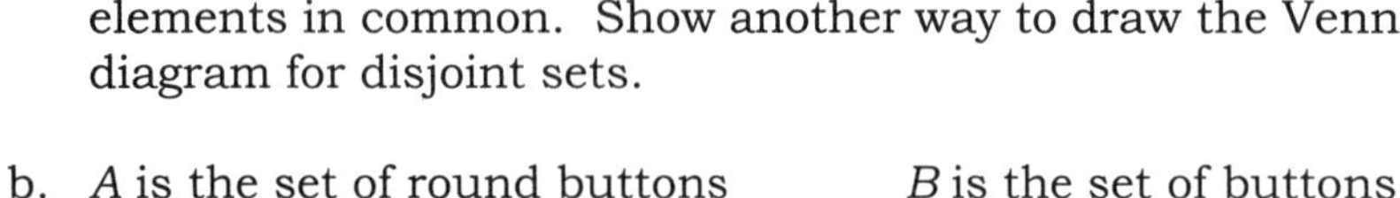

   b. *A* is the set of round buttons   *B* is the set of buttons with two holes

   What do you observe about the relationship between these sets?

   A and *B* are said to be **intersecting sets** since they have some elements in common.

c. A is the set of buttons with 4 holes B is the set of square buttons with 4 holes

What do you observe about the relationship between these sets?

In this case, all of the elements in set B are also elements of set A. In such a case, we say set B is a **subset** of set A, written $B \subseteq A$. Show another way to draw the Venn diagram when one set is a subset of another.

3. Next, consider several different ways of combining sets to form a third new set. For parts (a) – (d), let <u>A be the set of round buttons and B be the set of buttons with 2 holes</u>. Place all the buttons in the appropriate regions on Venn diagram #2 on Materials card 2.1.1.

   a. The set consisting of all common elements of sets A and B is called the **intersection** of sets A and B, written $A \cap B$. Describe the buttons in $A \cap B$. Then, shade the Venn diagram to the right corresponding to $A \cap B$.

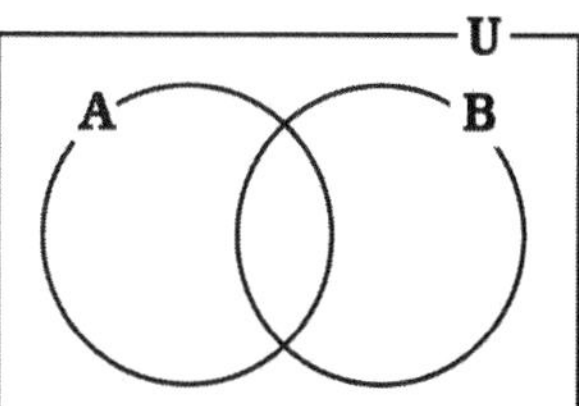

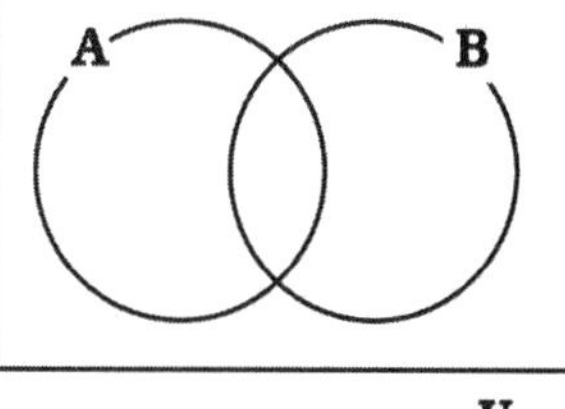

   b. Describe the buttons in the region corresponding to the shaded portion of the Venn diagram to the right.

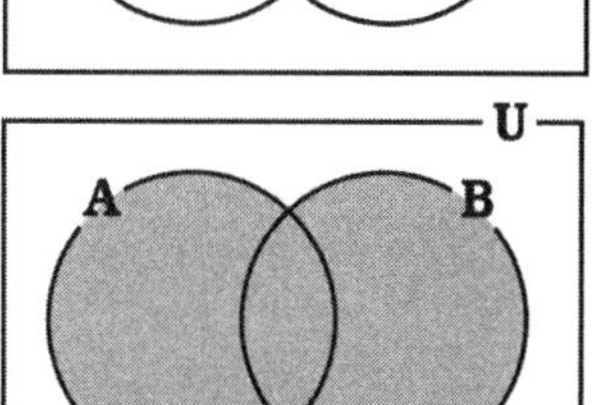

   These elements are in what is called the union of sets A and B, written $A \cup B$. The **union** of two sets is formed by putting the elements that are in A *or* in B *or* in both A and B together.

   c. Describe the buttons in the shaded region shown to the right.

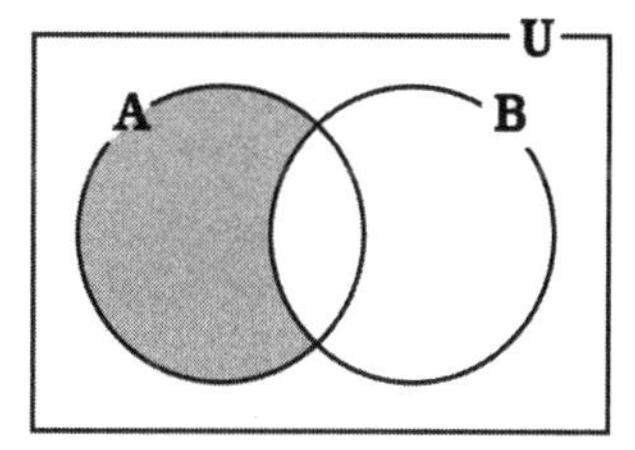

   These elements are in what is called the **set difference** of set B from set A (all elements in A that are not in B), written $A - B$.

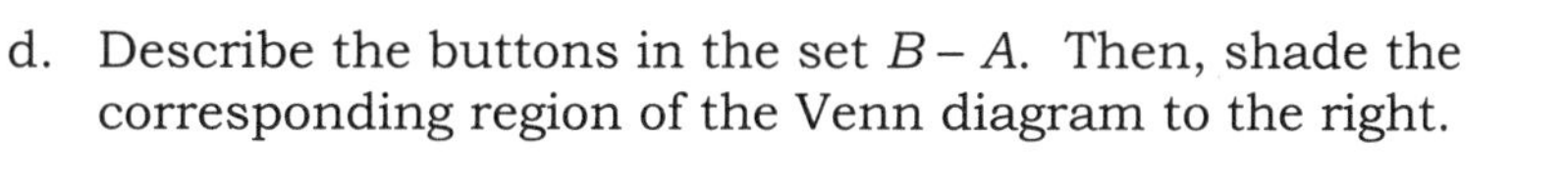

   d. Describe the buttons in the set $B - A$. Then, shade the corresponding region of the Venn diagram to the right.

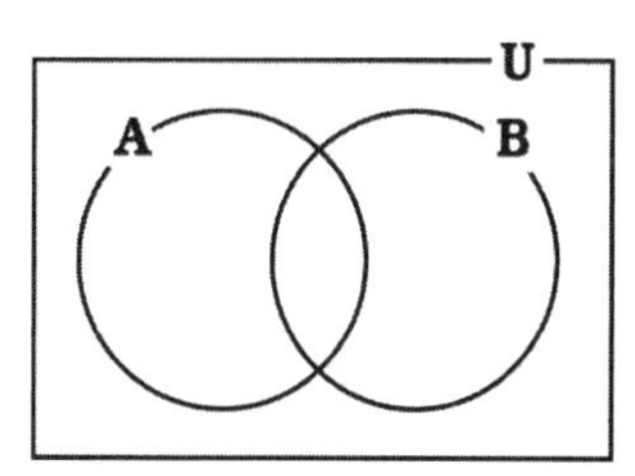

4. From the buttons on Materials Card 2.1.1, choose your own sets A and B. Then, describe the buttons in each of the following sets and shade a Venn diagram for each one.

A = ______________________ B = ______________________

a. $A \cap B$ b. $\overline{A \cup B}$ c. $B - \overline{A}$

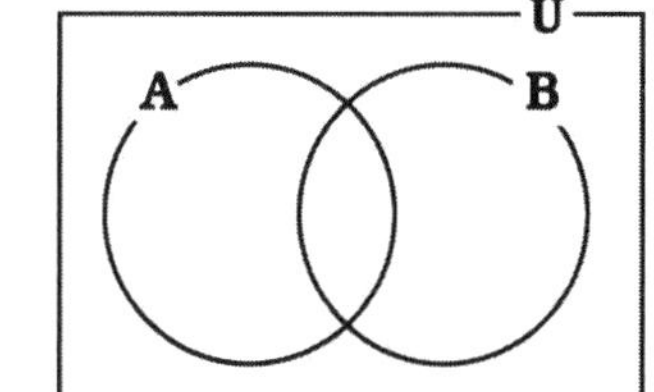

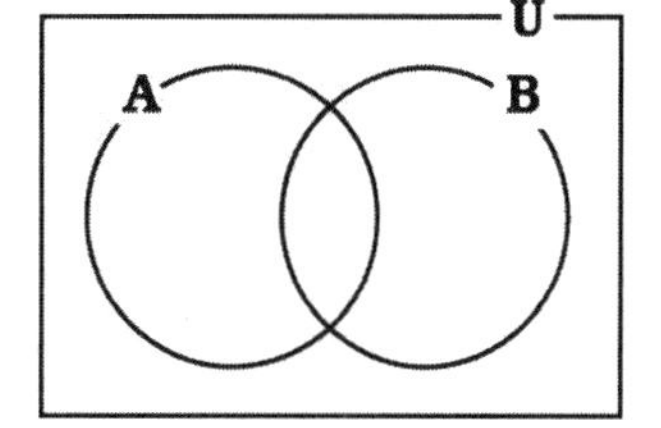

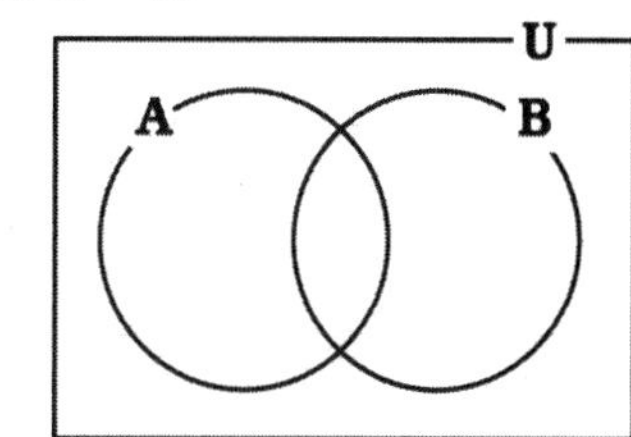

OBJECTIVE: **Use Venn diagrams to show relationships of sets of whole numbers**

You Will Need: **(eManipulative option: *Venn Diagrams*)**

The next activity shows several ways that Venn diagrams can be useful for displaying information and picturing relationships among sets of whole numbers.

1. The general three-set Venn diagram is shown to the right.

   For this problem, let the universal set be the set $U = \{2, 3, 4, 6, 8, 9, 10, 12, 14, 15, 16, 18, 21, 22, 25\}$.

   Set A = multiples of 3
   Set B = multiples of 2
   Set C = multiples of 4

   a. Write each of the numbers in the universe in the appropriate region of the Venn diagram.

   b. Notice that there are two regions in the diagram that have no elements. Can you think of numbers to put into the set U that would fit in these empty regions? Why or why not?

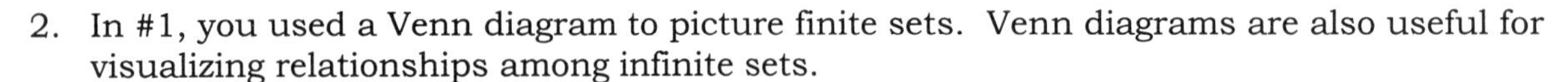

2. In #1, you used a Venn diagram to picture finite sets. Venn diagrams are also useful for visualizing relationships among infinite sets.

   a. As an example, suppose the universal set U is the set of *all* whole numbers. Let A be the set of all multiples of 2, B the set of all multiples of 4 and C the set of all multiples of 8. Determine if the Venn diagram at the right represents sets A, B, and C correctly. Explain your reasoning.

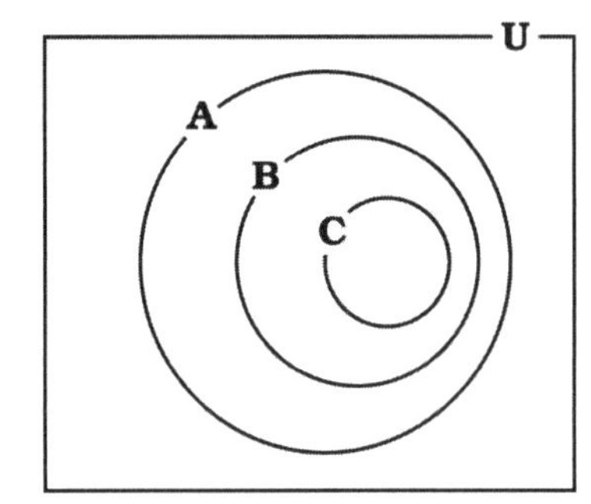

   b. How would you describe the set $\overline{A}$? Shade the Venn diagram to show this set.

   c. Sketch two copies of the Venn diagram in part (a). Shade one of them to show $A \cap B$ and the other to show $A \cup B$. What do you notice about these sets?

3. For this problem, suppose the universal set U is the set of all whole numbers. Let A be the set of all multiples of 2 and B the set of all multiples of 3. For each of the following sets, shade a Venn diagram. Then describe the elements of the shaded region and give a few examples of the elements. The first one is partially done for you as an example.

   a. $\overline{A \cap B}$ It can be helpful to shade an intermediate diagram to find the final shaded region. First, shade $A \cap B$ in the diagram on the left, then shade the complement of this set in the other Venn diagram:

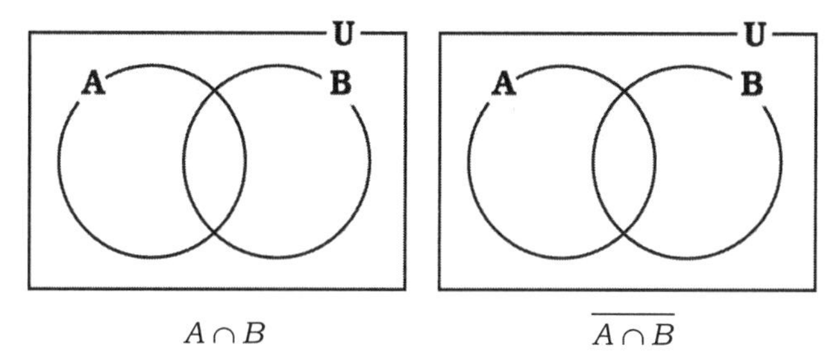

$\overline{A \cap B}$ is the set of whole numbers that are not multiples of both 2 and 3. For example, 2, 3, 5, 8, 9, and 16 are some of the elements of $\overline{A \cap B}$. Write each of these numbers in the appropriate region of the diagram for $\overline{A \cap B}$.

   Can you think of another way to describe the elements of $\overline{A \cap B}$?

   b. $\overline{A} \cup \overline{B}$

   c. $\overline{A \cup B}$

   d. $\overline{A} \cap \overline{B}$

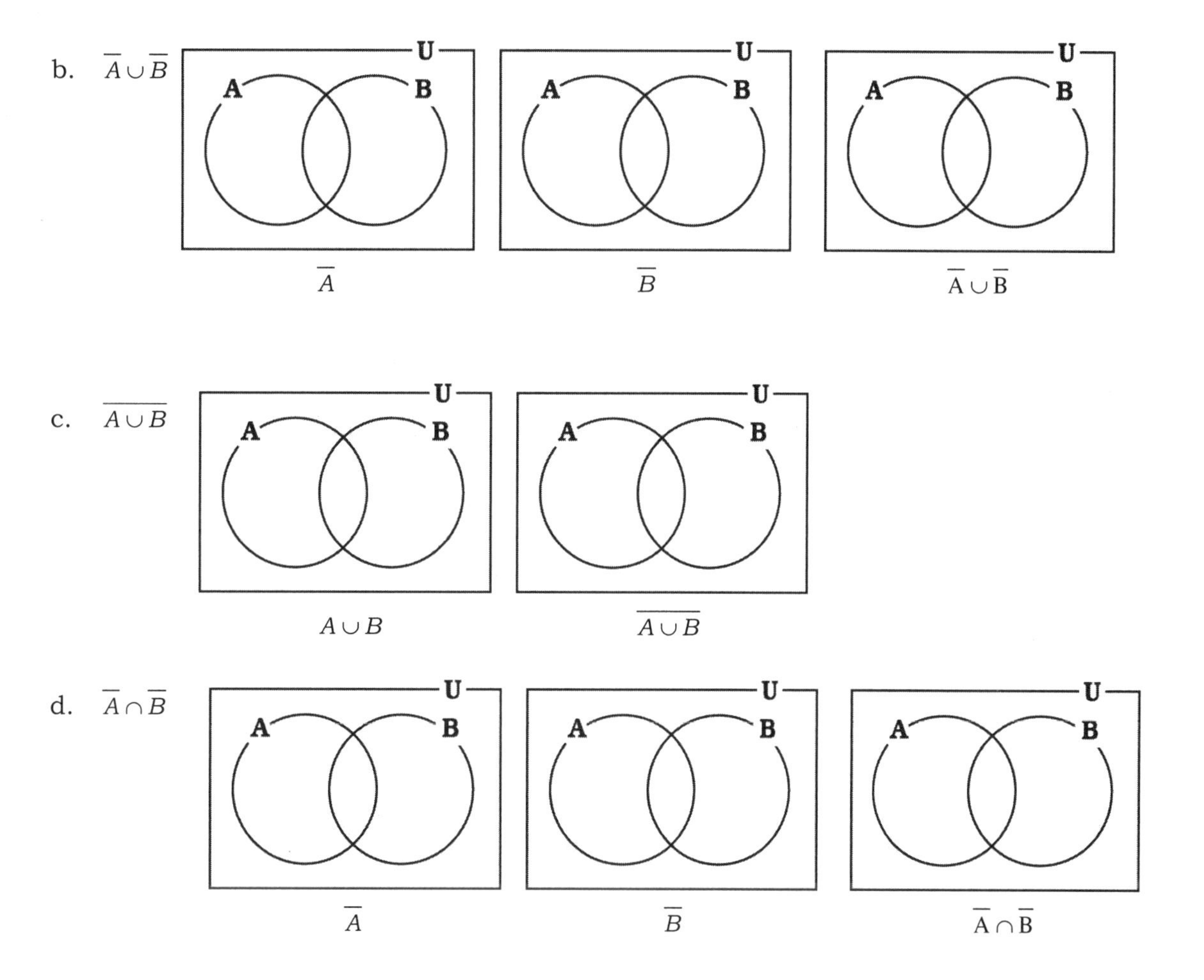

   e. What do you notice about the sets you shaded in parts (a)- (d)?

OBJECTIVE: **Investigate regrouping and place-value in base three**

You Will Need: **Materials Card 2.3.1 and a handful of paperclips or other objects**

The numeration system we use today is called the Hindu-Arabic numeration system. Making sense of the way our numerals represent numbers requires an understanding of place value and regrouping. In the next activity, you will use a common hands-on model, called base pieces, to investigate these concepts. To make the experience more like that of your future students, you will be working in bases other than ten with which you are already familiar.

1. The pieces pictured to the right and on Materials Card 2.3.1 are called **base-three pieces.**

   What relationship do you observe between the different pieces?

   They are all multiples of 3

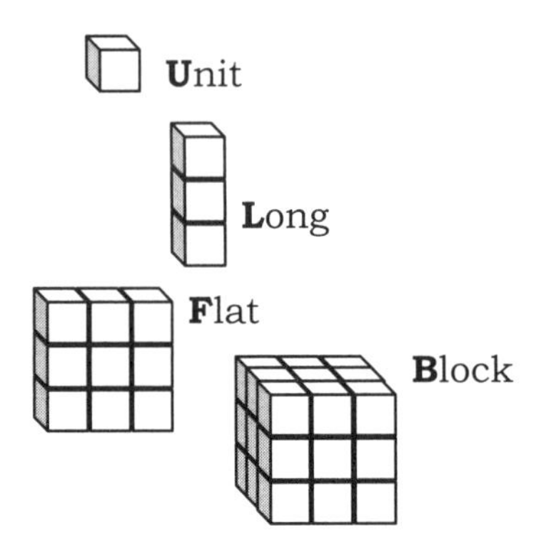

2. Place a pile of paperclips (or other small objects) in front of you. Now make a pile of unit cubes with the same number of units as paperclips.

   a. Exchange (or **regroup**) your base-three pieces until the number of paperclips is represented by as few base-three pieces as possible. Record the final number of pieces of each type in this chart:

B	F	L	U

   b. Repeat part (a) , using a different number of paperclips, and record the number of pieces on this chart:

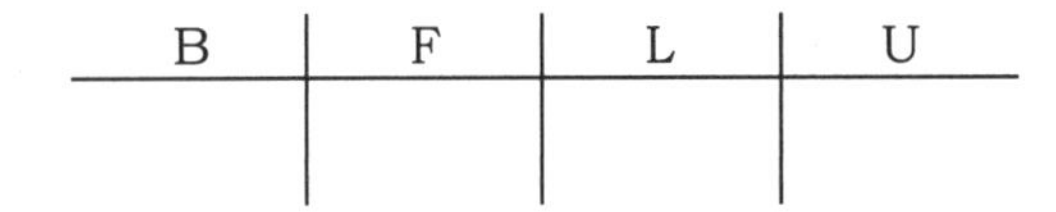

B	F	L	U

   c. When recording the number of base three pieces in a collection that has the fewest pieces possible, what digits are allowed in any single column? ______________Why?

3. Charts like those in #2 show how to write numerals in base three. For example, the numeral that describes the pieces pictured to the right is 1022_{three}(one block, zero flats, two longs, two units).

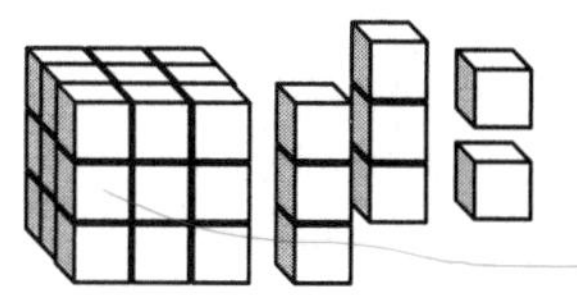

   Write a base-three numeral for each set of base-three pieces shown here.

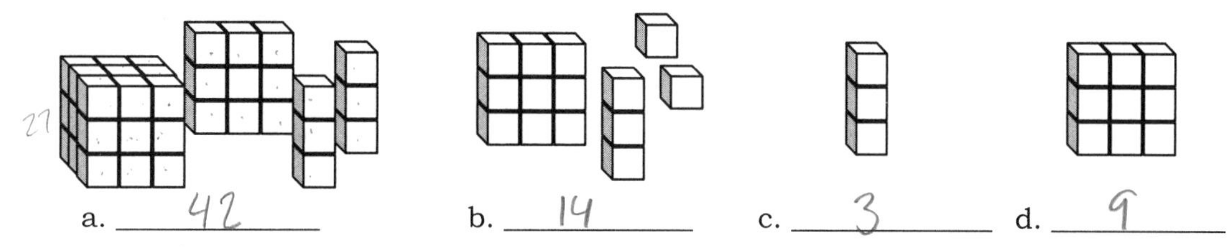

   a. ________ b. ________ c. ________ d. ________

4. The charts you used in problem #2 were labeled with the base-three piece for each column.

   a. In the following chart, give the **place value** of each column. That is, re-label each column with the *base-three* numeral for that base-three piece.

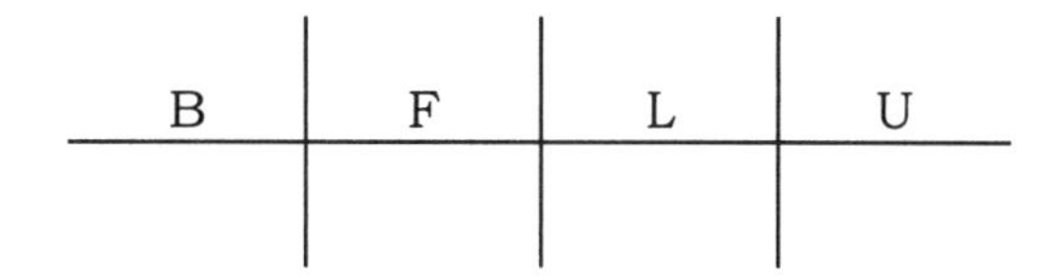

B	F	L	U

   b. How would the base-three place values you wrote in part (a) be expressed using base-*ten* numerals? ________________

5. Using the place values you wrote in problem #4, we can write numbers in **expanded notation**. This is when you decompose a number into pieces that show the meaning of the numeral. Consider this example, which shows 112_{three} in expanded notation.

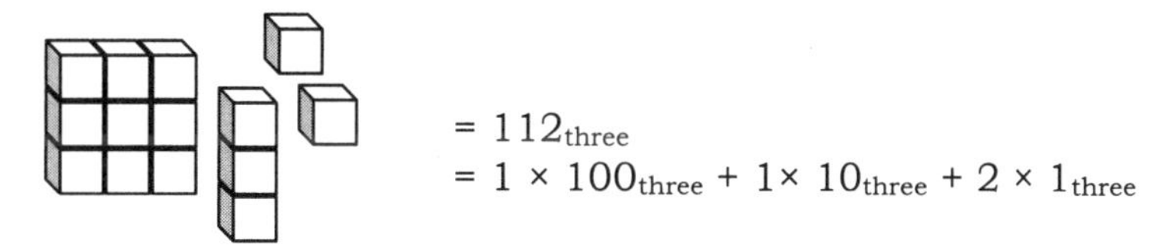

$= 112_{three}$
$= 1 \times 100_{three} + 1 \times 10_{three} + 2 \times 1_{three}$

   Write each of the following numerals in expanded notation:

   a. 201_{three} = ________________________________

   b. 2010_{three} = ________________________________

6. Now think about base five.

   a. Sketch the base-five pieces.

   b. What are the base-five place values, using base-five numerals? ________________

   c. How would the base-five place values you wrote in part (b) be expressed using base-*ten* numerals? ________________

OBJECTIVE: **Change numerals from one base to another**

You Will Need: **Base-three pieces (Materials Card 2.3.1)**

In the next activity, you will develop methods for converting numerals from one base to another. Solving problems like these will help you gain a deeper understanding of our own base-ten numeration system.

1. Each of the following numerals represents a number in base three. Represent each number using your base-three pieces from Materials Card 2.3.1. Then, write each numeral in base ten.

   a. 1201_{three} b. 2122_{three} c. 1221_{three}

2. Draw pictures to represent the following numbers with base five pieces. Write each numeral in base ten.

   a. 1234_{five} b. 2312_{five} c. 1341_{five}

3. For problems #1 and #2, did you count each unit one-by-one to determine the base-ten numeral? ________

   a. Describe a process that does not involve counting that you could use to change a numeral from a given base to base ten. If you need to, try more examples.

   b. What role does place value play in the process you wrote for part (a)?

4. Next, consider the reverse process – converting from base ten to another base. Take 14 units from Materials Card 2.3.1. Make exchanges until you have a collection with as few base three pieces as possible. What base-three numeral represents the 14 units?

   ____

5. Beginning with the following number of units, draw pictures to show the process of converting to base three. Try to discover an efficient procedure for doing this.

   a. 19 b. 25 c. 34

6. Rewrite each of the following base ten numerals as a base five numeral. For this problem, try converting *without* using drawings of base pieces. Show your process.

   a. 39 b. 78 c. 156

7. Describe a process for changing a base ten numeral to another base. In your description, include the role of place value in your process.

OBJECTIVE: **Investigate the definition of function using arrow diagrams**

You Will Need: **(eManipulative option: *Function machine*)**

The concept of function is one of the most fundamental in all of mathematics. The formal definition of function is complex. Introducing the definition of function through arrow diagrams can be helpful in visualizing all the parts of this definition.

Definition: A **function** is a correspondence between a first set of objects, the domain, and a second set of objects, the codomain, with the property that to each element in the domain there corresponds exactly one element in the codomain.

Notice that there are many aspects of this definition. When thinking about a function, a person must keep in mind the domain, the codomain, the correspondence (or relationship) between the elements of the sets, and the property that *every* element of the domain set correspond to exactly one element in the codomain. It is often possible to organize all of this information into a diagram called an **arrow diagram**.

Here's an example:

Arrows begin at an object in the domain and point to their corresponding element in the codomain.

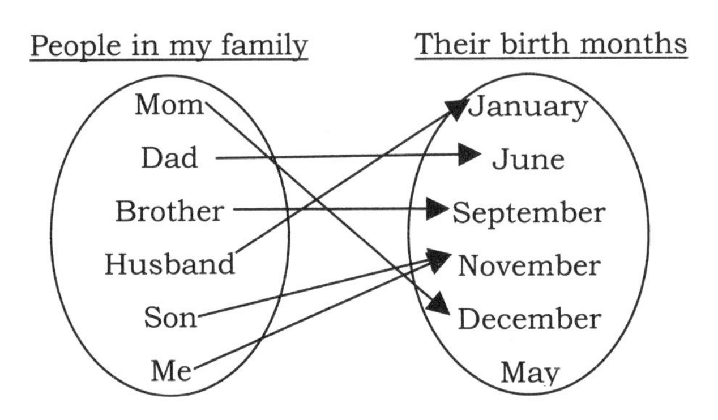

1. Refer to the arrow diagram in the example to answer each of the following questions:

   a. Identify the domain and codomain in the example.

   b. Is the correspondence displayed in the arrow diagram a function? Use the definition of function to explain your answer.

2. Now consider the correspondences displayed in the following arrow diagrams. In which of these is a function displayed? Explain your answers. That is, either show that the correspondence fits the definition of function or, if a correspondence is not a function, explain why not.

a.
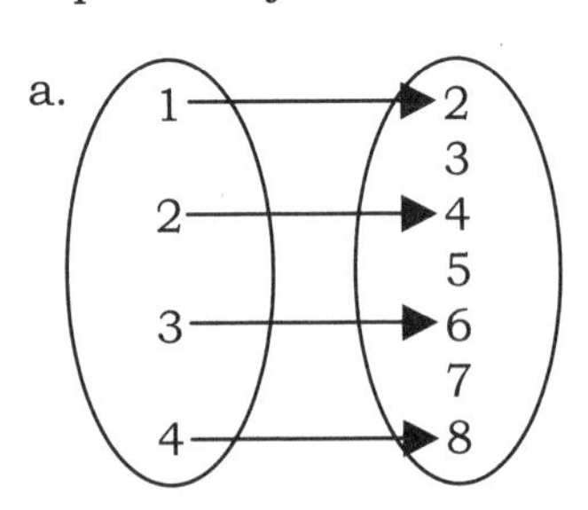

b.
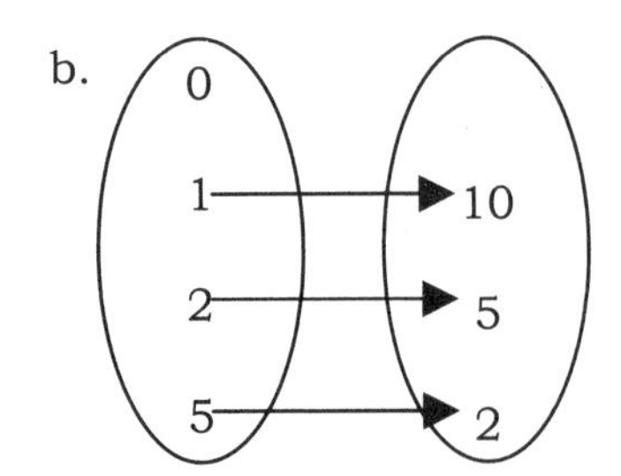

c.
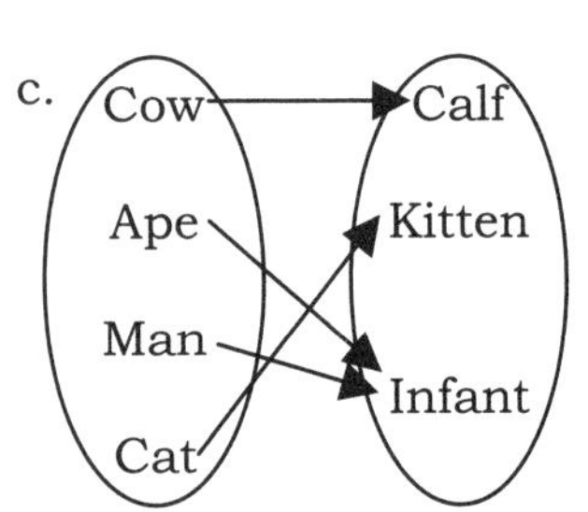

d.

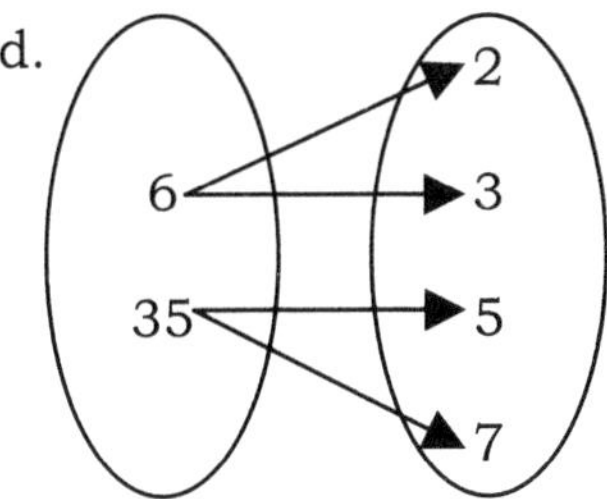

e.

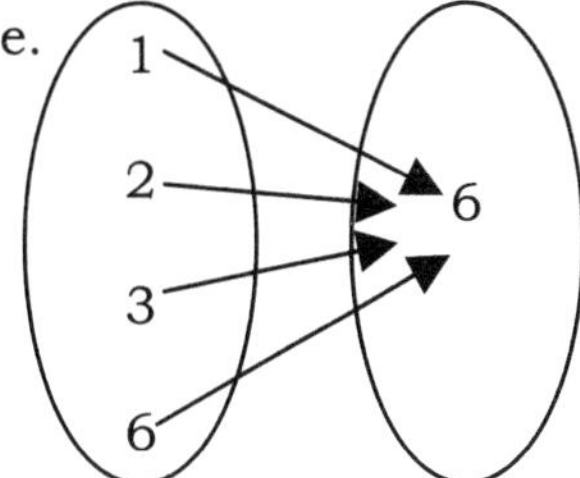

f.

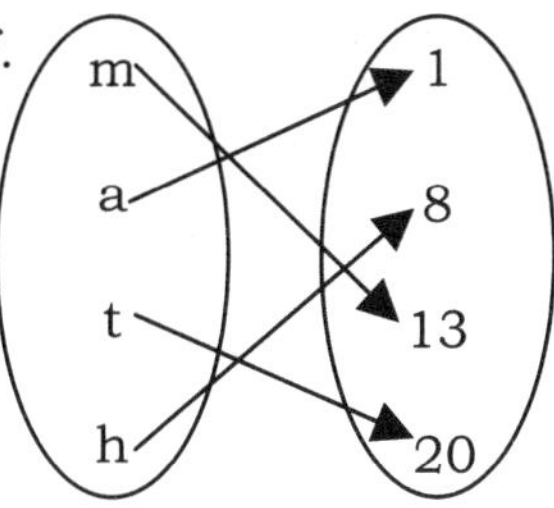

3. You may have noticed that for some of the arrow diagrams in problem #2, it is possible to describe the relationship between the elements of the domain with the elements of the codomain in words. For example, one could describe the correspondence in part (a) like this: "Each number in the domain corresponds to its double."

   Keeping in mind that not every correspondence is a function, try using a word description to describe each of the remaining correspondences displayed in the arrow diagrams in problem #2.

   b.

   c.

   d.

   e.

   f.

4. Now compare your answers to #3 with the answers of a classmate. Do the two of you agree? If not, how can you explain the differences?

5. Describe how a person could tell, just by looking at an arrow diagram, whether a correspondence is a function.

6. Under what circumstances do you think it is impossible to actually *draw* an arrow diagram, for a function?

   a. Write down an example of a function for which you could not draw the arrow diagram.

   b. For the example you gave in part (a), can you imagine what the arrow diagram would look like if you could draw it? Describe it in words.

EXERCISE

HIDDEN MESSAGE

To find the hidden message, follow these directions:

- Choose any numeral in the left-hand column.
- Find its corresponding base ten numeral in the right-hand column
- Put the letter in the blank with the corresponding problem number in the message below.

NUMERALS	BASE TEN	
1. 112_{twelve}	U	73
2. 320_{four}	T	84
3. 100010_{two}	S	31
4. 11111_{two}	R	158
5. 111_{five}	P	65
6. 111000_{two}	N	196
7. 200_{seven}	M	13
8. 211_{five}	L	69
9. 70_{eight}	I	27
10. 343_{five}	H	43
11. 124_{eight}	F	106
12. $2T_{twelve}$	E	98
13. 2120_{three}	C	49
14. 51_{six}	B	34
	A	56

___ ___ ___ ___ ___   ___ ___ ___   ___   ___ ___ ___ ___ ___ !

12 8 4 10 5   9 1 7   2   3 13 6 14 11

CONNECTIONS TO THE CLASSROOM

1. An elementary student conducted a survey of all 25 students in his class. He questioned them about their activities and interests in three areas. As part of a class presentation on the results of his project, the student wants to make a Venn diagram to use as a visual aid. His rough draft is shown to the right.

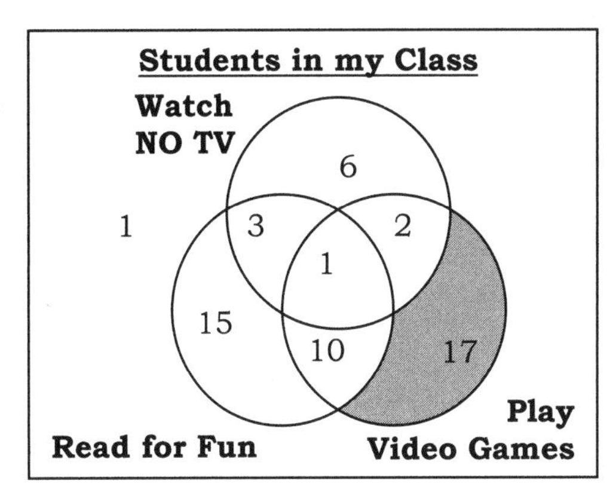

   a. The student has placed numbers of students in each region. How can you tell at a glance that the student is missing an important concept involved in Venn diagrams?

   b. In the Venn diagram above, write in what you believe should be the correct number for each region. Explain your reasoning. Then, show how you could check that your numbers are correct.

   c. In words, describe the set represented in the shaded region of the student's Venn diagram.

      Use set notation to write a name for this shaded region. ________________

      Why do you suppose the student emphasized this region of the diagram?

   d. What mistake do you think the student likely made in placing the numbers in the diagram? Did the student get any of the numbers correct? Which one(s)? Why do you suppose the student got these numbers correct?

2. Counting the blocks in a pile on her desk, a student says "...twenty-six, twenty-seven, twenty-eight, twenty-nine, twenty-ten, twenty-eleven, ..." How is the student's reasoning? Would most understand this student if she used this terminology in public? What could a teacher do to guide this student to more appropriate terminology?

3. a. A student writes 21 for 201 and 423 for 400,203. Why do you think the student is doing this? Show how to use hands-on models presented in this chapter to help the student see there is a difference between 201 and 21.

   b. Another student writes 201 for 21 and 400,203 for 423. Why do you think the student is doing this? How would you help this student?

MENTAL MATH

Write the base five numeral that represents the number of base-five pieces in each group shown below. What are the next two numerals in the counting sequence?

a. 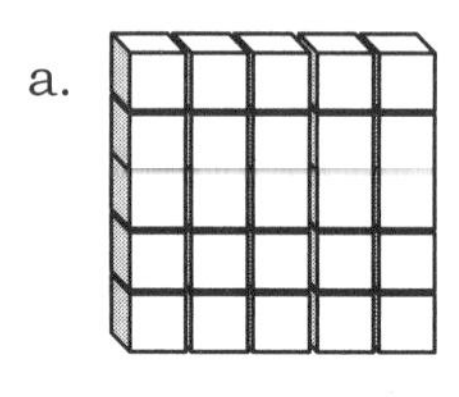 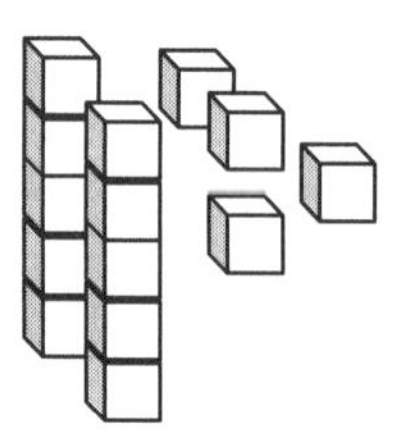

b. 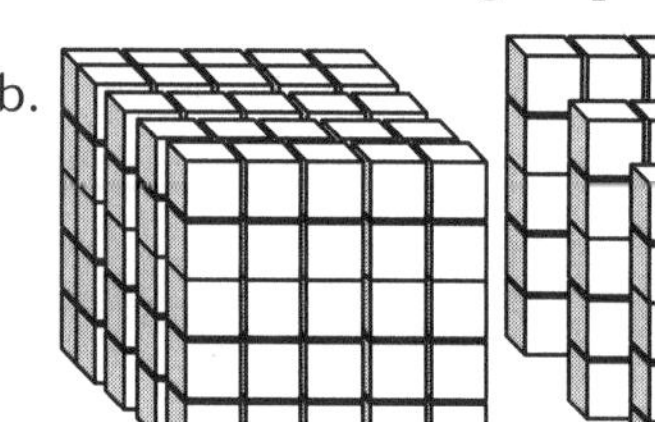 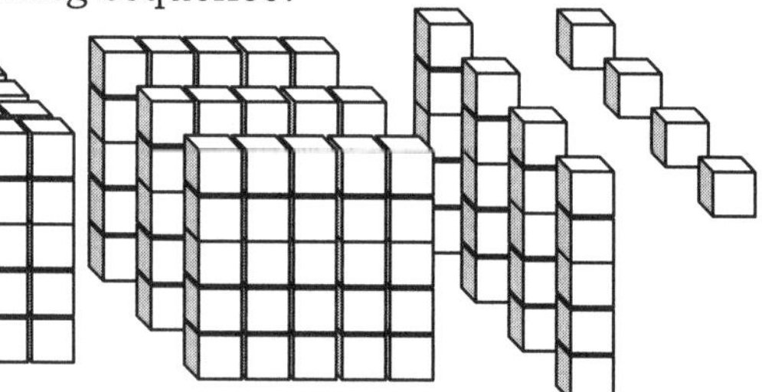

If these two groups were combined, what number would be represented?

DIRECTIONS IN EDUCATION
Manipulative Materials
When I Do, I Learn

Manipulative materials provide the opportunity for teachers to combine the child's natural curiosity and social interest with content learning. Lessons carefully structured to illustrate mathematical content through manipulatives allow children to become involved, investigate and persist at activities in a stimulating and challenging environment.

What are manipulatives?

- Manipulative materials are concrete, "hands on" models which may be used to illustrate mathematical concepts.
- Manipulative materials are objects which appeal to a variety of senses – visual, aural, or tactile – and which capture the attention of the learner.
- Manipulative materials are objects which **students** can touch, move and rearrange – the materials may come from the students' own environment or may be designed specifically to represent mathematical ideas.

Reasons to use manipulatives:

- Manipulative materials can be used to make the abstract world of mathematics meaningful because they bridge the gap between concrete thinking and the abstract content of mathematics.
- Students, particularly children, need physical involvement in order to add new ideas to their cognitive structure. Concrete experiences are necessary before the learner can think pictorially or abstractly.
- Manipulation leads to understanding and abstraction. Students who have had the opportunity to see and manipulate a variety of objects have clearer mental images and can represent abstract ideas more completely. Experiencing ideas to be learned through action can lead students to the use of associate symbols.
- Conceptual learning is maximized when learners are exposed to concepts through a variety of physical contexts.
- Research indicates that mathematical achievement is enhanced through the use of manipulatives.
- Manipulatives are fun!

Criteria for selection of manipulative materials:

- Sufficient materials should be provided to allow individual student manipulation.
- The materials should be a clear representation of the mathematical idea, appropriate for the students' developmental level, and as versatile as possible.
- The materials should be made to withstand normal use and handling by children and must also be safe for children to use.
- To increase student motivation and interest, the materials should be bright, precisely constructed and aesthetically pleasing.
- While still clearly portraying the concepts being taught, materials should be simple to operate and manipulate.
- Teachers should look for materials which can be distributed and collected with a minimum of time and which can be easily stored in the classroom.
- The initial cost as well as maintenance and replacement charges need to be considered. Also of concern will be the versatility of and life durability of the materials.
- Teacher-made manipulatives may often be as effective as commercial ones.

Guidelines for the use of manipulatives:

- Teachers act as guides, asking frequent questions, probing and extending the student's understanding.
- A variety of manipulative materials should be provided.
- Provide time for free exploration when introducing new manipulatives.
- Be aware of irrelevant details of the manipulatives which may distract students.

As you think about manipulative materials, ask yourself:

1. What materials can I think of to represent fractions and their operations? number concepts? geometry and measurement concepts?
2. What materials are appropriate for primary grades? upper grades? middle school?
3. What are good sources of manipulatives?

3 Whole Numbers: Operations & Properties

THEME:

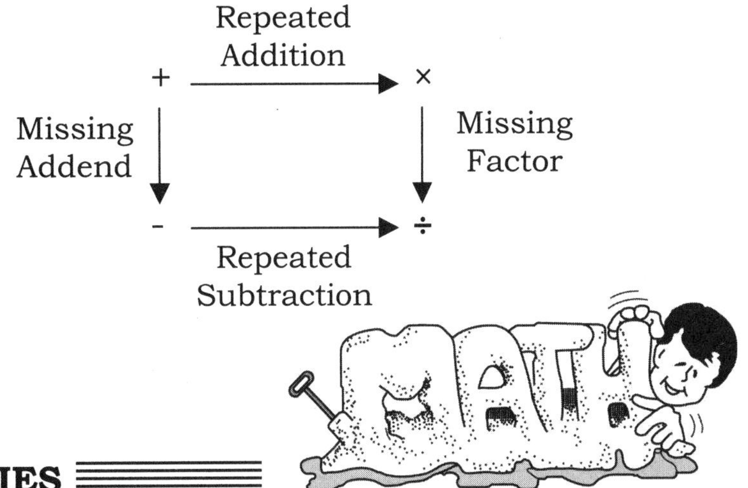

HANDS-ON ACTIVITIES

The four basic operations of addition, subtraction, multiplication, and division are concepts central to mathematics. To become proficient at using these operations, students must achieve a deep understanding of the meaning of each operation and the connections between them, as well as the properties each operation possesses. Simply memorizing facts is not enough to prepare students for solving non-routine problems in which they must be able choose the appropriate operation.

As the activities in this chapter illustrate, each operation can be viewed in different ways using various models. For example, consider the following three ways students may think about the division problem 15 ÷ 5. Can you see the connection each student makes to one of the other four operations in each case?

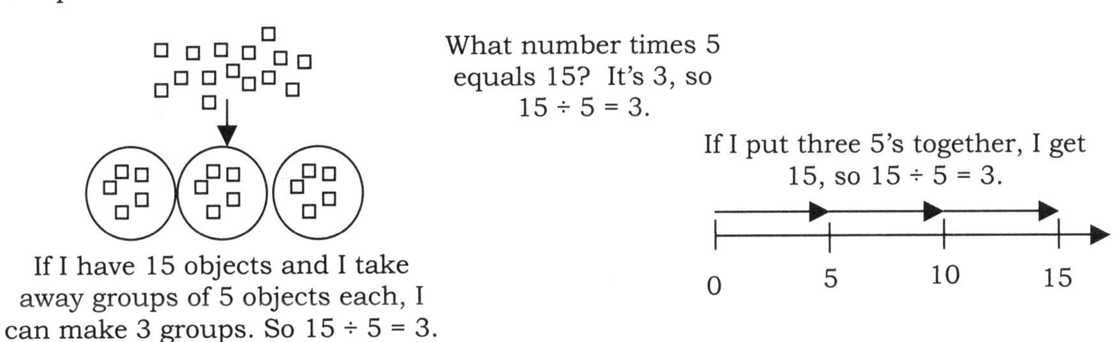

The focus of this chapter is on the meanings of the four operations. In contrast, the focus of Chapter 4 will be on computational procedures, called algorithms, which allow us to efficiently find answers to problems involving operations.

OBJECTIVE: **Use a measurement model to discover properties of addition**

You Will Need: **Materials Card 3.1.1**

In this activity, you will use centimeter strips, a popular hands-on manipulative, to model addition. Centimeter strips can be made of paper, such as those on your materials card or a colorful wooden version may be purchased in classroom sets. Centimeter strips are versatile and useful for modeling topics throughout the elementary, middle and high school curricula, from addition and subtraction, to fractions, to factoring polynomials.

1. Henry has 2 meters of rope and Ahmed has 6 meters of rope. To find out how much they have between the two of them, one could imagine putting their lengths of rope end to end and measuring the result. This is an example of the **measurement model** for adding whole numbers.

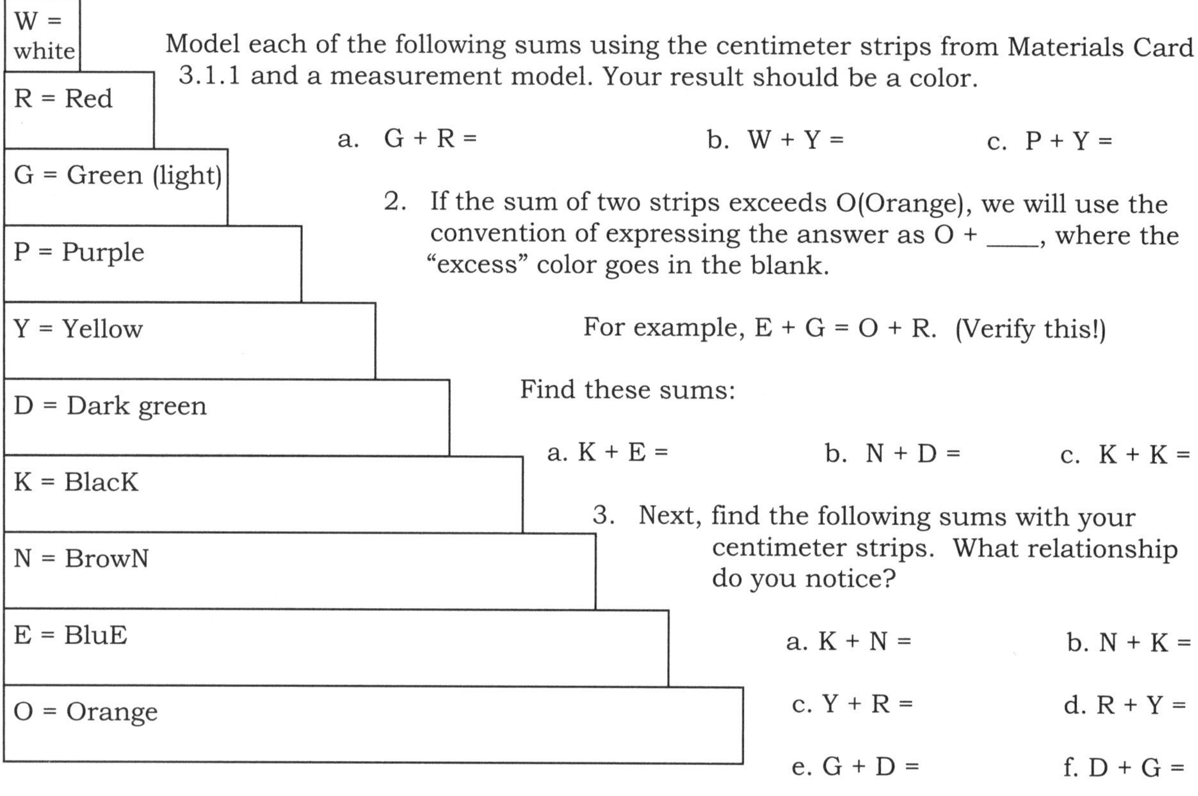

Model each of the following sums using the centimeter strips from Materials Card 3.1.1 and a measurement model. Your result should be a color.

a. G + R = b. W + Y = c. P + Y =

2. If the sum of two strips exceeds O(Orange), we will use the convention of expressing the answer as O + ____, where the "excess" color goes in the blank.

For example, E + G = O + R. (Verify this!)

Find these sums:

a. K + E = b. N + D = c. K + K =

3. Next, find the following sums with your centimeter strips. What relationship do you notice?

a. K + N = b. N + K =

c. Y + R = d. R + Y =

e. G + D = f. D + G =

4. a. Fill in the portion of this table above the dotted line.

b. Use your discovery in problem #3 to complete the rest of the table.

c. Write an equation for the property of addition that helps you to simplify your task in part (b). This is called the **commutative property** of whole number addition.

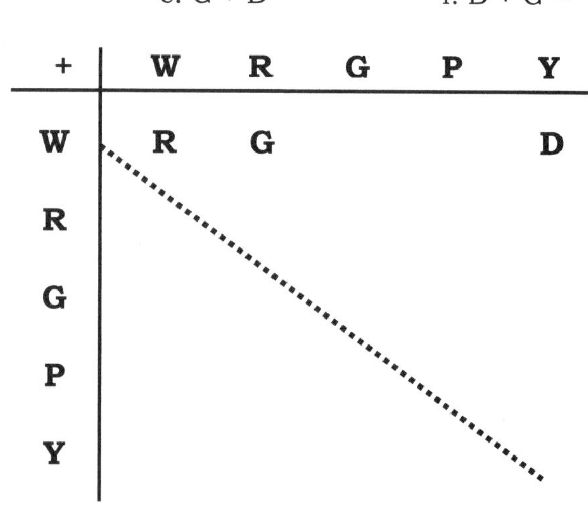

+	W	R	G	P	Y
W	R	G			D
R					
G					
P					
Y					

5. a. Next, use your centimeter strips to find the sum W + R. To this result, add Y. We write this as (W + R) + Y where the parentheses tell you which part to add first. In summary, (W + R) + Y = _______.

   b. Now consider W + (R + Y). Which sum do you find first? _______ Record your result: _______ How does this result compare with part (a)?

   c. Compute the following sums using your centimeter strips.

   (P + R) + G =   P + (R + G) =

   What do you notice?

   d. Write a general statement or equation that summarizes your findings in parts (a)-(c). This is called the **associative property** of whole number addition.

OBJECTIVE: **Represent subtraction using a set model**

You Will Need: **Materials Card 3.1.2**

The operation of subtraction can be approached in several different ways, as the problems in this activity illustrate. Having an understanding of each of these approaches is important when trying to decide which operation is appropriate for solving a given problem.

1. Using the squares from Materials Card 3.1.2, demonstrate how you could model the following problems concretely. In each case, draw and label a picture that shows your method. Then, write a mathematical expression and answer the problem.

   a. Jacques has 15 postage stamps. If he gives 6 stamps to his cousin, how many stamps will he have then?

   b. Quinn has 8 crackers and 5 pieces of cheese to put on the crackers. How many crackers will not have cheese on them?

   c. Huyen wants to collect the trading cards of all 12 members of the Chicago Bulls. If she now has 7 cards, how many more does she need to collect?

   d. Troy has some baseball trading cards. If he collects 8 more, he will have 17 cards. How many cards does he now have?

2. Describe the similarities and differences between the problems in #1.

3. Some subtraction problems can be thought of using the "Take Away" approach, where from a set of objects, a subset is taken away. Which problem(s) in #1 can be approached in this way? Explain.

4. Another approach to subtraction is the "Missing Addend" approach, where one finds the number c that must be added to a known number b to obtain the result a. In other words, c is the missing addend in the equation $a = b + c$. Which problem(s) in #1 can be solved using this approach? Explain.

5. Look again at problem #1b. The take-away approach cannot be applied directly, since the set of cheese slices is not a subset of the set of crackers. However, if we first make a comparison by matching each cheese slice with a cracker (as shown in the following diagram), then one of the two subtraction approaches can be applied.

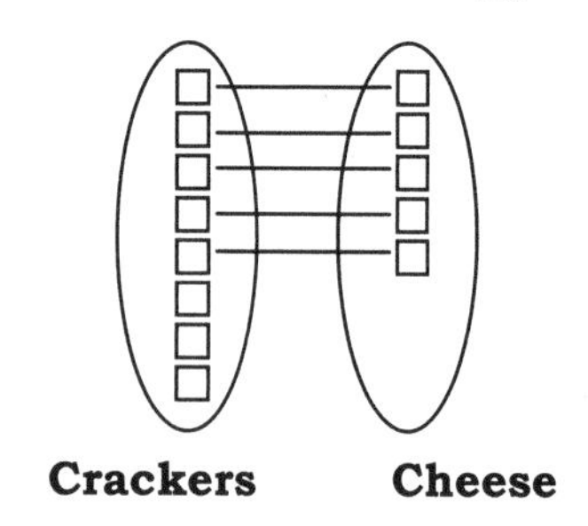

Now the take-away approach is appropriate because we think, "Take away 5 crackers from the set of 8 crackers. The 3 remaining crackers do not have cheese." Explain how the missing-addend approach could also be applied to solve this problem.

6. Write three more word problems that have $9 - 5 = 4$ as the solution. Write one take-away problem, one missing-addend problem, and one problem for which a comparison must be made before applying one of these two approaches. In each case, draw and label a picture of the squares from Materials Card 3.1.2 to show how to model the problem concretely.

Take away: Missing Addend: Comparison:

OBJECTIVE: **Investigate the relationship between subtraction and addition**

You Will Need: **Centimeter strips from Materials Card 3.1.1 (eManipulative option: *Number Line Bars*)**

The missing addend approach to subtraction can help students learn subtraction facts because it shows how addition and subtraction are related. The idea that addition and subtraction are inverse operations is important, not only for learning basic subtraction facts, but also in developing the concepts of algebra.

1. The centimeter strips to the right show how to model the missing-addend approach to find the difference E – D.

   E = BluE

   D = Dark green

   ?

   a. That is, what must be added to D to obtain the result of E?_____

   b. Use your centimeter strips to solve the following, using the method in part (a).

   Y + _____ = O   K + _____ = O + R   D + _____ = O + P

   c. Rewrite each of the addition statements in part (b) as an equivalent subtraction statement.

   d. Rewrite each of the following subtraction problems in its missing-addend equivalent and find the solution.

   N – R = _________   K – D = _________   (O + W) – Y = _________

2. The table to the right is the one you should have constructed in Activity 3.1.1. Using the missing addend approach and this table, find the following differences. Show your process.

+	W	R	G	P	Y
W	**R**	**G**	**P**	**Y**	**D**
R	**G**	**P**	**Y**	**D**	**K**
G	**P**	**Y**	**D**	**K**	**N**
P	**Y**	**D**	**K**	**N**	**E**
Y	**D**	**K**	**N**	**E**	**O**

   a. N – Y = _____   b. O – Y = _____

   c. K – P = _____   d. D – P = _____

   Use your centimeter strips to check your answers.

3. In problem #2(a), you used the fact that Y + G = N to find the result N – Y = G. Also, since G + Y = N, we know that N – G = Y. These two addition equations and two subtraction equations together are called **a four-fact family**. Write the four-fact family for each of the remaining parts of #2.

   b.   c.   d.

OBJECTIVE: **Discover properties of multiplication**

You Will Need: **Squares from Materials Card 3.1.2**

The next activity shows two ways to visualize the concept of multiplication, first using a set model and then with a rectangular array. These models can be useful for seeing the connection between multiplication and addition, as well as for discovering properties of multiplication.

1. As part of a classroom project to make a rug for the reading corner, three students in your class have each brought in four carpet squares. To model this situation, use the squares from Materials Card 3.1.2 to make 3 groups of 4 squares.

   What *addition* problem is represented by these sets of squares? repeated addition
   Another interpretation of these sets of squares is the product 3×4. This example shows that the operation of multiplication is simply repeated addition.

   That is, the product of any two whole numbers *m* and *n*, where $m \neq 0$,
   $m \times n =$ ______________________

2. How can you use repeated addition to make sense of a product like 5×0? What is this product?

   so it is 0+0+0+0+0 = 0

3. Using 24 of the squares from Materials Card 3.1.2 and a set model, show four multiplication facts for 24. In other words, find four pairs of **factors** of 24. Record a drawing for each *or* write a description of your sets of squares.

   2x12 3x8 1x24 4x6

4. If the 3 groups of 4 squares in problem #1 are pushed together, a rectangle is formed. This shows the product 3×4 using a **rectangular array approach**.

   4 columns
   3 rows

   Sketch a rectangular array of squares to represent the product 2×5:

5. Suppose that the three students in problem #1 are joined by a fourth student. In order to make the same-sized rug, how many carpet pieces will each of them need to bring to school? ________

   a. Draw a *set* model to illustrate your answer.

   b. Describe the difference between your drawing in part (a) and the one shown in problem #1.

c. Sketch the rectangular array that represents this situation. How does it look different from the one shown in problem #4?

d. Although the sets and rectangular arrays you drew to represent 3×4 and 4×3 *look* different, the total number of squares in each of these is 12. Write a general statement or equation that summarizes this result.

What is the name of this property for whole number multiplication?

6. A group of students decides to make 4 rugs, each 2 squares by 3 squares. We will represent this as $4 \times (2 \times 3)$, where the product in parentheses represents the number of squares. Make a drawing of rectangular arrays to model this situation.

How many squares will they use altogether? $4 \times (2 \times 3) =$ ________

Another group decides to make 3 rugs, each 4 squares by 2 squares. This could be represented by $3 \times (4 \times 2)$ or, using the commutative property, as $(4 \times 2) \times 3$. Make a drawing of rectangular arrays to model this situation.

How many squares will this group use altogether? $(4 \times 2) \times 3=$ ________

Write a general statement or equation that summarizes your findings above.

What is the name of this property for whole number multiplication?

7. Another group of students has two rug portions, one that has 2 rows and 5 columns and another with 2 rows and 3 columns. To determine how many squares they have altogether, one student suggests finding out how many are in each portion and then adding the results together. That is, $(2 \times 5) + (2 \times 3) =$ ________.

A second student suggests they could put the portions together to form a larger rectangular rug. How many rows are in the final rug? ________ How many columns? ________ How does the number of columns relate to the original portions that were combined?

What is $2 \times (5 + 3)$?

Write a general statement or equation that summarizes your findings above.

This is called **the distributive property of multiplication over addition** for whole numbers.

OBJECTIVE: **Use a set model and a measurement model for division**

You Will Need: **Squares from Materials Card 3.1.2**

Just as addition and subtraction are inverse operations, so are multiplication and division. Also, just as multiplication and addition are related, so are division and subtraction. In this activity, you will use both a set model and a measurement model to illustrate these connections between the four operations.

1. A teacher wants to group his 24 students for a cooperative learning activity. Using the squares from Materials Card 3.1.2 to represent the students, demonstrate how you could divide the students into groups as described below. Record a drawing of your result. Then, write a mathematical expression and explain how to obtain each answer from your drawing.

   a. 4 students per group

   24÷4 – 6 there will be 6 groups of 4 children in order to allow all the children to be in equal groups

   b. 3 students per group

   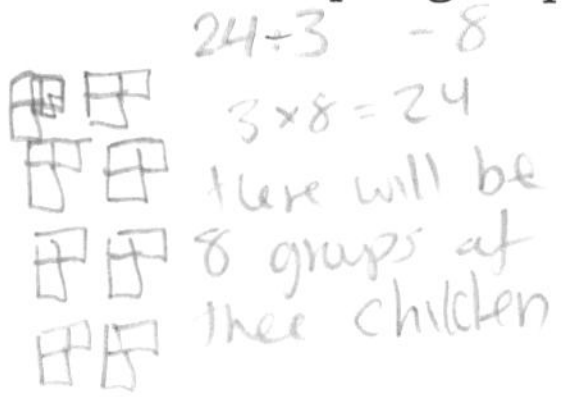

   c. 5 students per group

   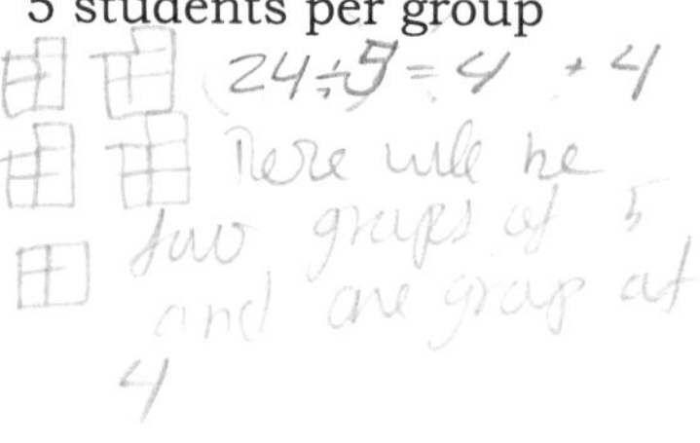

2. For an activity at recess, the teacher wants to group his 24 students into groups of equal size as described below. Using squares, demonstrate how you could divide the class. Record a drawing of your result. Then, write a mathematical expression and explain how to obtain each answer from your drawing.

   a. 4 groups

   b. 3 groups

   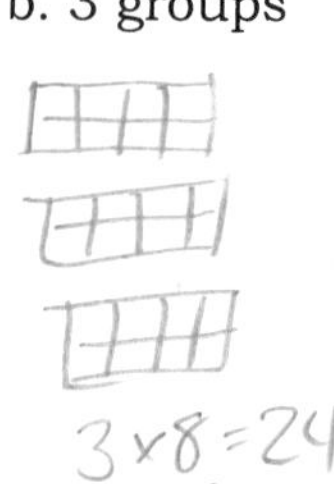

   c. 5 groups

3. Compare the two situations in problems #1 and #2. What are the similarities? What are the differences?

   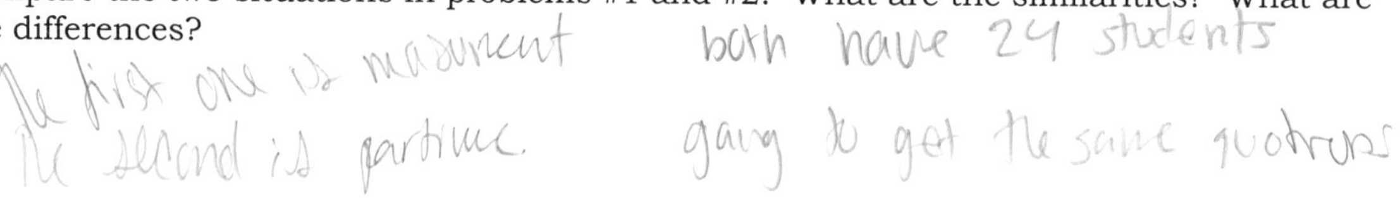

4. In problem #1, you used the **measurement concept** of division because you measured out groups of a certain size and obtained the answer by counting the number of groups formed.

   In problem #2, you used the **partitive concept** of division because you partitioned the objects into a certain number of groups and obtained the answer by counting the number of objects in each group. This concept of division can often be visualized by thinking of dealing cards. You "deal" the squares, one at a time, into a given number of piles until you use all your squares.

Now suppose the students in an elementary class have collected 45 sample squares to make the rug for the reading corner. If the students decide to make a rug with 5 rows of squares, how many squares will there be in each row? Describe how you could determine the answer using the measurement approach. Then describe how you could determine the answer using the partitive approach.

for partitive you would start the five rows and just keep adding one at a time till you run out of rugs

45 ÷ 5 = 9

5. In problems #1 and #2, you used a set model to represent each situation. A number line can also be used to model measurement division.

   a. Let the number line shown below represent 20 kilometers of highway. Create a story problem that would represent 20 ÷ 5 from the *measurement* perspective. Explain how you could obtain the result from the number line.

   -5 -5 -5 -5

   0 5 10 15 20

   you could fit 20 cars on the high way. There were 5 of each color, how many color cars where there?

   There was a high way that was 20 kilometers. There was a gas station every 5 kilometers. How many gas stations are there.

   b. Explain how the number line model you used in part (a) can help to show the connection between division and subtraction.

   you repeate subtraction, working backwards

6. A student asked to find 16 ÷ 2 explained his thought process like this: "Two times what number is sixteen? It's eight, so sixteen divided by two is eight." missing factor

   a. Explain the connection the student is making between multiplication and division.

   they are using missing factor

   b. This student's method could be described as a "missing factor" approach to division. Use a similar method to find the following quotients. Write a sentence to explain your thought process.

   24 ÷ 4 4×6=24 24÷6=4

   0 ÷ 4=0 0×4

   24 ÷ 0 undifined

   Summarize what the last two examples illustrate about division with zero.

7. When a division problem produces a remainder, the result of the division may be reported in various ways depending on the physical situation involved. Give a real-life situation when the result of 23 ÷ 5 would be the number given below.

   a. 4 — an estimation low

   b. 5 — estimating high

   c. 4 r3 — giving an exact answer

   d. 4 3/5 (assuming fractions) — giving answer using fractions

OBJECTIVE: Develop laws of exponents

You have seen that multiplication is repeated addition and division can be viewed as repeated subtraction. In the same way, the concept of exponent can be used to simplify many multiplication problems. In the next activity, you will apply definitions to develop laws for simplifying exponent expressions.

1. If a and m are whole numbers, where $m \neq 0$, then how do you write a^m as a product?

   a×m

   In this expression, the number a is called the **base** and m is called the **exponent**.

2. Complete the following to discover a rule for *multiplying factors with the* *same base*:

   By definition, 3^4 means 3·3·3·3 ______ and 3^2 means 3·3 ______, so $3^4 \times 3^2$ means (3·3·3·3)·(3·3) ______. Write this expression with a single exponent: 3^6 ______.

   Use what you discovered in the previous example to complete this rule.

Let a, m, and n be any whole numbers where m and n are nonzero. Then $a^m \times a^n =$ a^{m+n}

3. Use your rule from #2 to find the following products. Write your answer with a single exponent.

   a. $5^6 \times 5^3 =$ 5^9 ______

   b. $8^2 \times 2^5 =$ 2^{11} ______
   $2^3 \cdot 2^3$ 2^6

   c. Why can't you apply your rule directly when finding the product in part (b)?
   because the base's are not equal

4. Complete the following to discover a rule for *dividing two numbers with the* *same base*:

   Find the missing factor in the following multiplication statement: $5^3 \times$ 5^4 ______ $= 5^7$. This shows that $5^7 \div 5^3 =$ 5^4 ______.

   What shortcut does this example suggest?
   subtraction of the exponents

   Complete this rule.

Let a, m, and n be any whole numbers where $m > n$ and a is nonzero. Then $a^m \div a^n =$ a^{m-n}

5. Use your rule from #4 to find the following quotients. Write your answer with a single exponent.

   a. $7^9 \div 7^3 =$ 7^6 ______

   b. $9^3 \div 3^5 =$ 3 ______
   $3^2 \cdot 3^2 \cdot 3^2$
   $3^6 \div 3^5$

6. Complete the following to discover a rule for *multiplying factors with the same exponent*:

   By definition, 2^3 means $2 \cdot 2 \cdot 2$ and 5^3 means $5 \cdot 5 \cdot 5$, so $2^3 \times 5^3$ means $(2 \cdot 2 \cdot 2)(5 \cdot 5 \cdot 5)$. Write this expression with a single exponent: $(2 \cdot 5)^3$. What property of whole-number multiplication did you need to use for this? less than and multiplcation of whole #'s

   Use what you discovered in the previous example to complete this rule.

   > Let a, b, and m be any whole numbers where m and n are nonzero. Then
   > $a^m \times b^m = (a \cdot b)^m$

7. Now try this example to discover a rule for *raising a power to a power.*

   By definition, a^3 means $a \cdot a \cdot a$ so $\left(5^4\right)^3$ means $(5 \cdot 5 \cdot 5 \cdot 5)^3$. Use one of your previous rules to write this expression with a single exponent 5^{12}.

   Use what you discovered in the previous example to complete this rule.

   > Let a, m, and n be any whole numbers where m and n are nonzero. Then
   > $\left(a^m\right)^n = a^{m \cdot n}$

8. To discover *a rule for zero as an exponent,* first compute each of the following.

   $5^4 = 5 \cdot 5 \cdot 5 \cdot 5$ 625
   $5^3 = 5 \cdot 5 \cdot 5$ 125
   $5^2 = 5 \cdot 5$ 25
   $5^1 = 5$ 5

   As the power decreases by one at each step, what pattern do you notice in the results you found? Extend the pattern you found to find the next step.

   you divide each one by 5
   $5^0 = 1$

   Use what you discovered in the previous example to complete this rule:

   > Let a be any nonzero whole number.
   > Then $a^0 = 1$

   Notice in the statement of your rule, **0^0 is not defined**. To see why, consider these two patterns:

$4^0 = 1$	$0^4 = 0$
$3^0 = 1$	$0^3 = 0$
$2^0 = 1$	$0^2 = 0$
$1^0 = 1$	$0^1 = 0$
According to this pattern, 0^0 should be 1.	According to this pattern, 0^0 should be 0.

   Since these two patterns are not consistent, we say that 0^0 is undefined.

EXERCISE

Proper – T – Practice

Each of the following computations could be simplified by applying one of the properties listed at the right below. For each computation, identify the property or properties that could be used and place the code letter on the line in front of the computation.

____ 1. (96 + 56) + 44

____ 2. (56 × 29) + (56 × 71)

____ 3. } 132 + (51 + 68)

____ 4.

____ 5. 4 × (250 × 29)

____ 6. 21 + (39 + 0)

____ 7. (121 × 49) – (21 × 49)

____ 8. } 8 × (57 × 125)

____ 9.

____ 10. (56 × 1) × 4

11. (46 × 27) + (54 × 27)

Properties of Whole-Number Operations

(C) commutativity for addition

(M) associativity for addition

(E) identity for addition

(H) commutativity for multiplication

(A) associativity for multiplication

(I) identity for multiplication

(T) distributivity for multiplication over addition

(S) distributivity for multiplication over subtraction

Now unscramble these 11 code letters to identify a subject that is an art and a tool, as well as a science.

___ ___ ___ ___ ___ ___ ___ ___ ___ ___ ___

CONNECTIONS TO THE CLASSROOM

1. A student asked to find the sum of the first four counting numbers, 1, 2, 3, and 4, showed the following work:

$$1 + 2 = 3 + 3 = 6 + 4 = 10$$

   a. Is it clear to you from these equations what the student did to obtain her answer? Is her answer correct? Was her thinking likely correct?

   b. Is this student's use of an equal sign incorrect? Explain.

   c. Show another way to write out the process for this problem that makes proper use of equals signs.

2. Before they learn their basic subtraction facts, many students use informal strategies such as adding on, counting up, or counting down. Using concepts from this chapter, describe each of the following students' methods, by name if possible.

   a. One student solved the subtraction problem 9 – 6 by counting down from 9 on his fingers like this: "I start with 9 fingers. Eight, seven, six, five, four, three. When I put down six fingers, I have three left, so nine minus six is three."

   b. Another student solved the subtraction problem 9 – 6 by counting up from 6 on his fingers like this: "I start with six fingers up. Seven, eight, nine. I had to put three more fingers up to get to nine, so nine minus six is three."

   c. One student solved the subtraction problem 9 – 6 by counting down on her fingers from 9 to 6 like this: "I start with nine fingers up and put down fingers until I reach six. Eight, seven, six. I had to put down three fingers so nine minus six is three."

3. A student asks, "I know how to use the distributive property of multiplication over addition for this problem: $2 \times (3 + 4)$. I can do $(2 \times 3) + (2 \times 4)$. I think it works the same way if I have multiplication in the parentheses, like $2 \times (3 \times 4) = (2 \times 3) \times (2 \times 4)$? Is the student correct? Explain your answer.

4. A student says, "I know what to do when we raise a product to a power. The rule is $(a \times b)^m = a^m \times b^m$. Does this work when I'm adding inside the parentheses, like this: $(a + b)^m = a^m + b^m$?" How could you help this student discover whether this is a law of exponents?

MENTAL MATH

In each maze below, trace a path from the start to the triangle so that the sum of the numbers in the squares that you pass through is equal to the result in the triangle at the end. You do not have to pass through all of the squares but you cannot pass through any square more than once.

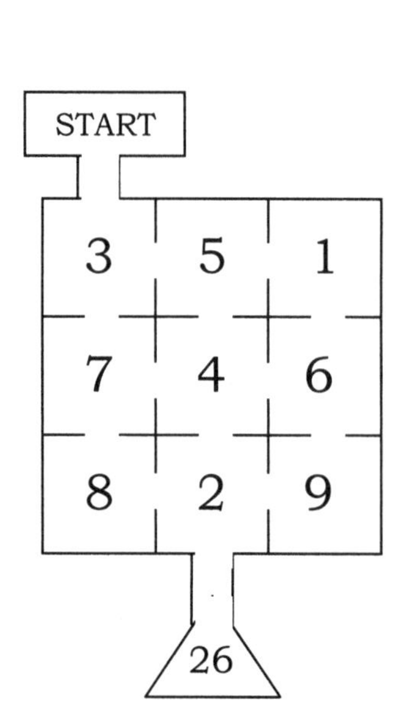

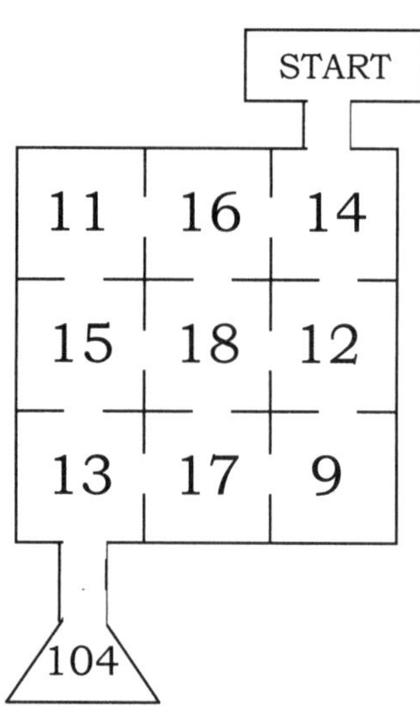

DIRECTIONS IN EDUCATION
Developmentally Appropriate Practices
I Think I Can!

In his book, *Insult to Intelligence*, Frank Smith suggests that young students enter school as "explorers". They are eager to join the "Learners' Club" – to be accepted by the learners and to be like the learners. By the time these children enter school they are already accomplished learners, having attained an understanding of the language and culture within which they live. To maintain this positive attitude toward learning, schools must match the activities provided to the developmental needs of children.

For young children, the partitioning of school subjects into discrete subsets of knowledge further divided into discrete skills and facts is not developmentally appropriate. A focus on skill acquisition is frequently based in correcting student failure rather than celebrating student success. Because the domains of a child's development are interwoven, a sense of failure in one domain (cognitive) may also invade into the development of the others (physical, social, emotional). It is essential for teachers of primary grade children to understand developmentally appropriate practices which extend learning, and also foster feelings of success and a lifelong love of learning.

What are some components of a developmentally appropriate classroom for young children?

The National Association for Education of Young Children (NAEYC) has identified guidelines for developmentally appropriate practices:

- Curriculum goals are designed to promote physical, social, emotional, and intellectual growth. Both curriculum and instruction are responsive to individual differences with a focus on the development of self-esteem, and positive attitudes toward learning.

- Instruction is designed to encourage the "explorer" in every child. The environment is created with materials which are concrete, real, and relevant to children's lives. These materials are arranged in ways to invite interest and participation.

- Interaction with peers in small cooperative groups promotes communication while the children engage in work or play.

- Math content is integrated with other relevant projects as children use math through exploration, discovery, and solving meaningful problems.

- Social skills are developed which enable the child both to interact appropriately with others and to become a self-manager.

- Teachers build on children's internal motivation to make sense of the world and to acquire competence.

- Parents are involved as partners in the learning process. This involvement enhances the home/school transition for students and allows parents to reinforce and support the learning process at home.

- Evaluation of student work is done by the student in combination with the teacher and/or peers. The purpose of evaluation is to guide students to see alternatives, improvements, and solutions.

- Student progress is reported to parents in narrative form based on observational records rather than as letter or numerical grades.

- Children are allowed to progress through the curriculum at their own appropriate pace. The curriculum is adapted to meet the needs of the child. The child is not asked to adapt to the demands of the curriculum.

As you think about developmentally appropriate practices, ask yourself:

1. Do I have an adequate understanding of the development that typically occurs in the lives of primary grade children?

2. How can I plan instruction which addresses required curriculum yet provides for individual student differences?

3. Am I aware of instructional material and manipulative equipment which will meet the needs for concrete exploration and involvement by the learner?

4. What traits might by typical of a lifelong learner? What can I do in the classroom to develop and enhance these traits?

5. How can I focus on individual accomplishments rather than on comparison against the group or against and established norm?

6. At what age can symbols begin to replace concrete objects as students think about problems?

4 Whole-Number Computation — Mental, Electronic, and Written

THEME: Understanding Algorithms

HANDS-ON ACTIVITIES

As a teacher, you will help your students learn many algorithms. In mathematics, an **algorithm** is a step-by-step procedure for performing a computation. Algorithms play a vital role when you perform paper and pencil calculations. Additionally, they may help you when you compute or estimate in your head and perhaps when you use a calculator.

When you understand algorithms and can explain how they work, you will be better able to diagnose student error patterns, as well as help your students find and explain their mistakes. You will also be able to recognize when a student's different way of calculating is correct. Through the activities in this chapter, you will take a deeper look into algorithms, standard and nonstandard.

Consider these different methods for computing the division problem 320 ÷ 5:

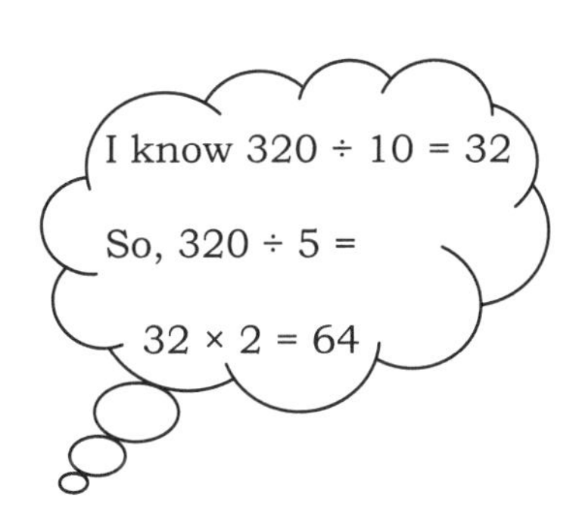

Can you explain why this mental process works?

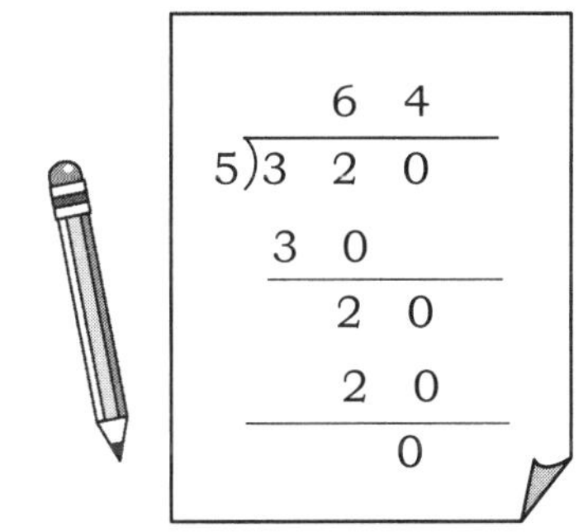

How would you help a student make sense of this familiar process?

What was the error?

OBJECTIVE: **Investigate and Use Mental Math Techniques**

In many day-to-day situations, computing with paper and pencil or a calculator is not practical or possible. Developing the flexibility to compute mentally takes practice. The first activity helps you investigate a few common techniques for doing mental computations.

1. Compute each of the following mentally and write a sentence describing how you thought about each one.

   a. 14 + 29 + 36

   79 – 36 + 14 = 50, 50 + 29 = 79

   14 & 36 are compatible numbers so I added them

   b. 25×101

   275 multiplied 25×100 then added 25

   Distributive

2. Some mental calculations are easier if pairs of compatible numbers are combined and properties of whole number operations (such as the commutative property) are applied.

   a. **Compatible numbers** are *pairs* of numbers that are easy to add, subtract, multiply or divide. In problem #1a, you may have noticed that 14 and 36 *when added* are compatible numbers. Explain why.

   b. Are 14 and 36 compatible numbers when multiplied? Why or why not? Give an example of a pair of 2-digit numbers that are compatible with respect to multiplication.

   c. When computing the sum 14 + 29 + 36 mentally, one could use two properties of whole number addition to group compatible numbers. Identify the properties that were used.

$$14 + (29 + 36) = 14 + (36 + 29) = (14 + 36) + 29 = 50 + 29 = 79$$

3. Identify the property that could be used to do these calculations mentally. Do the calculations in your head and record your thought process for each one.

   a. 25×101

   275
   Distributive addition

   b. 19×21-19×11

4. Use compatible numbers and properties of whole number operations to do the following computations in your head. Record your thought process.

   a. 5×36×2

   b. 82 + 15 + 18 + 41 + 55

5. Another common technique for mental calculation is the method of **compensation**. As an example, consider the subtraction problem 47-18. Since it is easier mentally to subtract 20 than it is to subtract 18, one could first add 2 to 18. Next, to *compensate* for adding 2 to 18 in this subtraction problem, 2 must also be added to 47. The thought process could be explained this way:

$$47 - 18 = (47 + 2) - (18 + 2) = 49 - 20 = 29$$

   a. Write a sentence explaining why the answer to the original subtraction problem 47 - 18 is the same as the answer to the reformulated problem 49 - 20. Sketch a number line to accompany your explanation.

   b. When asked to mentally compute the difference 75 – 27, one of your students does the following. Is this method correct? Would it be appropriate to call this a compensation method? Why or why not?

   $75 - (27 - 2) = 75 - 25 = 50$, then subtract 2 from the answer to get $50 - 2 = 48$

   c. Would it be appropriate to call the following method a compensation method?

$$47 - 18 = (47 + 3) - (18 + 3) = 50 - 21$$

   Would you recommend using this method as shown? Why or why not?

6. Each of the following problems has been changed to make the numbers easier to compute mentally. Explain how you must change the other number in each pair to compensate for the change made in the first number. Find the answer to the original problem using compensation. Record your thought process in each case.

   a. $127 + 38$
   Add 3 to 127.

   b. 14×4
   Divide 14 by 2.

   c. $144 \div 18$
   Divide 144 by 2.

7. Using compensation do the following computations in your head. Record your thought process in each case.

   a. $163 - 46$

   b. $57 + 24$

   c. 28×25

   d. $1700 \div 50$

OBJECTIVE: **Use Computational Estimation Techniques.**

You Will Need: **A calculator**

Often we need to make a quick calculation that does not have to be exact. An estimate is a "ballpark" figure and, for a given problem, there may be several different estimates. Estimation is especially important for developing "number sense" and predicting the reasonableness of answers when using a calculator. Three common techniques are contained in the next activity.

1. Sometimes when estimating it is enough to know an interval (or range) that will contain the answer. As an example, consider the following estimation procedure.

   Find a range estimate for 472 + 718.

   Low estimate: 400 + 700 = 1100 High estimate: 500 + 800 = 1300
   The answer is between 1100 and 1300.

   Explain how these estimates were obtained.

2. In the following problems, a number and a missing number blank are given on the left. Using estimation, fill in the blank with numbers so that the result will fall within the given range. Check your estimates with a calculator and adjust as necessary.

	Start	Range (inclusive)	List estimates tried
a.	36 + _____	(80, 90)	____________________
b.	374 - _____	(235, 245)	____________________
c.	27 × _____	(1200, 1300)	____________________
d.	134 × _____	(2000, 2500)	____________________
e.	856 ÷ _____	(20, 30)	____________________
f.	2406 ÷ _____	(90, 120)	____________________

3. If an estimate is needed, we can round the numbers *before computing* to replace complicated numbers by simpler numbers. Each number may be rounded up, rounded down, or rounded to the nearest specified place value, depending on the context of the problem. Round each of the following to the specified place value.

   a. 1058 (nearest thousand) __________
   b. 435 (nearest ten) __________
   c. 1245 (nearest hundred) __________

   d. Use a number line to explain why the answer to part c is *not* 1300.

4. Use rounding to estimate each of the following. Show your process in each case.

   a. 624 - 289 - 132 ≈ _______ b. 47 × 28 ≈ _______ c. 472 ÷ 46 ≈ _______

5. Rather than rounding to a specified place value, one can round either up or down to compatible numbers, as shown in the next example.

$$3418 \div 84 \approx 3600 \div 90 = 40$$

Round to compatible numbers to estimate each of the following.

a. $74 \times 98 \approx$ ________________ b. $56 \times 18 \approx$ ________________

c. $248 \div 46 \approx$ ________________ (Can you do this one in two different ways?)

Explain why the descriptor 'compatible numbers' is appropriate in each of the above.

6. Use estimation to answer the following questions.

a. Estimate the total travel cost. Tell which estimation technique you used and why you chose it.

Gas: $16.28, Hotel: $57.83, Food: $25.57

b. Estimate the total cost of 3 items that cost $8.43 each.

c. When might you estimate like this?

Actual prices are: $6.39, $3.75, and $5.98.
The total will be $7 + $4 + $6 ...about $17, maybe a little less.

d. In determining an evacuation zone, a scientist must estimate the distance that lava from an erupting volcano will flow. Should she overestimate or underestimate? Explain.

e. Li is sent to the donut shop with $5.00 to buy 5 dozen donuts for the office staff. He notices that the price is listed as $1.20 per dozen, but that a special is advertised for $.25 off each dozen when you buy 5 dozen or more. Will he be able to buy 5 dozen of the donuts? Li lives in Oregon, where there is no sales tax. How much will he pay the salesclerk? Discuss when an estimate would be appropriate in this scenario and when an exact calculation is necessary.

OBJECTIVE: **Explore Features of your calculator**

You Will Need: **A calculator**

Various calculators have different features. The next activity helps you explore some of these features.

1. a. Enter the digits 1, 2, 3, etc. until the display is full. How many digits does the display hold? What happens if you try to enter more digits?

   b. Enter a change of sign [+/-] or [+ ↔ -]. Does that alter the number of digits displayed?

   c. Clear the display ([ON/C] or [ON/AC] should do that). Next, enter a decimal point followed by the digits 1, 2, 3, etc. until the display is filled. How many digits to the right of a decimal point will your calculator hold?

   d. What is the largest positive number your calculator will hold in its display? What is the smallest positive number?

2. You have already seen that [ON/C] or [ON/AC] is one way to clear the calculator display. This "All Clear" key is generally the most comprehensive way to clear the calculator because it clears more than just the display. Other keys that may appear on your calculator include [CE], meaning "clear entry" [CE/C] or [Backspace].

   a. Using those keys which appear on your calculator, try to find ways to make the following corrections without starting again.

Actual Problem	Your Mistaken Keystrokes	Correction
256 × 719	256 × 716	
538 × 923	538 × 623	
762 × 516	762 + 516	

   b. For those keys on your calculator, summarize what each clears.

3. Does your calculator have a memory? Look for keys like [M+], [x↔M], or [STO].

   a. To store a number, enter a number, say 100, and press [M+], [x↔M], or [STO]. What does your display look like?

   Is there an M displayed to indicate there is a nonzero number in memory?

   Enter the number 55 and store it. What is now displayed?

   b. To recall the number stored in memory, press [MRC], [MR], or [RCL].

   Next, do some simple calculations such as 25 [×] 32 [=] and 478 [+] 918 [=].

   Now recall the memory. Has it remembered the number that was in memory?

c. Eventually you will probably want to start a new computation in memory and/or clear out memory entirely (so that the M is no longer displayed). Find at least two ways to clear the memory on your calculator. Describe the ways.

d. After entering the following, predict what has been stored in memory. Check your predication by pressing recall.

3 [×] 5 store	Prediction: _____	Check: _____
3 [×] 5 [=] store	Prediction: _____	Check: _____

e. Begin a new computation by entering 60 into memory. Then enter the number 45 and press [M+] or [SUM]. Now recall the contents of memory. What has happened?

f. Enter the number 15 and press [M-] or press [+/-] and [M+]. What happens to the contents of memory?

g. Compute the following using the memory: $(5 \times 6) + (10 \times 13) - (7 \times 9)$

4. How would you find 2^7 using your calculator?

Enter the base, in this case 2, and then press the exponent key on your calculator. Look for [y^x] or [^]. Next enter 7. What does your calculator display show? Do you have to press [=] to see the answer?

Does your calculator have [x^2]? If so, list the keystrokes you need when finding 14^2 using this key.

OBJECTIVE: **Explore operations of your calculator**

You Will Need: **A calculator**

To understand how and when to use calculators, it is important to know how they operate and what their capabilities and limitations are.

1. a. Enter each of the following keystroke sequences on your calculator and describe what happens.

[AC]	[+]	3	[=]	[=]	[=]	____________
3	[+]	[=]	[=]	[=]	[=]	____________
3	[+]	3	[=]	[=]	[=]	____________
5	[+]	3	[=]	[=]	[=]	____________

If any of these produce a sequence counting by 3, your calculator has an automatic constant function. If not, you may have to use 3 [+] 3 [+] 3, etc.

b. How can you get your calculator to reproduce the following sequences? List the keystroke sequences.

5, 10, 15, 20, . . .   8, 16, 24, 32, . . .   3, 8, 13, 18, . . .

c. How can you get your calculator to count by the even numbers? By the odd numbers?

2. a. Record the results your calculator gives to the following expressions:

4 [+] 6 [×] 3 [=] ________   12 [×] 2 [-] 3 [×] 5 [=] ________

b. Does your calculator give the results 30 and 105?
Some do. To get these results, in what order did the calculator perform the operations? (Watching the display as you slowly enter the operations may give you a clue.)

Calculators that perform operations in this manner, namely as the operations are entered, have what is called **arithmetic logic**.

c. Does your calculator give the results 22 and 9? ________If so, in what order were the operations performed to get these results?

Calculators that perform operations in this manner, namely according to the usual mathematical convention for *order of operations*, have **algebraic logic**.
To avoid confusion, mathematicians have developed this convention for the order of operations: within the innermost parentheses, calculate exponents, then perform multiplication and division from *left to right*, and lastly perform addition and subtraction from *left to right.*

A mnemonic device (or memory aid) often used to help students remember the order of operations is: "Please excuse my dear aunt Sally." Make up your own sentence that could be used as a mnemonic device for the order of operations and write it here.

d. Does your calculator have arithmetic logic or algebraic logic?

e. Since parentheses have the highest priority, calculators often have [( ] and [ )] keys. Insert parentheses as needed in the following expressions to obtain the result shown.

$12 \div 3 + 3 \times 5 = 19$ $25 + 20 \div 5 - 2 \times 4 = 21$ $3 + 2^2 + 12 \div 2 \times 3 = 25$

$12 \div 3 + 3 \times 5 = 10$ $25 + 20 \div 5 - 2 \times 4 = 1$ $3 + 2^2 + 12 \div 2 \times 3 = 9$

$12 \div 3 + 3 \times 5 = 35$ $25 + 20 \div 5 - 2 \times 4 = 60$ $3 + 2^2 + 12 \div 2 \times 3 = 33$

$25 + 20 \div 5 - 2 \times 4 = 28$

f. How could you obtain the previous results without using parentheses (on a calculator with either arithmetic or algebraic logic)?

3. Sometimes you will try to do an operation that your calculator cannot do. What happens, for example, when you enter the following?

12 [÷] 0 [=] ________

What is wrong with the operation you tried to do?

Which clearing key on your calculator do you need to use to clear out the result?

4. In Activity 4.1.3, you discovered how many digits can be displayed on your calculator. What happens when the results of a computation are either too large or too small? Perform the following calculations to find out.

   98989898 [×] 19191919 [=] ________   0.0000003 [×] 0.0000005 [=] ________

   a. Does your calculator give an error message for overflow or underflow?

   b. Does your calculator give answers like [1.8998 15] and [1.5 -13]?
   These answers are expressed in scientific notation and represent the product of a decimal number between 1 and 10 (10 not included) and a power of 10. For example, [1.8998 15] means 1.8998×10^{15} or 1,899,800,000,000,000 and [1.5 -13] means 1.5×10^{-13} or 0.00000000000015. Notice that these are only approximate results to the original problem.

   c. How could you use your calculator to find the exact answer for 98,989,898 × 19,191,919? Explain your steps using what you know about place value and the usual multiplication procedure.

OBJECTIVE: **Explore and reinforce concepts using mental math and a calculator**

You Will Need: **A calculator and a partner**

It is fun and challenging when students practice computing mentally and then use a calculator to check their work. Try the following games with a partner.

1. The following are sequences of numbers. Mentally decide what operation and number needs to be put in the blanks to obtain the number given on the next line. The first blanks have been filled in as an example. Use a calculator to verify your solutions.

a.			b.		
52361	–	2000	78	__	______
50361	__	______	70	__	______
50360	__	______	7	__	______
50300	__	______	70000	__	______
503	__	______	70200	__	______
6503			7020		

c. Make up some similar sequences and have your partner do them.

2. Play the following game with your partner or in a larger group. If desired, a 10-second time limit can be established for responses.

   The first player enters a two-digit number, say 37, and passes the calculator to the second player. The first player then calls out a new two-digit number, say 53. The second player must obtain the new number by adding to or subtracting from the number displayed, here by adding 16. The second player then passes on the calculator and calls out a new number.

3. One person enters a number into the calculator, say 75,198, and hands it to the other person, asking them to change one of the digits, say 5, to a 0 by subtracting an appropriate number. That person identifies verbally what is being subtracted. Exchange roles and repeat. As a variation, you could ask the other player to change the chosen digit to something other than 0, say change the 5 to a 2.

4. One player selects a "secret" two-digit number. He or she then performs a sequence of operations, known to both players, and hands the result to the second player. The object for the second player is to determine the original number.

   For example, ___ [×] 2 [=] [+] 15 [=] [÷] 3 [=] 19.

   (Note: You might be able to omit [=] depending on your calculator's logic.)

5. Using the digits 2, 3, 4, 6, 9, complete the following exercises so that you have the largest possible answers and the smallest possible answers. Check with a calculator and revise if necessary.

   a. Largest possible answers

   □□ + □□□ =

   □□ × □□□ =

   □□□□ × □ =

   □□□ - □□ =

   b. Smallest possible answers

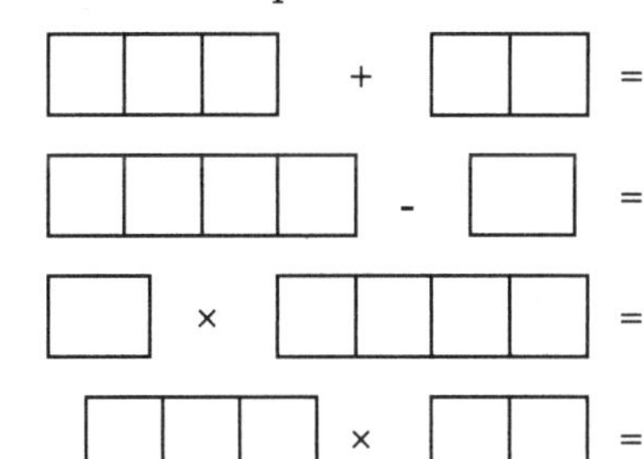

c. Choose 5 different digits and repeat parts (a) and (b).

OBJECTIVE: **Explore the standard multiplication algorithm using base ten pieces**

You Will Need: **Materials Card 4.2.1 (Physical Manipulative option: *Base Ten Pieces*; eManipulative option: *Rectangle Multiplication*)**

In mathematics, **algorithms** are step-by-step procedures you follow such as when you add a column of numbers or do a long division problem. The following activity will help you gain further insight into 'how-to-do' multiplication using our standard algorithm.

1. You will recall from Chapter 3 that a rectangular array was used to model multiplication. For each of the following rectangles, give the factors that are multiplied and the product.

   a.  b. 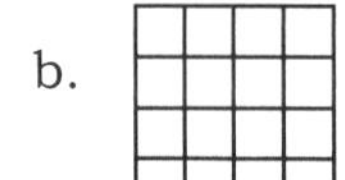

2. A rectangular array can also model the product of larger numbers. Consider the following rectangle made up of base ten pieces (flats, longs and units).

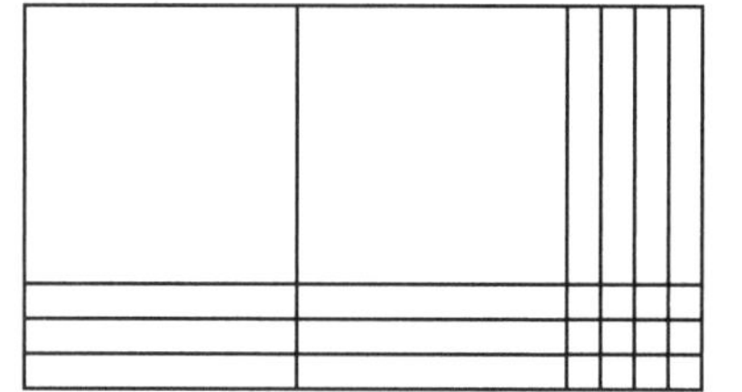

   a. Label the dimensions of this rectangle.

   b. What multiplication problem is modeled here?

   c. When the pieces representing the area are combined what number does the area represent? Explain.

3. If we separate the tens and units of the two factors, and separate the pieces similarly, four rectangles are formed.

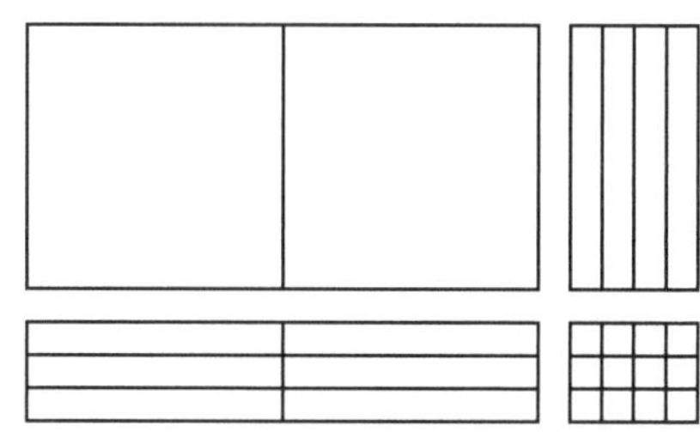

   a. Label the dimensions and area of each of these rectangles.

   b. In the *intermediate* multiplication algorithm we write four partial products, as shown here. Explain how the algorithm is related to these rectangles.

$$\begin{array}{r} 24 \\ \times\ 13 \\ \hline 12 \\ 60 \\ 40 \\ 200 \\ \hline 312 \end{array}$$

   c. Write out the *standard* pencil-and-paper algorithm for 24 × 13. Then make a sketch of base piece rectangles that model the partial products in this algorithm. Clearly label your drawing.

4. Sketch rectangles to compute the following products. In each case, show how the rectangles relate to the standard multiplication algorithm.

   a. 52×14 b. 26×34 c. 17×24

ACTIVITY 4.2.2

OBJECTIVE: **Investigate the algorithm for long division using base ten pieces**

You Will Need: **Base ten pieces from Materials Card 4.2.1 (Physical Manipulative option: *Base Ten Pieces*; eManipulative option: *Rectangle Division*)**

The usual long division algorithm involves many steps and is complex. In the next activity, we will use base ten pieces to look at a concrete model that parallels the pencil-and-paper algorithm.

1. Consider the division problem $23 \div 2$.

   a. Write out the usual pencil-and-paper long division algorithm for this problem. As you do it, record your thinking next to your work. That is, what thought do you have at each step?

   $2\overline{)23}$

   b. Using the fewest possible base ten pieces, represent 23. Now divide those pieces into two equivalent piles. Do you have any leftover units? Draw a sketch of this.

   Each pile has _____ longs and _____ units, with _____ units left over. The number of leftover units is called the **remainder**.

   $23 \div 2 =$ _____ longs_____ units, remainder _____ or $23 \div 2 = 11$, remainder 1.

   c. How does the method in part (b) relate to the standard long division algorithm shown in part (a)?

d. Now write out the long division algorithm for the 23 ÷ 3. As you do this problem, write out your thought process.

e. Divide the base ten pieces representing 23 into 3 parts. Make a sketch of this and show any remainder.

f. Why is this problem more difficult than 23 ÷ 2? Discuss this difficulty both in terms of the algorithm and the base piece model.

2. An intermediate algorithm for long division can help students to make sense of the standard algorithm. As an example, consider the division problem 653 ÷ 4. Model each step with your base ten pieces as described. Compare your model to each step of the intermediate algorithm shown to the right.

Represent 653 using as few of your base-ten pieces as possible.

Beginning with the largest pieces, put the greatest number of those pieces you can into 4 groups. What is the value of the pieces you have in each group? _____ What is the total value of the pieces you used to make these 4 groups? _____ What is the value of the pieces that are left? _____

$$\begin{array}{r} 1 \\ 4\overline{)653} \\ \underline{400} \\ 253 \end{array}$$

Make any necessary exchanges so you can repeat this process with the next largest pieces. Put the greatest number of those pieces you can into 4 groups. What is the value of the pieces you have put in each group at this step? _____ What is the total value of the pieces you used at this step? _____ What is the value of the pieces that are left? _____

$$\begin{array}{r} 16 \\ 4\overline{)653} \\ \underline{400} \\ 253 \\ \underline{240} \\ 13 \end{array}$$

Exchange as needed and repeat the process with the last pieces. What is the value of the pieces you put in each group at this step? _____ What is the total value of the pieces you used at this step? _____ What is the value of the pieces that are left? _____

$$\begin{array}{r} 163 \\ 4\overline{)653} \\ \underline{400} \\ 253 \\ \underline{240} \\ 13 \\ \underline{12} \\ 1 \end{array}$$

In the margin at the right, write out the standard long division algorithm for this problem. How does this intermediate algorithm differ from the standard division algorithm?

Write out the intermediate algorithm shown above for 653 ÷ 3. Model this with your base pieces, if you need to.

3. The following shows the first step of the standard algorithm along with the base ten pieces for that step.

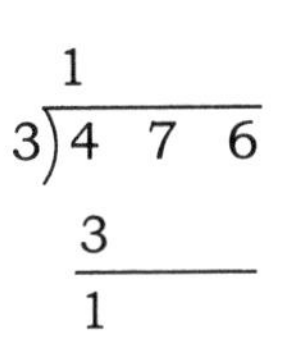

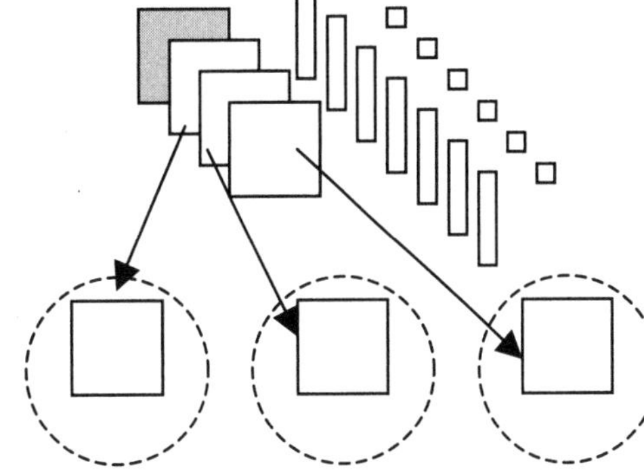

Think: "1 flat in each of the 3 groups. I used 3 flats with 1 flat left over."

a. In terms of the base-piece model, what does the "3" written beneath the 4 represent? Why do we write 3 in this position?

b. In terms of the base-piece model, what does the "1" written beneath the 3 represent?

c. What do we do with the one flat that is unused?

d. The next step in the algorithm is "Bring down the 7." How is this represented with the base ten pieces?

e. Finish this division problem and show each step as in the example.

OBJECTIVE: **Use a chip abacus to add and subtract in base three**

You Will Need: **Materials Card 4.3.1 (eManipulative option: *Chip Abacus*)**

The concepts of place-value and operations with whole numbers were introduced in Chapter 3 using base ten pieces. In the next activity, you will use a chip abacus, a more abstract model, to examine addition and subtraction algorithms. You will work in base three so you can experience the challenges your future students may face when learning these concepts in base ten.

For this activity, you will use a **chip abacus** which can be made out of paper or cardboard. The columns represent the place values – depending on the base. We represent numbers by putting chips in the appropriate columns.

Example: 345 is represented as

100	10	1
• • •	• • • •	• • • • •

on a base ten abacus.

Notice that the chips used in each column are the same size and shape. They do not represent different values due to their shape and size as did longs and flats. The chips represent different values based on which column they are placed in.

1. a. How would the columns of a base three abacus be labeled, using *base three* numerals? Sketch the *base three piece* associated with each of these place values.

   b. How are these base three place values expressed using *base ten* numerals?

   c. What is the largest number of chips that can appear in any single column of a base three chip abacus? Why?

2. Write the base three numerals that are represented by these diagrams.

1000_3	100_3	10_3	1
	• •	•	• •

____

1000_3	100_3	10_3	1
•		•	• •

____

3. Sketch a chip abacus to represent each of these numerals.

   a. 120_{three} b. 12_{three} c. 1002_{three}

4. A chip abacus can be used to represent addition problems, such as $112_{three} + 212_{three}$. First represent both numbers (addends) on the abacus.

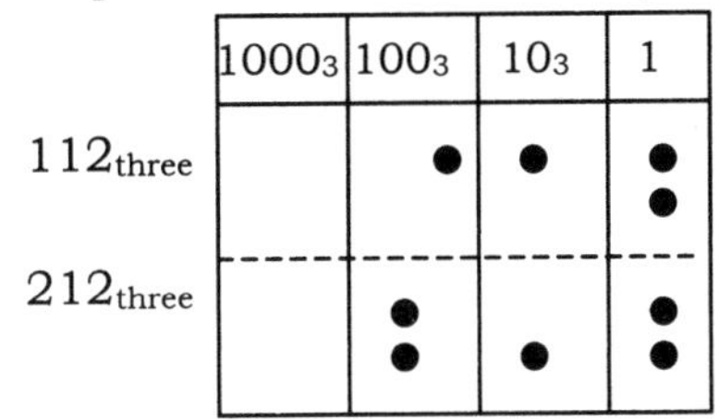

To compute the sum we combine the chips in each column and make exchanges as needed.

a. Show all of the necessary exchanges.

b. In part (a), you likely worked column by column from right to left. Repeat the addition process only this time begin with the left-hand column and work to the right until all exchanges have been made. Did you get the same answer?

c. Does it matter which column you begin in? Is addition easier if we always work from right to left? Explain.

d. Use the chip abacus to find the sum $112_{three} + 212_{three}$. ________________

e. Now try this addition problem. Compute the sum on your chip abacus and show each exchange. Be sure to label your drawings so it is clear how you obtained the answer using the abacus.

$122_{three} + 202_{three} + 12_{three} =$ ________________

5. The chip abacus can also be used to represent a subtraction problem, such as $122_{three} - 12_{three}$.

a. In this example, we use the *take away approach* to subtraction.

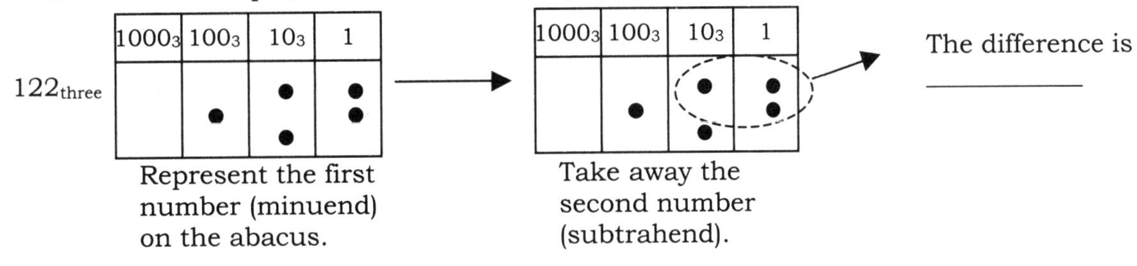

b. Now show how to use a chip abacus to model the difference $122_{three} - 12_{three}$ using the *comparison approach* to subtraction. How are the diagrams you draw different from the diagrams in part (a)?

6. Next consider the subtraction problem $211_{\text{three}} - 12_{\text{three}}$.

   a. What chips must be removed from the abacus to show the difference $211_{\text{three}} - 12_{\text{three}}$?

1000_3	100_3	10_3	1
	● ●	●	●

   b. Explain why we cannot subtract 12_{three} from 211_{three} directly.

   c. Starting in the units place show all the exchanges that must be made.

1000_3	100_3	10_3	1
	● ●	●	●

   d. Lastly, show the take away process for each column. Record the difference here.

7. Compute the following *without* using your abacus.

   a.
$$\begin{array}{r} 1\,2\,2\,1_{\text{three}} \\ +\ 2\,1\,2\,2_{\text{three}} \\ \hline \end{array}$$

   b.
$$\begin{array}{r} 2\,2\,1\,1_{\text{three}} \\ -\ 1\,2\,1\,2_{\text{three}} \\ \hline \end{array}$$

   Did you use an algorithm for these? If so, how is the algorithm related to how you usually add or subtract?

OBJECTIVE: **Use the chip abacus to multiply in base three**

Your Will Need: **Chips and Chip Abacus from Materials Card 4.3.1 (eManipulative option: *Chip Abacus*)**

A chip abacus can also be used to look more closely at the operation of multiplication. As you do the next activity, think about the algorithm *you* use to do multiplication.

1. a. Show 121_{three} on your abacus:

   b. Put two *more* 121_{three}'s on your abacus and make all necessary exchanges.

1000_3	100_3	10_3	1

   c. Compare the result in part (b) with 121_{three}. How are they alike? How are they different?

   d. Since we can think of multiplication as repeated addition, the result obtained in part (b) was three times 121_{three}, or $10_{\text{three}} \times 121_{\text{three}}$. Can you suggest a shortcut for multiplying any number (in base three) by 10_{three}? by 100_{three}?

2. Find the product $2_{three} \times 102_{three}$ by putting 102_{three} on your abacus twice and then making exchanges. Record a sketch of this.

   $2_{three} \times 102_{three} =$ __________

3. Now see if you can use what you learned in problems 1 and 2 to finish the following computation. Identify the two *properties* that can help you simplify.

   $20_{three} \times 102_{three} = (2_{three} \times 10_{three}) \times 102_{three} =$

4. a. The product $21_{three} \times 102_{three}$ can be computed using the following process. Identify the property that is used in this sequence of equations and finish the problem to find the product. Check this problem with your chip abacus if you need to.

   $21_{three} \times 102_{three} = (1_{three} + 20_{three}) \times 102_{three} = (1_{three} \times 102_{three}) + (20_{three} \times 102_{three}) =$

   b. How is this related to the usual multiplication algorithm?

5. Use a process similar to that in problem 4 to find these products:

   a. $12_{three} \times 111_{three} =$ __________

   b. $21_{three} \times 111_{three} =$ __________

   c. $21_{three} \times 201_{three} =$ __________

   d. $12_{three} \times 120_{three} =$ __________

OBJECTIVE: **Use a lattice to add in base three**

The algorithms we usually use are not the only algorithms for performing operations. An alternative to the standard algorithms for addition and multiplication is called the lattice method. Exploring non-standard algorithms can help students to better understand and appreciate the standard procedures.

1. Examine this addition procedure for the problem $122_{three} + 221_{three}$:

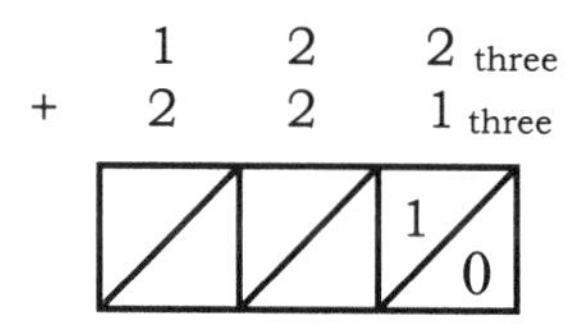

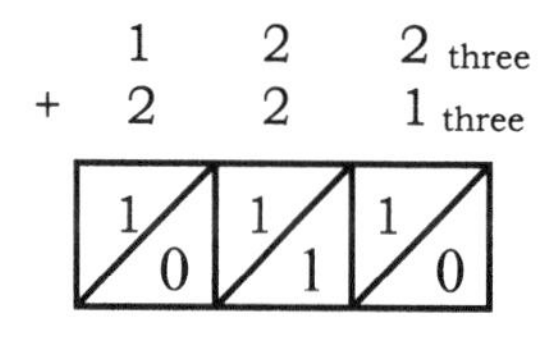

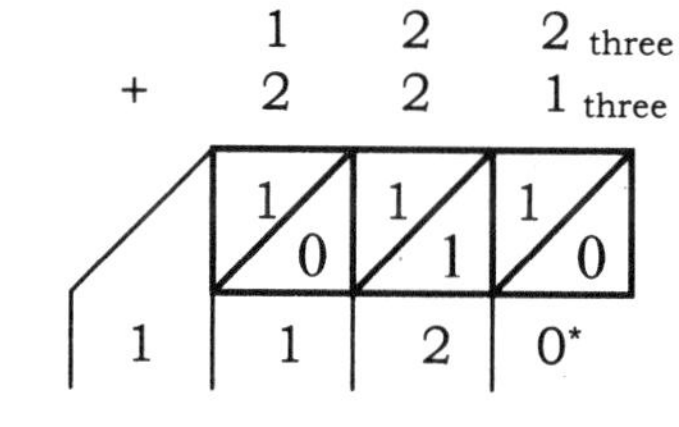

* This sum was obtained by adding down the diagonals.

a. Try these addition problems using the lattice method. Remember that no subscript indicates base ten.

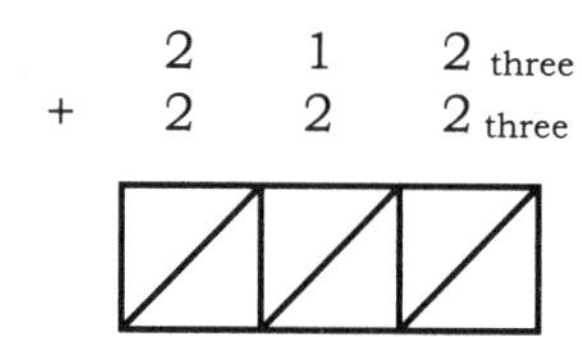

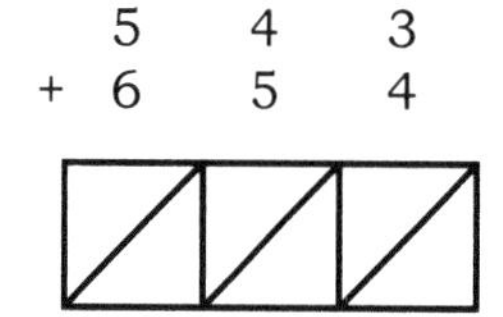

b. Describe how this method relates to the standard addition algorithm. How is it different? How is it similar?

2. Next examine this multiplication procedure for 34×56:

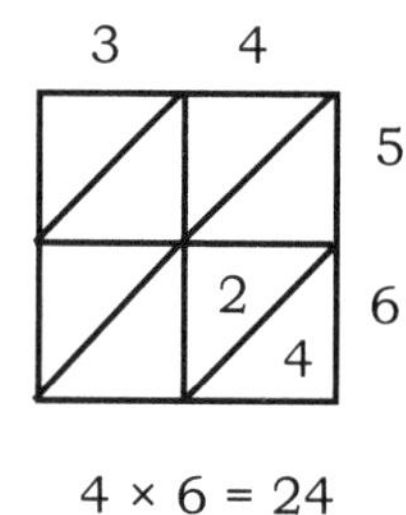

$4 \times 6 = 24$

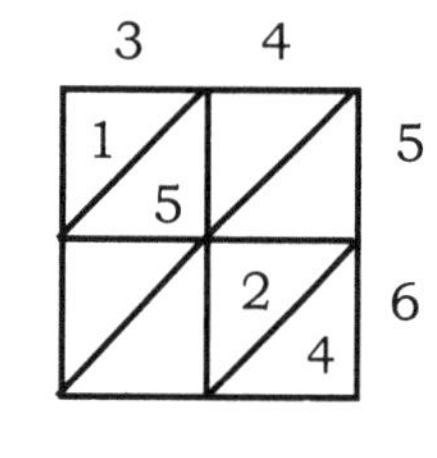

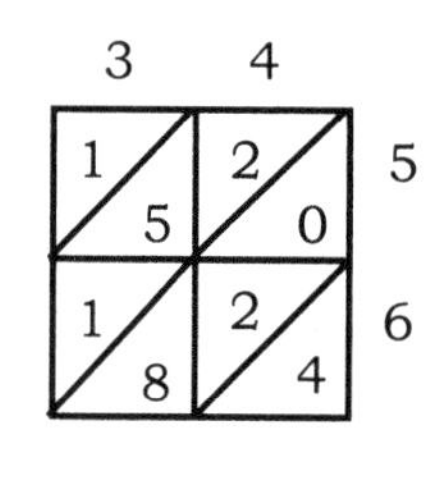

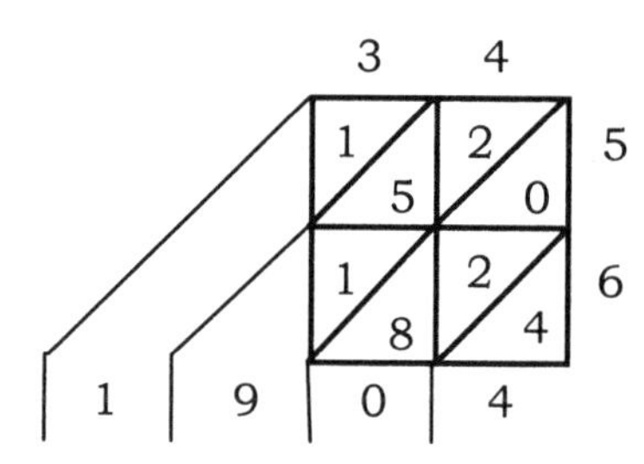

By adding down the diagonals and regrouping, we see that $34 \times 56 = 1904$.

a. Try the lattice method for the following multiplication problems:

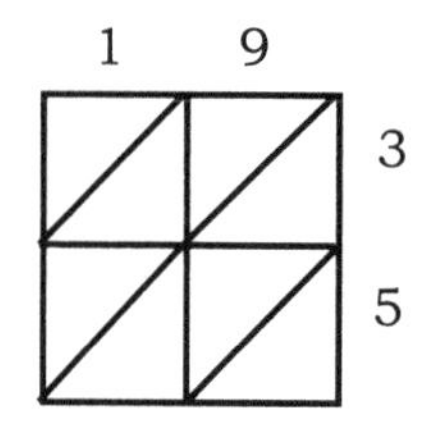

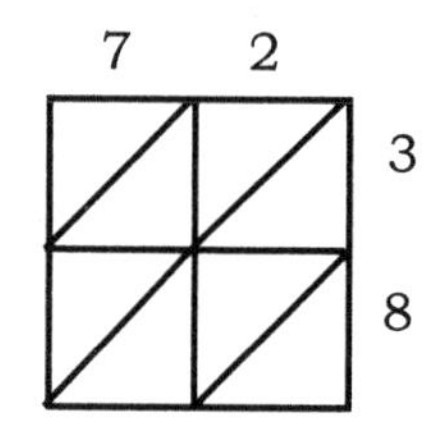

b. Use the lattice method with 3-digit numbers to compute the product 327 × 408. It is left to you to draw the lattice.

3. Use the notion of place value to explain *why* the lattice method for multiplication works.

EXERCISE

Want Ads

Needed: Doctors who know how to operate.
Tools Available: 3, 5, 6, 7, 9
Skills Required: +, – , ×, ÷, and ( )

First, choose three different tools to complete the following operations. Remember the order of operations. There may be several correct answers.

9 + 6 – 5 = 10	□ × □ ÷ □ = 2
□ × □ – □ = 9	□ × □ – □ = 15
□ + □ × □ = 21	(□ + □) × □ = 60

Now that you are getting proficient, you choose four different tools as well as any of the operations (don't forget parentheses can be used). Fill in the triangles (numbers given above) and place operations and parentheses between the triangles to make a true equation.

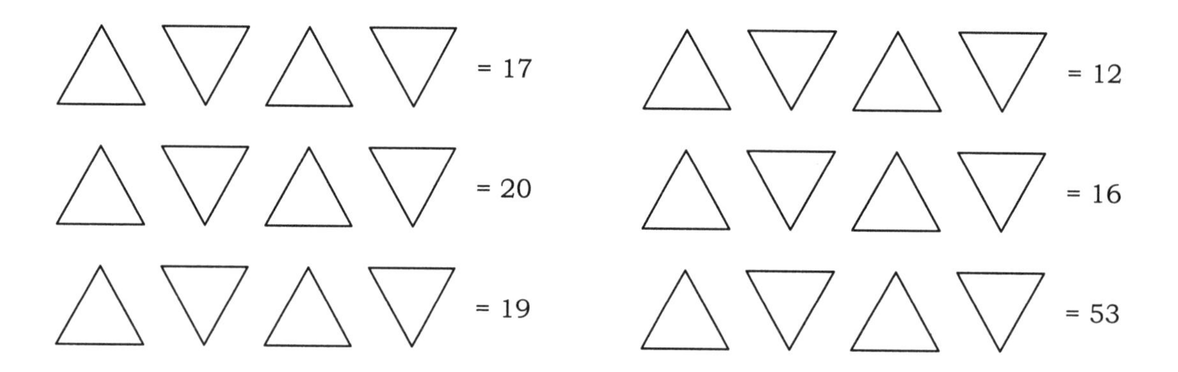

CONNECTIONS TO THE CLASSROOM

1. When asked to mentally compute the sum 35 + 41, a student described his thought process as follows:

   "I first add 30 and 40 to get 70. Next I add 5 and 1 to get 6. So the answer is 70 + 6 or 76."

   a. Describe this student's method in a step-by-step fashion.

   b. Write out this solution as a series of equations from left to right showing what properties were used at each step.

2. A student asks "How can I use my calculator to find the whole number quotient and remainder when 6,289,214 is divided by 92,365?"

   Without simply telling the student which steps to follow, explain how you could help this student. Can you think of some questions you might ask the student to help lead them to their own understanding of this process?

3. When asked to compute $3 + (2^2 + 8) \div 2 \times 3$, a student showed the following work:

$$3 + (2^2 + 8) \div 2 \times 3 = 3 + 12 \div 6 = 3 + 2 = 5$$

   Is this student's work correct? If not, what are some questions that you can ask the student to elicit a correct solution?

4. a. The following is a sample of a student's paper. Find the error pattern in the work and complete the last two subtraction problems as this student would using this method.

$$\begin{array}{r} 1\,\overset{6}{\not7}\,{}^{1}9 \\ -\quad 2\;6 \\ \hline 1\;4\;1\;3 \end{array} \qquad \begin{array}{r} 2\,\overset{3}{\not4}\,{}^{1}7 \\ -\quad 2\;5 \\ \hline 2\;1\;1\;2 \end{array} \qquad \begin{array}{r} 2\,\overset{4}{\not5}\,{}^{1}6 \\ -\quad 3\;9 \\ \hline 2\;1\;7 \end{array} \qquad \begin{array}{r} 3\;7\;2 \\ -\quad 4\;8 \\ \hline \end{array} \qquad \begin{array}{r} 4\;8\;6 \\ -\quad 7\;2 \\ \hline \end{array}$$

   b. What instructional procedures and/or hands-on materials might you use to help the student with this problem?

5. When asked to solve these long division problems, a student showed the following work:

$$\begin{array}{r} 37\ R\ 1 \\ 4\overline{)1229} \\ \underline{12} \\ 29 \\ \underline{28} \\ 1 \end{array} \qquad \begin{array}{r} 92\ R\ 3 \\ 7\overline{)6317} \\ \underline{63} \\ 17 \\ \underline{14} \\ 3 \end{array} \qquad \begin{array}{r} 86\ R\ 2 \\ 6\overline{)4838} \\ \underline{48} \\ 38 \\ \underline{36} \\ 2 \end{array}$$

a. Show how to use compatible number estimation to check the reasonableness of each of the student's answers.

b. Describe the error pattern this student is using. Why might he be using such a procedure?

c. Show how you could use base ten pieces to help this student understand his errors.

MENTAL MATH

Find the correct path to the calculator.

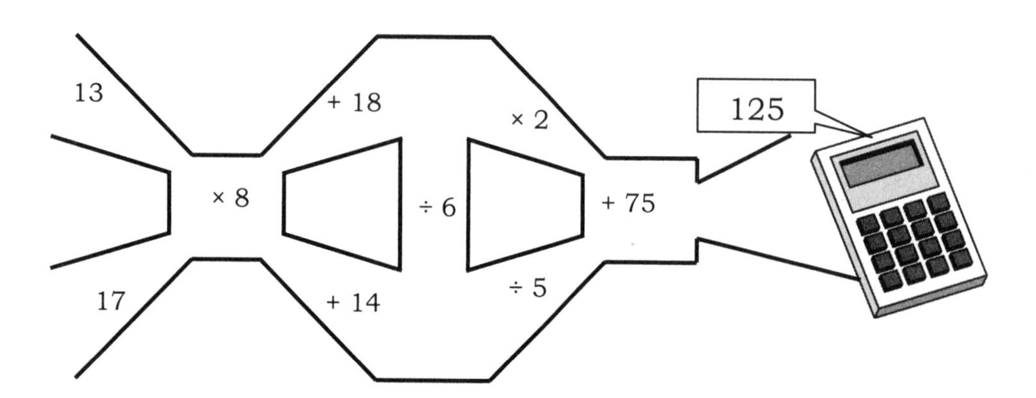

DIRECTIONS IN EDUCATION
Estimation/Mental Math
It's About...

Estimation and the ability to perform mental computation are useful, practical skills. We all make estimates each time we answer questions such as the following:

- How long will it take to drive to the airport?
- Do I have enough cash to buy everything I need?
- How much food should I prepare for dinner?
- How much money should I budget for extras?
- How many people were at the ball game?
- How much wallpaper should I order?

Some studies have found that adults use estimation on 75% of the non-occupational calculations they perform. With increasing reliance on calculators for exact computation, the skill of estimation to predict or assess the result displayed on the calculator is essential.

Much of classroom instruction in mathematics is focused on exact computation and on arriving at the "right answer". Students often receive little or no instruction and practice in the skills of estimation and approximation. Should it simply be left to chance that students will learn effective techniques to use mathematics in the ways common to mainstream society?

Estimation is the process used in answering such questions as, "About how many...?" or "About how much...?" **Mental math** is the process of performing exact computations mentally.

Reasons to teach estimation and mental math:

- They are basic skills in the application of mathematics.
- They provide a form of self-checking skills to take into the real world.
- They comprise an experiential background from which errors can be explained.
- Number sense is reinforced and enhanced through practice in estimation.
- Experience with estimation helps students to understand mathematical concepts.
- Estimation encourages flexibility in reasoning about numbers.
- Students develop a "feel" for very large or very small number and for measurements through experiences with estimation.
- Estimation enhances problem-solving skills and the ability to use certain problem-solving strategies.
- Estimation activities aid in the development of a positive attitude toward mathematics by helping to eliminate the "right answer syndrome".

What kinds of estimation skills should be taught?

- Skills dealing with measurement:
  - Ability to use units of measure.
  - Ability to measure a given object using a given unit.
  - Ability to match an object or entity to a given measurement.
  - Ability to check estimates by using measuring devices.

- Skills dealing with relative size:
  - Understanding of the concepts of great-than, less than and betweenness.
  - A "feel" for numbers and orders of magnitude.

- Skills dealing with questions of computation:
  - Understanding place value which permits leading-digit estimation.
  - Rounding skills used in conjunction with addition, subtraction, multiplication, and division.
  - Averaging skills which provide for grouping of similar numbers in a problem.
  - Using a reference point to determine in an answer will be over or under that specified point.
  - Finding a range of answers. This skill is particularly useful in assessing the reasonableness of results.

How can you help your students become good estimators and mental math calculators?

- Encourage students to think mentally first.
- Incorporate mental-math and estimation techniques in all the discipline area and in everyday situations that occur in the classroom.
- Use mental-math and estimation techniques frequently.
- Promote creativity and independent thinking in your classroom.

The one remaining ingredient required to insure the success of efforts to include estimation and mental math in the mathematics program is that of **teacher commitment.** Time must be devoted to such study on a regular basis. The use of estimation and mental math must be interwoven with the use of pencil-and-paper skills. The teacher should learn to model these skills for the class and to encourage the use of these skills by students.

As you think about estimation and mental math, ask yourself:

1. How often and in what settings do I use estimation or mental math?

2. Do I rely on paper-and-pencil skills when a mental calculation might be more efficient?

3. Can I think of ways to incorporate estimation or mental math in my lessons?

5 Number Theory

THEME: Composing and Decomposing with Primes

HANDS-ON ACTIVITIES

Number Theory is the study of attributes of counting numbers. This branch of mathematics provides a variety of rich problems that can help students deepen their number sense.

Some of the terms used to describe relationships among counting numbers are "factor", "multiple", and "divisor". In this chapter, you will connect many of the ideas you have studied previously to these new, yet related, concepts. For example, in Chapter 3, you saw that two of the ways to model multiplication problems are with rectangular arrays and as repeated addition.

The rectangular array below shows 3 × 4 = 12. To describe this relationship, we say 3 is a factor of 12. What is another factor of 12?

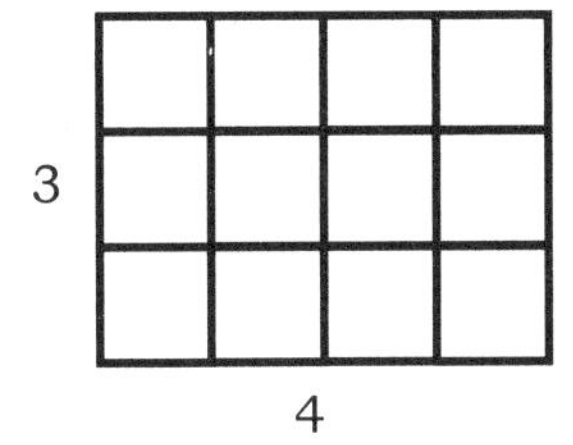

4	4	4

Another way to view multiplication is as repeated addition. This model shows three strips of length 4 units placed end-to-end. Since the total length is 12 units, we say that 12 is a multiple of 4.

When solving problems in Number Theory, it is often important to classify counting numbers as prime or composite and, as you will see in the activities in this chapter, there are several methods for doing so. One such method involves the use of the Sieve of Eratosthenes, which allows us to find prime numbers in a given range. Two other important Number Theory concepts are greatest common factor and least common multiple. When you want to find the greatest common factor or least common multiple of two counting numbers, it can be helpful to write composite numbers as a product of prime numbers. These will be key ideas in your study of fractions in Chapter 6.

OBJECTIVE: **Investigate factors, primes and composites**

You Will Need: **Squares from Materials Card 3.1.2**

In this first activity, you will re-visit the rectangular array model that was introduced in Chapters 3 and 4. Being able to visualize multiplication and division in this way will lead naturally to the concepts of factors and divisors, as well as the definitions of prime and composite numbers.

1. Use the squares from Materials Card 3.1.2 for the following problems.
   a. Form all the possible rectangles that can be made up of exactly 12 squares. Record sketches of your rectangles.

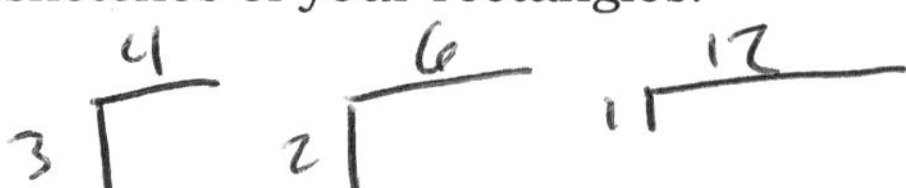

   b. Repeat part (a), using 13 squares. -prime cause only one type

   c. How are the dimensions of the rectangles you formed above related to the number of squares in each rectangle?

   d. One of the rectangles you formed in part (a) had dimensions 6 squares by 2 squares. Since $6 \times 2 = 12$, we say that 6 and 2 are **factors** of 12.

   List all the factors of 12: ____________ List all the factors of 13: ________

   e. Since 12 objects can be arranged in a rectangular array with 6 rows (or columns) we say, "6 **divides** 12", denoted $6 \mid 12$. Sketch an array made up of squares from the materials card to explain why 5 does not divide 12.

   f. Another common way to express the relationship "6 divides 12" is to say "6 is a divisor of 12." What are all the divisors of 12?

   g. How are the factors of 12 and the divisors of 12 related? What is the difference in how we think of these two terms?

2. Find all the factors of each of the following numbers by making rectangular arrays with your squares from the materials card. Record the factors below.

   a. 7 ____________ b. 9 ____________

   c. 19 ____________ d. 20 ____________

   e. 25 ____________ f. 36 ____________

3. List the numbers in problem #2 for which you could make only one rectangle: ______________ These are examples of **prime numbers**, they have exactly two factors. List the numbers in problem #2 for which you could make more than one rectangle: ______________ These are examples of **composite numbers**, they have more than two factors. Explain why the number 1 is neither prime nor composite.

4. a. Which numbers in problem #2 have an odd number of factors? ______________

   b. What is special about one of the rectangular arrays you made for each of these numbers?

   c. Use your observation in part (b) to explain why every perfect square will have an odd number of factors.

5. When trying to find all the factors of a number, it can be helpful to list them systematically.

   a. To find all the factors of 64, you could begin with the smallest factor of 64, namely 1. Then, find its "partner", 64. The next factor, 2, is paired with 32. Continue listing the factors of 64 in this way, until you have found them all.

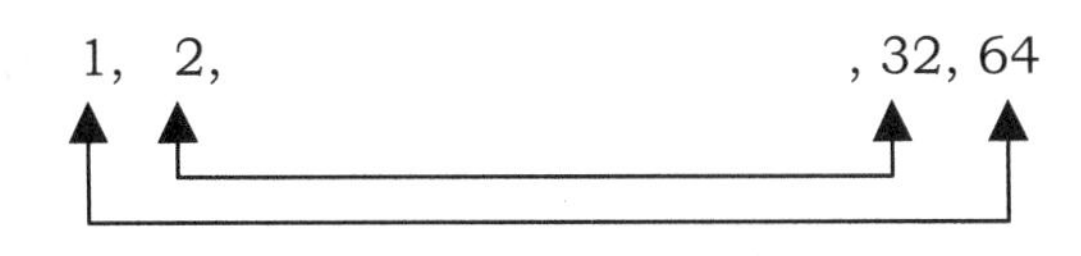

   b. When you get to 8, you can stop checking for factors. Explain why.

   c. How is the "cross-over point", 8, related to the number 64?

   d. List the factors of 150 using the method in part (a). How does computing the square root of 150 help you to solve this problem?

OBJECTIVE: **Investigate prime factorization**

You Will Need: **(eManipulative option: *Sieve of Eratosthenes*)**

Decomposing a composite number into a product of primes can help students gain a sense for the "building blocks" of the number. Representing numbers in this way is the basis for many of the concepts of Number Theory. When writing prime factorizations, it will be helpful if you are familiar with the first several primes. To this end, first consider the Sieve of Eratosthenes, a procedure for finding primes between 1 and any chosen number.

1. The following example shows how to find all the primes less than 100.
   a. On the grid, begin by crossing out 1. Then circle 2 and cross out every second number after 2. What is a factor of each of the numbers you have crossed out at this step? _____ Next, circle 3 and cross out every third number after 3, even if you have already crossed it out. What is a factor of each of the numbers you crossed out at this step? _____ Continue in this manner until all the numbers in the grid are either circled or crossed out.

Sieve of Eratosthenes

1	2	3	4	5	6	7	8	9	10
11	12	13	14	15	16	17	18	19	20
21	22	23	24	25	26	27	28	29	30
31	32	33	34	35	36	37	38	39	40
41	42	43	44	45	46	47	48	49	50
51	52	53	54	55	56	57	58	59	60
61	62	63	64	65	66	67	68	69	70
71	72	73	74	75	76	77	78	79	80
81	82	83	84	85	86	87	88	89	90
91	92	93	94	95	96	97	98	99	100

   b. Why do you suppose this is called a "sieve"?

   it colects all the prime numbers and all the crossed out

   c. The numbers in the grid, except 1, that are crossed out are composite numbers since having a prime factor caused each of them to be crossed out. The circled numbers in the grid are the primes less than 100. List them here.

2. It is sometimes useful to write a composite number as a product of primes, called its **prime factorization**. Factor trees can help you do this. As an example, consider the following factor trees for 12.

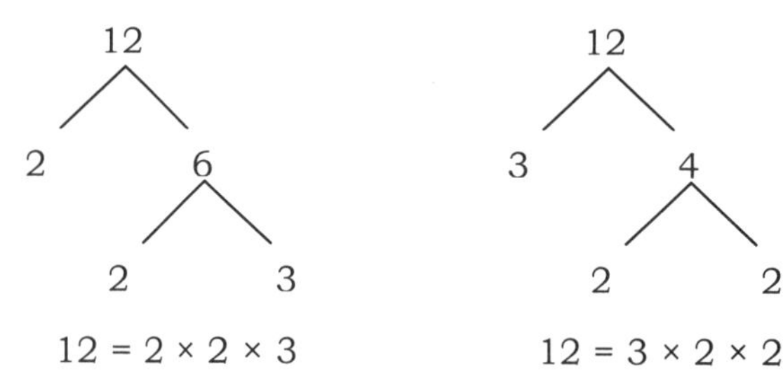

   a. Compare the two ways of writing the prime factorization of 12 shown above. How are they different? How are they alike?

b. Express each of the following composite numbers as a product of primes, using a factor tree.

14 16 150 252

3. In Activity 5.1.1 #5(d), you found all twelve of the factors of 150. Explain how you could systematically find these twelve factors using only the prime factorization of 150, $2 \times 3 \times 5^2$.

4. Yolanda wants to write the prime factorization of 127.

a. She begins by checking whether the smallest prime, 2, is a factor of 127. How can she decide whether 2 is a factor of 127, without having to divide 127 by 2? That is, what is a **test for divisibility** by 2?

b. Next, Yolanda checks whether 127 is divisible by the primes 3, 5, 7, and 11. Show how to do this. If you use any tests for divisibility, describe these.

c. Why doesn't Yolanda need to check 127 for divisibility by any composite numbers, like 6?

d. How does Yolanda know that 11 is the largest prime she must check as a possible factor of 127??

e. What can she conclude about 127?

OBJECTIVE: **Model GCF and LCM**

You Will Need: **Centimeter strips from Materials Card 3.1.1**

There are several different methods for finding the greatest common factor (GCF) and the least common multiple (LCM) of two numbers. Starting with a hands-on method may help students build a mental picture of the meaning of these two concepts. This mental picture, in turn, can be helpful for understanding other more efficient methods for computing GCF and LCM.

1. Take out your centimeter strips from Materials Card 3.1.1. For this activity, assign a value of 1unit to the white centimeter strip.

   In this case, what are the values of the other strips?

   Red = ____ Green (light) = ____ Purple = ____ Yellow = ____ Dark green = ____

   blacK = ____ browN = ____ bluE = ____ Orange = ____

2. Lay out these centimeter strips:

6		
2	2	2

   a. What product is modeled here? ____ × ____ = ____
   This shows that ____ is a factor of ____.

   b. How would you show 3 is a factor of 6? Record a sketch of the centimeter strips you use.

   c. Use your centimeter strips to illustrate *all* the factors of 16 and *all* the factors of 24. Don't forget that 1 and the number itself are factors. For example, sixteen 1 rods placed end-to-end shows 1 is a factor of 16. Record sketches to show your process.

   d. Notice that you can put eight 2 rods end-to-end to make a total of 16. Also, you can put twelve 2 rods end-to-end to make 24. Thus, 2 is a **common factor** of 16 and 24. What are the other common factors of 16 and 24? ____________

   e. What is the largest of the common factors you wrote in part (d)? _____ This is called the **greatest common factor** of 16 and 24, denoted GCF(16, 24).

3. Use your centimeter strips to find the greatest common factor of each of the following pairs of numbers. Record your answers below.

   a. GCF(8, 16) = _____ b. GCF(8, 9) = _____

4. If a is a factor of b, what is GCF(a, b)? _____ Explain your answer.

5. The following example shows 2, 4 and 6 which are the first three **multiples** of 2.

2

2	2

2	2	2

a. Use your centimeter strips to find the first eight multiples of 4. Record sketches to show your process.

Multiples of 4: _________________________________...

b. Use your centimeter strips to find the first six multiples of 6. Record sketches to show your process.

Multiples of 6: _________________________________...

c. Examine the multiples of 4 and 6 that you wrote in parts (a) and (b). What numbers are multiples of both 4 and 6? _______________ These are the first few **common multiples** of 4 and 6.

d. What is the smallest of the common multiples you wrote in part (c)? _____ This is called the **least common multiple** of 4 and 6, denoted LCM(4, 6).

6. Use your centimeter strips to find the least common multiple of each of the following pairs of numbers. Record your answers below.

a. LCM(8, 16) = _____

b. LCM(4, 9) = _____

7. If a is a factor of b, what is LCM(a, b)?_____ Explain your answer.

8. When will LCM(a, b) = $a \times b$? Explain your answer.

OBJECTIVE: **Use prime factorization to find GCF and LCM**

You Will Need: **(eManipulative option: *Factor Tree*)**

In the next activity, you will see how prime factorization can be used to determine the greatest common factor and the least common multiple of two numbers. This method is convenient when the numbers involved are large.

1. a. Determine the prime factorization of each of the following numbers and circle the common factors.

   84 = ____ × ____ × ____ × ____

   90 = ____ × ____ × ____ × ____

   The prime factors common to 84 and 90 are ____ and ____.

   b. A student says, "3 is greater than 2, so 3 is the largest factor common to 84 and 90." How could this student obtain a larger factor of both numbers?

   c. What is the GCF(84, 90)? _____

2. Use prime factorization to find the greatest common factor of each of the following pairs of numbers. Show your process in each case.

   a. GCF(54, 63) = _____ b. GCF(12, 35) = _____ c. GCF(315, 525) = _____

3. The first few multiples of 84 are: $84 = 1 \times 84$, $168 = 2 \times 84$, $252 = 3 \times 84$, ...

   Notice that 84 is a factor of each of its multiples. Thus, any multiple of 84 must contain all the prime factors of 84. With this in mind, consider the following method for "building up" the least common multiple of 84 and 90.

   The LCM(84, 90) must contain all the factors of 84, so start by putting $2 \times 2 \times 3 \times 7$ in the LCM. In order to be a multiple of 90, the LCM(84, 90) must also contain all the prime factors of 90. What factors of 90 must be put in with these? (Remember, choose these so the LCM remains as small as possible while still having all the factors of both numbers.)

   $84 = 2 \times 2 \times 3 \times 7$

   $90 = 2 \times 3 \times 3 \times 5$

   LCM(84, 90) =

   $2 \times 2 \times 3 \times 7 \times$ ____________

4. Use prime factorization to find the least common multiple of each of the following pairs of numbers. Show your process in each case.

   a. LCM(54, 63) = _____ b. LCM(12, 35) = _____ c. LCM(315, 525) = _____

EXERCISE

What do Alexander the Great and Smokey the Bear have in common?

DIRECTIONS: Find the greatest common factor or least common multiple for the numbers given on the left. Draw a straight line connecting the black square (■) next to each problem with the ■ next to its answer. Each line will cross a letter. Wherever the number for that letter appears in the code at the right, fill in the letter.

GCF(10, 35)	■	■	9
GCF(21, 36)	■	■	3
GCF(24, 64)	■	■	6
GCF(21, 84)	■	■	5
GCF(150, 375)	■	■	48
LCM(3, 9)	■	■	75
LCM(10, 12)	■	■	42
LCM(24, 16)	■	■	15
LCM(15, 9)	■	■	21
LCM(6, 7)	■	■	45
GCF(12, 18, 30)	■	■	30
GCF(15, 45, 90)	■	■	8
LCM(5, 7, 15)	■	■	60
LCM(5, 6, 15)	■	■	105

Letters: A, V, T, E, B, H, L, D, M, Y, I, O, N, S

9 – 48 – 6 – 15

75 – 8 – 9 – 48

48 – 5 – 3 – 6

9 – 48 – 6

105 – 5 – 45 – 6

45 – 60 – 42 – 42 – 21 –6

30 – 5 – 45 – 6

CONNECTIONS TO THE CLASSROOM

1. A student says, "I know that 15 divides 30 and 15 divides 45, so that must mean that 15 divides 75."

   a. How is this student reasoning?

   b. Will the student's method always work? That is, is the following always true: if a divides b and a divides c, then a divides $(b + c)$? Explain your answer using hands-on models presented in this chapter.

c. This student's method can be helpful when you want to *mentally* determine whether one number divides another. For example, how could you determine that 24 divides 288 using this students method? Do this problem in your head and then show your thought process.

2. Kathy says, "2 divides 12 and 3 divides 12, therefore 2 × 3 = 6 divides 12." Al says, "3 divides 12 and 3 divides 12, therefore 3 × 3 = 9 divides 12."

   a. For each of these students' statements, determine whether it is correct or incorrect. Use the prime factorization of 12 to explain your answers.

   b. Kathy wants to use her method to determine whether 20 divides 720. What convenient factors of 20 could she use in order to make this problem easy to do mentally? Do the problem in your head and record your thought process.

3. When computing the least common multiple of several pairs of numbers, a student notices a pattern:

   LCM(4, 7) = 28 LCM(8, 9) = 72 LCM(5, 11) = 55

   He says, "I discovered a shortcut for finding the least common multiple of two numbers. You just multiply the numbers together."

   a. When will this student's shortcut work?

   b. If the student tries to find the LCM(8, 10) using his shortcut, he will get 80. Using the prime factorization method presented in this chapter, what answer would the student get? Show the process.

   c. After using prime factorization to find the LCM(8, 10) = 40, the student says, "If you divide 80 by 2, you get 40." What relationship has the student noticed; that is, in the context of what you have learned in this chapter, how is 2 related to 8 and 10?

   d. How would the student revise his original shortcut so that it applies to finding the LCM of any pair of counting numbers? Choose another pair of numbers and try the new shortcut.

MENTAL MATH

OBSERVE THIS PATTERN:

Begin with one hexagon. =1

Surround it with hexagons. 2 3

Count the number of hexagons. + 2 = 7

The sum is prime.

Draw another ring of hexagons:

- What is this sum?
- Is it a prime?
- What will be the sum for the next figure?
- Will it be a prime?
- If you continue the pattern, adding 2 more rings, will the sum be a prime?

DIRECTIONS IN EDUCATION
Technology in the Math Classroom
What's New

Most of today's students have access to calculators and computers. The question is no longer, "Should students be allowed to use calculators and computers in the math classroom?", but "What are appropriate uses of calculators and computers in the classroom?" To insure that **every** student has access to such technology, mathematics instruction at all levels must incorporate the tools of current technology.

The National Council of Teachers of Mathematics, in the *Curriculum and Evaluation Standards for School Mathematics*, has recommended that:

- Appropriate calculators should be available to all students at all times.
- A computer should be available in every classroom for demonstration purposes.
- Every student should have access to a computer for individual and group work.
- Students should learn to use the computer as a tool for processing information and performing calculation to investigate and solve problems.

How should calculator usage impact mathematics instruction?

- Use of calculators as a tool to explore the patterns and concepts of mathematics.
- More emphasis on estimation and determining reasonableness of the result.
- Less emphasis on complex paper-and-pencil algorithms.
- More time spent analyzing **how** to solve problems.

What is necessary to facilitate appropriate instruction with calculators?

- The opportunity to choose whether to use a calculator, mental math, or paper-and-pencil algorithms and an understanding of when each is most appropriate.
- An understanding of the potential for error with calculators and skills for assessing the reasonableness of results obtained on the calculator.
- An awareness that facility with basic skills, including number sense and skills in mental computation and estimation, enhances calculator use.
- The opportunity to view the calculator as one of many tools available to the student of mathematics.
- The opportunity to solve real world problems with numbers that exceed the students' paper-and-pencil skills.
- The opportunity to use calculators in testing situations.
- An environment that encourages calculator usage.

How can computers be used to enhance mathematics instruction?

Classroom computer instruction may be of several distinct types: computer-assisted instruction, environments for forming and testing conjectures, exploration, drill and practice, and computer applications.

- Computer-assisted instruction (or programmed learning) may be interesting and motivational for some students on a limited basis. It is often of the drill and practice nature and is frequently limited to objective "right/wrong" responses. While this type of computer use may be appropriate for limited outcomes, it does little to enable students to see the rich potential for applications of computers in the world of mathematics.

- Applications of computers in the classroom may take on many distinctly different roles including:
  - ❖ The use of commercially (or teacher) prepared software such as spreadsheets to collect and manage data which may be used in problem solving or in the study of statistical concepts.
  - ❖ The use of commercially prepared software such as LOGO or the Geometric Supposer to explore concepts, to form and test conjectures, and to solve problems.
  - ❖ Student-generated software which provides solutions to single problems or to categories of problems – some facility with programming language may be required.

- Classroom use of computers should serve to illustrate:
  - ❖ The power of the computer to perform repeated calculations in short periods of time, particularly in performing simulations.
  - ❖ The versatility of the computer as a mathematical tool.
  - ❖ The applications of mathematics on the computer in the real world.
  - ❖ The role of mathematics in enhancing the utility of computers in other fields such as science.

As you think about using technology in your classroom, ask yourself:

1. How proficient am I in the use of calculators and computers?
2. Am I familiar with software and/or instructional materials which facilitate the use of calculators and computers in my classroom?
3. How can I keep my skills up-to-date as technology continues to advance?
4. How can I keep informed of new developments in the use of technology?

6 Fractions

THEME: **Understanding Fractions and Their Operations**

HANDS-ON ACTIVITIES

One of the main ideas of the elementary mathematics curriculum is the concept of fraction. A fraction, by definition, is a number that has the form a/b, where a and b are whole numbers and b is non-zero. But fractions are much more than one number "over" another. These numbers are used to describe part-to-whole relationships, and they give us a way to name quantities that are not possible only with the use of whole numbers.

As a teacher, you will need to be able to answer questions such as, "Why is 1/2 the same as 3/6?" or "Why is 3/4 greater than 2/3?" or "Why do I need a common denominator to add fractions?" Merely teaching your students rules for performing operations with fractions will not be enough. Rules can be easily forgotten. But if they learn the *why* behind the rules students will be able to reconstruct the rules, if they forget them. Building a deep understanding of fraction concepts and operations will help your future students to compute confidently and efficiently with fractions.

As you will see in this chapter, there are many connections between the whole-number operations you have studied and operations with fractions. While there are many similarities, fraction operations also involve many new procedures and algorithms. In the following activities, you will use some familiar and some new hands-on materials to investigate fractions and their operations.

Fractions are a part of everyday life from measuring 5/16 of an inch using a tape measure . . .

. . . to buying 1/2 of a gallon of milk . . .

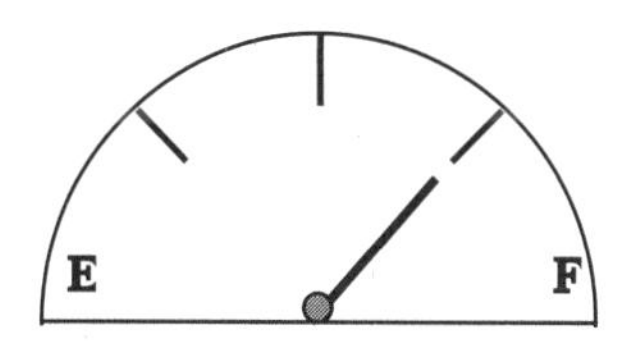

. . . to figuring out if 3/4 of a tank of fuel is enough to finish your car trip.

OBJECTIVE: **Use models to represent fractions**

You Will Need: **Materials Cards 6.1.1A and 6.1.1B (eManipulative options: *Parts of a Whole, Visualizing Fractions & Naming Fractions*)**

One meaning of fractions is that they represent the number of equivalent parts being considered out of the unit amount. Using an area model can be helpful for visualizing this idea. Two of the common area models used in elementary classroom are circles divided into parts and shaded fraction strips. You will use both models in this activity and in several of the activities later in the chapter.

1. Take out the whole circle from Materials Card 6.1.1A. Assign a value of 1 unit to the circle.

   a. Find the two pieces of equal size that will exactly cover the circle. Each of these pieces is called one-half of the circle, written 1/2, since it is one of two equal-sized parts comprising the whole circle. Label each of these.

   b. If you are sharing a pizza with a friend who cuts the pizza as pictured to the right and gives you piece B, have you received 1/2 of the pizza? Why or why not?

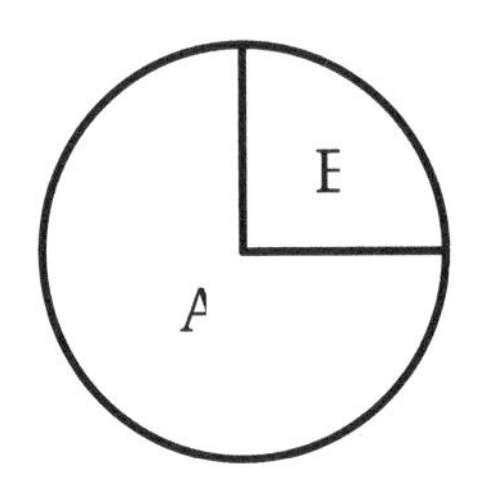

   c. Find the three pieces on the materials card that exactly cover the circle. What do we call these pieces? _______ Label each of them.

   d. Continue in this manner until all the pieces are labeled. What are the fractions represented by the other pieces? ______________________

2. When writing a fraction such as 1/3, we see that two numbers are involved, each having a particular meaning.

   a. In 1/3, 3 is called the **denominator**. What does the 3 represent?
   the amount of times the whole is evenly divided

   b. In 1/3, 1 is called the **numerator**. What does the 1 represent?
   The number of equal parts being considerd

   c. What does the fraction 2/3 represent?
   2 of 3 equal parts of a whole

3. Now take out the fraction strips from Materials Card 6.1.1B. With this model, assign a value of 1 unit to each of the fraction strips. For each of the following, identify the fraction represented by the shaded portion of the fraction strip.

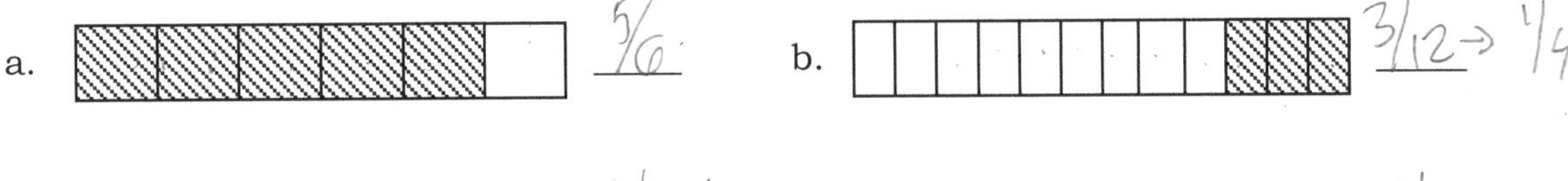

4. Find the fraction strips that represent the following fractions. Describe what you are looking for in each case.
   a. 7/12 b. 4/6 c. 1/4 d. 9/12

5. Sketch and shade fraction strips to model each of the following fractions.

   a. 1/9

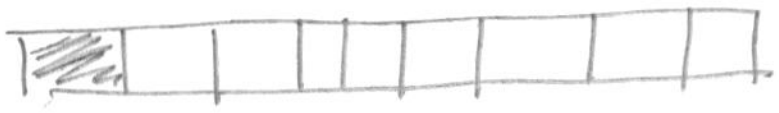

   b. 4/5

6. If you were given a fraction strip, that did not have any subdivisions marked like the one pictured below, how would you determine the fractional amount of the bar that is shaded?

   use a ruler to find the equal parts

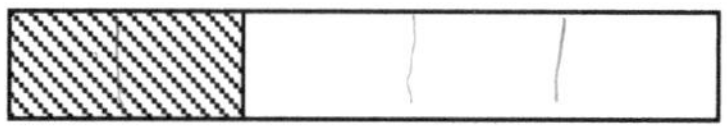

7. A fraction whose numerator is greater than its denominator is called an **improper fraction**.

   a. For example, in the fraction 5/4, the denominator tells you the unit should be divided into 4 equal-sized parts, as shown at the right. What do you need to draw in order to show 5 of these 1/4 pieces? Make a sketch of shaded fraction strips to show 5/4.

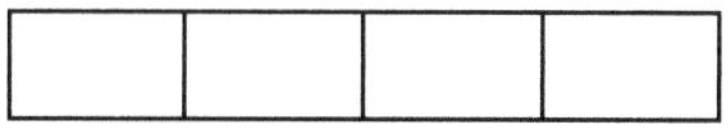

   b. Another way to write 5/4 involves thinking of the number as having a whole number part and a fraction part. How would you do this? Use your drawing of fraction strips in part (a) to explain your answer.

   1 (one whole strip) 1/4 (whats left over) = 1 1/4

   c. Your answer in part (b), read "one and one-fourth" is called a **mixed number** since it has a whole number part and a fraction part. What does the word "and" indicate in the name of this fraction?

   including

   d. Sketch fraction strips to represent each of the following improper fractions. Then write the mixed number for each one.

   13/5 = 2 3/5

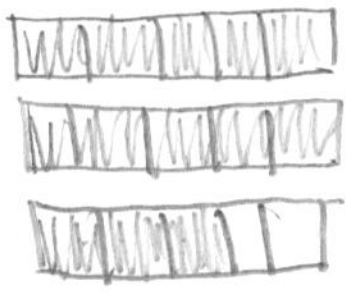

   11/3 = 3 2/3

   e. In order to write 13/5 as a mixed number, a student follows a procedure she learned in class, but she doesn't understand why. How can you help this student make sense of her method?

   "First, I divide the numerator by the denominator and I get 2, remainder 3. Then I write my answer as $2\frac{3}{5}$ where the quotient, 2, is the whole-number part and the remainder, 3, goes over the denominator."

OBJECTIVE: **Use different amounts as the unit**

You Will Need: **Centimeter strips from Materials Card 3.1.1**

Some examples of the everyday uses of fractions are: 1/2 of a sandwich, 3/4 of the students in a class, and 2/3 of a teaspoon of baking soda. Notice each of these fractions is *of* some amount; this amount is called the unit. As these examples illustrate, the unit can be an object, a set of objects, or a quantity. In previous chapters, you used centimeter strips to model addition as well as LCM. This versatile manipulative can also help students visualize the unit involved in a fraction.

1. Use the following abbreviations for the colors of the centimeter strips:

   W = White, R = Red, G = Green (light), P = Purple, Y = Yellow, D = Dark green,
   K = BlacK, N = BrowN, E = BluE, O = Orange

   For this problem, let O represent the unit. Two Y's can be placed end-to-end to make the same total length as O, as shown below.

O	
Y	Y

   a. Y is what fraction of O? ________

   b. With O as the unit, use your centimeter strips to find the value of each of the following and write the fraction in the space below. Make a sketch of your strips.

   W = ________ R = ________ E = ________

2. Now let N be the unit. What color represents each of these fractions? Record a sketch of your strips for each one.

   a. 1/2 = ______________ b. 1/8 = ______________

   c. 1/4 = ______________ d. 5/4 = ______________

   e. 3/4 = ______________

3. Using two O strips as the unit, find the value of the following strips.

   a. Y = ________ b. P = ________ c. R = ________

4. Complete the following statements with a color.

   a. W is 1/2 of ______________ b. R is 1/2 of ______________

   c. ______________ is 1/2 of N d. ______________ is 1/2 of D

   e. Why do these different colors all represent 1/2?

5. Complete the following equations with a fraction.

   a. R = _______ of D b. R = _______ of N

   c. R = _______ of Y d. R = _______ of G

   e. How can R represent all these different fractions?

OBJECTIVE: **Represent equivalent fractions**

You Will Need: **Circle pieces and fractions strips (Materials Cards 6.1.1A and 6.1.1B) (eManipulative option: *Equivalent Fractions*)**

In the first two activities, you looked at several ways to represent fractions. The next activity will address the concept of **equivalent fractions** – fractions that represent the same amount, relative to the unit.

1. First, take out the circle pieces from Materials Card 6.1.1A. Find a piece that is 1/3 of the circle. Now find two equal-sized pieces that exactly cover the 1/3 piece. What is the value of each of these pieces? _______ You can say, "1/3 is equivalent to two 1/6 pieces." This is written as 1/3 = 2/6.

   a. Find a piece that is 1/2 of the circle. Find three different ways to cover the 1/2 piece using equal-sized pieces. Make sketches to record your process and record the equivalent fractions you find.

      1/2 = _______ 1/2 = _______ 1/2 = _______

   b. Use the same procedure as in part (a) to find a different fraction that is equivalent to each of the following fractions. Record a sketch in each case.

      2/3 = ______ 1/4 = ______ 3/3 = ______ 3/4 = ______ 5/6 = ______

c. Why is the last problem in part (b) difficult with these pieces?

2. Another way to model equivalent fractions is with shaded fractions strips.

   a. What fraction is represented by the shaded portion of the fraction strip shown to the right? _______

   b. Divide each of the subdivisions of this fraction strip into 2 equal-sized pieces. Now what fraction is represented by the shaded portion? _______

   c. Are the two fractions you wrote in parts (a) and (b) equivalent? Explain your answer.

   d. When you subdivided the parts by 2, what happened to the number of parts represented by the denominator (i.e. the total number of parts)? What happened to the number of parts represented by the numerator (i.e. the number of shaded parts)? How are the operations of "dividing" and "multiplying" related in this situation?

   e. In the fraction strip shown to the right, what fraction is represented by the shaded region?_______

      Divide each subdivision into 3 equal-sized pieces. What fraction is now represented by the shaded region? _______ How have the numerator and denominator been changed?

   f. Summarize how you can write another fraction equivalent to a/b.

3. Using your fraction strips, group them into sets that represent equivalent fractions. Record the equivalent fractions. Using your conclusion in problem #2(f), verify the equivalence of the fractions you have listed.

OBJECTIVE: **Use reasoning to order fractions mentally**

You Will Need: **(eManipulative option: *Comparing Fractions*)**

Determining which of two fractions is larger can be done in a variety of ways, including finding a common denominator, converting to decimals, and by cross-multiplying. These methods are convenient when you have pencil and paper or a calculator. However, in many day-to-day situations, you need to make a *mental* comparison between two fractions. In these situations, being able to reason from your knowledge of the meaning of fractions can be a more efficient method of ordering fractions.

1. To determine the larger of the two fractions 4/5 and 3/5, one could visualize the shaded fraction strips for each fraction shown to the right. Since 4 shaded parts out of 5 equal-sized pieces is more than 3 shaded parts out of 5 equal-sized pieces, 4/5 is greater than 3/5. This is written 4/5 > 3/5.

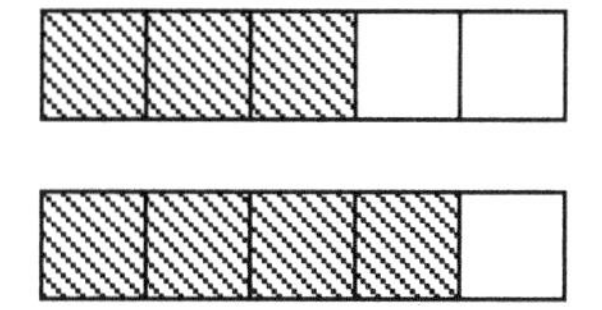

   a. Using the symbol for "is less than, namely '<' ", how would the relationship between 4/5 and 3/5 be written? Also write the relationship in words.

   b. In general, if two fractions have the same denominator, how do you know which one is the larger of the two?

   c. Order the fractions 7/10 and 9/15 by first finding a common denominator. Write an appropriate inequality relationship using one of the symbols "<" or ">".

2. Next, consider two fractions that have the same numerator, such as 4/5 and 4/7. One way to determine the greater of these two fractions is first to find a common denominator, then compare the numerators, as you did in problem #1. Instead, use reasoning to do this. Visualize the fraction bars for 4/5 and 4/7.

   a. How does your mental picture help you decide which fraction is larger? Use a description of the shaded fraction bars in your explanation.

   b. Complete this statement with "<" or ">": 4/5 ____ 4/7

   c. In general, if two fractions have the same numerator, how do you know which fraction is greater?

3. How can you determine which fraction is larger, 4/7 or 1/2, without finding a common denominator?

   a. Do this mentally and then record your thought process.

   b. Why is it easy to compare a fraction to 1/2?

4. When you want to order two fractions, this can sometimes be done by first comparing the two fractions to 1/2.

   a. For example, suppose a/b is less than 1/2 and c/d is greater than 1/2. Which of a/b and c/d is greater? Explain your reasoning.

   b. Use the method described in part (a) to order the fractions 7/13 and 5/11. Record your thought process.

   Complete this statement with "<" or ">": 7/13 ____ 5/11

   c. Another way that the fraction 1/2 can be helpful when ordering two fractions is by determining which fraction is closer to 1/2. For example, both 3/8 and 4/10 are less than 1/2. Use reasoning to decide which of these is closer to 1/2 and record your thought process.

   Complete this statement with "<" or ">": 3/8 ____ 4/10

   d. Use reasoning similar to that described in parts (a) - (c) to order each of the following pairs of fractions. Record your thought process in each case. Then complete each statement with "<" or ">".

   3/4 ____ 11/18 10/23 ____ 11/19

OBJECTIVE: **Model fraction addition**

You Will Need: **Circle pieces (Materials Cards 6.1.1A)**
(eManipulative option: *Adding Fractions*)

In the next activity, you will investigate the operation of addition with fractions. For now, suspend what you already know about adding fractions and try to think like a young student just being introduced to these concepts. Focus on the *why* behind the addition procedure, not just the *how*.

1. At a pizza parlor, you and a friend each order a mini pizza.
   a. You have one-sixth of your pizza left and your friend has two-sixths of his pizza left. Represent each of these amounts in the two circles shown at the right.

   b. To take the leftover pizza home, you combine it in a single container. Represent that amount on the third circle. How much pizza will you take home? Explain.

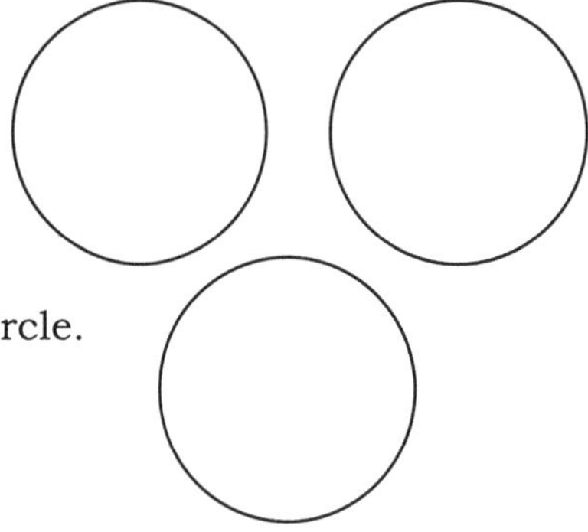

2. Represent the following problems with your circle pieces and find the solutions. Record a sketch of your pieces.

   a. $3/8 + 2/8 =$ _______   b. $1/6 + 5/6 =$ _______

3. Summarize how to add fractions with like denominators.

4. Suppose you have half of one pizza left and your friend has one-third of his pizza left. Represent the two amounts and the combined amount.

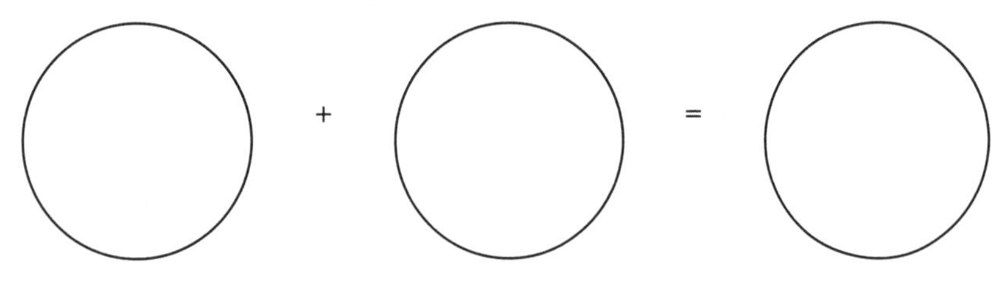

   a. Why can't you use the method you described in problem #3 to add these fractions?

   b. What equal-sized pieces can you find that will cover both the 1/2 and the 1/3 portions? ______How many of these pieces will cover the 1/2 portion? _____ the 1/3 portion? _____ How many does it take to cover the combination of the 1/2 and the 1/3 portions? _____ What is the $1/2 + 1/3$? ________

c. Make sketches of circle pieces and show how to find the sum of 3/4 and 1/6. Label your drawings so your process is clear.

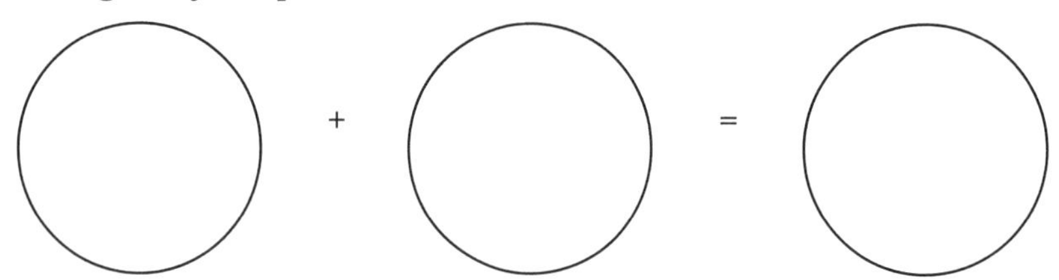

5. As you saw in problem #4, in order to add two fractions with different denominators, we first need to find equivalent fractions with the same denominator. Why do we need to do this?

6. When finding a common denominator of two fractions, the LCM of the two denominators is usually used. We call this the **lowest common denominator**. For each problem below, find the lowest common denominator and then determine the sum. Show all of your work.

   a. 3/5 + 4/7 =

   b. 4/18 + 5/30 =

   c. A student, who solved the problems in parts (a) and (b), noticed that the LCM of the denominators in part (a) is 5 × 7 = 35. She decided to do the problem in part (b) using 18 × 30 as the common denominator. Compute the sum in part (b) using this student's method. Do you get the same answer as before? How do you know?

   d. What are the advantages of using the student's method in part (c) for finding a common denominator of two fractions? Do you see any disadvantages?

7. Using ideas of fraction addition can help students make sense of the usual algorithm for rewriting a mixed number as an improper fraction. You may recall a procedure such as: "Multiply the whole number by the denominator and add the numerator to this product. Then write the result over the original denominator." For example, $2\frac{3}{5} = \frac{2 \times 5 + 3}{5}$.

   a. The mixed number $2\frac{3}{5}$ can be decomposed as the sum $2 + \frac{3}{5}$. Show how to find a common denominator in this sum.

   b. Now add the fractions you wrote in part (a). Write a sentence to explain how to do this.

   c. How does your work in parts (a) and (b) help to explain the usual procedure for changing a mixed number to an improper fraction?

OBJECTIVE: **Investigate subtraction of fractions**

You Will Need: **Fraction strips (Materials Cards 6.1.1B)**

Since addition and subtraction are inverse operations, most of the procedures for subtracting fractions are basically the same as those for adding fractions. However, some types of subtraction problems, those involving mixed numbers and regrouping, often pose a challenge for students.

1. Just as with whole numbers, the difference of two fractions can be modeled using the take away approach.

   a. To model the subtraction problem 5/8 – 3/8, begin by representing 5/8. Sketch a shaded fraction strip to show this.

   Represent the difference of 5/8 and 3/8 in your drawing showing 3/8 of the strip being taken away from the shaded part. What fraction of the strip remains?________

   Summarize how to subtract fractions with like denominators.

   b. Now consider the subtraction problem $4\frac{3}{5} - 2\frac{1}{5}$. To model this situation, begin with the fraction strips for $4\frac{3}{5}$ shown to the right. There are 4 whole units and three-fifths of another unit shaded. To model this subtraction problem with the take-away approach, what must be removed from this set? Write your answer in terms of fraction strips.

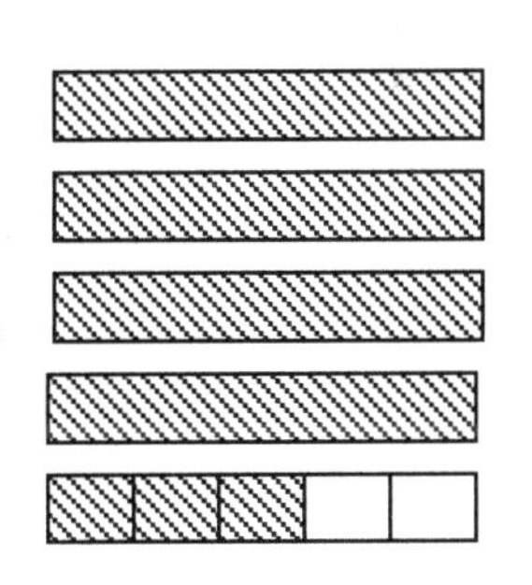

   Show the take away process in your drawing and find the difference: $4\frac{3}{5} - 2\frac{1}{5} =$ ________

2. When asked to find $4\frac{2}{5} - 2\frac{3}{5}$, a student sketches the fraction strips shown to the right and says, "You can't do this problem because 3/5 is bigger than 2/5. We don't have enough fifths to take away three of them."

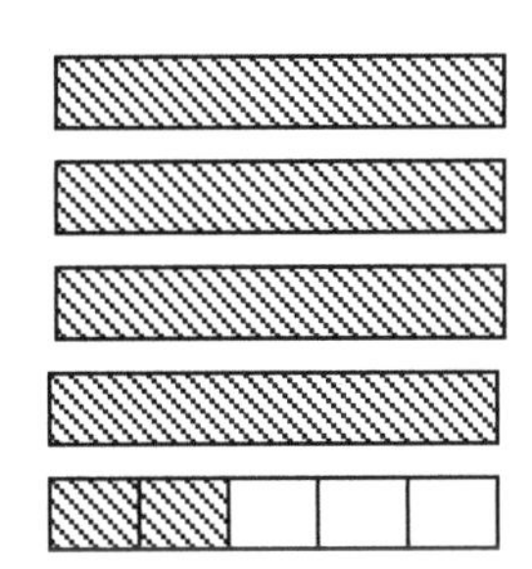

   a. It is possible to think of $4\frac{2}{5}$ in another way so that we can take away three 1/5 pieces. Show how you can alter the student's drawing to help her see that this problem is possible.

   b. What mixed number, equivalent to $4\frac{2}{5}$, does the model now show? ________

   Show the take away process in your drawing and find the difference: $4\frac{2}{5} - 2\frac{3}{5} =$ ________

3. When finding the difference $21\frac{7}{12} - 14\frac{11}{12}$, a student realizes she must first regroup since 11/12 is greater than 7/12. She begins to write out her process, and then gets stuck. She can't figure out what number to write in the missing numerator below.

$$\begin{array}{rcl} & 21\frac{7}{12} = 20\frac{}{12} \\ - & 14\frac{11}{12} = 14\frac{11}{12} \\ \hline \end{array}$$

   a. Without simply telling the student the correct number to write in the missing numerator, how could you help this student? Write some questions you could ask her to help her understand the concepts involved.

   b. Now finish the subtraction problem and find the difference.

4. Find each of the following differences and show your work.

   a. $\begin{array}{r} 14\frac{17}{30} \\ -\ 9\frac{23}{30} \\ \hline \end{array}$

   b. $\begin{array}{r} 104\frac{1}{7} \\ -\ 83\frac{3}{4} \\ \hline \end{array}$

   c. A student says, "I like to change mixed numbers to improper fractions when I do subtraction problems like these." Try this student's method for part (a) above and show your work.

   What are the advantages of using the student's method in part (c) for subtracting mixed numbers? Do you see any disadvantages?

OBJECTIVE: **Model fraction multiplication**

You Will Need: **Circle pieces (Materials Cards 6.1.1A)**

In previous chapters, you approached whole-number multiplication in several different ways; repeated addition was the first of these. In the next activity, fraction multiplication is introduced in the same way. Helping students make connections to what they already know about multiplication and addition of whole numbers, and addition of fractions can help them make sense of fraction multiplication.

1. Suppose three friends each order a mini pizza at a pizza parlor. They each have 1/4 of their pizza left.

   a. To model this situation, shade 1/4 of each of the three circles at the right. What repeated addition problem is modeled here?

   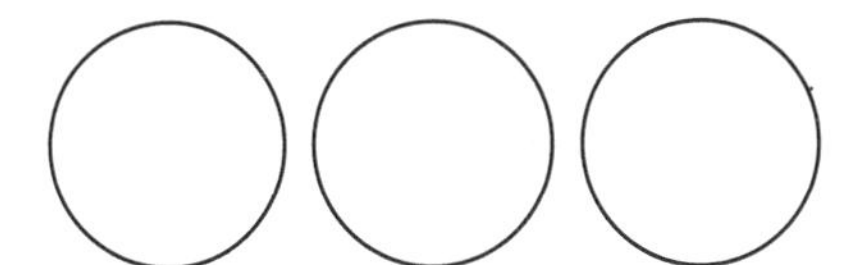

   To take their leftover pizza home, the three friends want to combine it in a single container. Rearrange the three one-fourth pieces on one circle.

   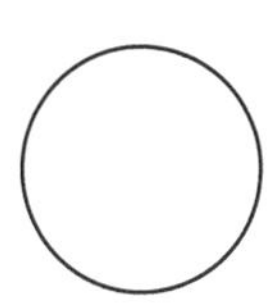

   What fraction of a whole pizza is shaded now? _______

   This shows 3 × 1/4 = _______

   b. Represent the following products and state the result. Label your drawing to make your process clear.

   2 × 1/6 = _______ 3 × 1/2 = _______

   c. Summarize how to multiply a whole number and a fraction.

2. Now suppose that two friends want to share three cookies equally. That is, they will each get one-half of the three cookies.

   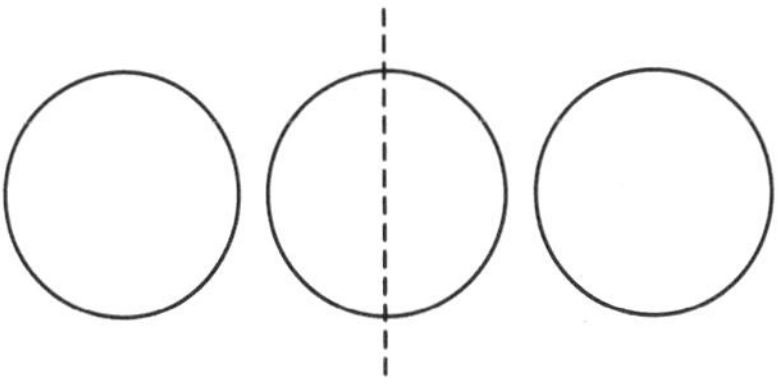

   The cookies are represented by the three circles shown to the right. A dashed line has been drawn to show how to divide the three circles into two equal-sized parts. Shade the part of the three cookies that one friend will get. The other friend will get the unshaded part. How many whole circles do you have shaded? _____ What part of another whole circle is shaded? _____ Taking one circle to be a unit, write the mixed number that represents how many shaded circles you have. ________ If you consider the original three circles *as your unit*, what fraction represents the shaded pieces? _______

   This shows 1/2 *of* 3 = 1/2 × 3 = _______ How does your drawing above differ from your drawing for 3 × 1/2 in problem #1(b)? How are they alike?

3. Circle pieces can also be used to model the product of two fractions.

   a. Shade 1/3 of the circle shown to the right.

   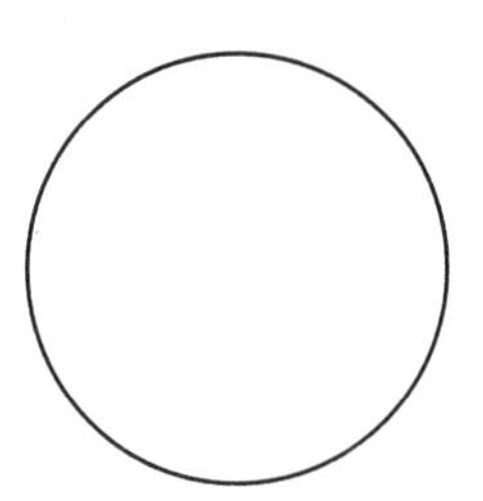

   What sized piece would cover half of the shaded portion?_______

   In other words, 1/2 *of* 1/3 = 1/2 × 1/3 = _______

   b. What pieces from the materials card do you need in order to cover 2/3 of 1/2 of the circle? Make a sketch of your pieces and label it.

   2/3 of 1/2 = 2/3 × 1/2 = _______

   c. Use your circle pieces to represent the following products. Record a sketch for each one.

   1/2 × 3/4 = _______ 1/3 × 3/6 = _______

OBJECTIVE: **Investigate the algorithm for fraction multiplication**

You Will Need: **4 square pieces of scratch paper (4" by 4") (eManipulative option: *Multiplying Fractions*)**

Another approach you used to represent multiplication of whole numbers was the rectangular array model. This model can be extended to help students visualize the product of two fractions and the usual algorithm for multiplying them.

1. For this entire activity, a square will represent one unit.

   a. Consider the following rectangles, made up of unit squares. For each one, give the whole number factors and the product.

   b. Follow the steps below to model 1/3 × 1/4.

Take one square piece of scratch paper and fold it lengthwise into quarters as illustrated.

Fold in half

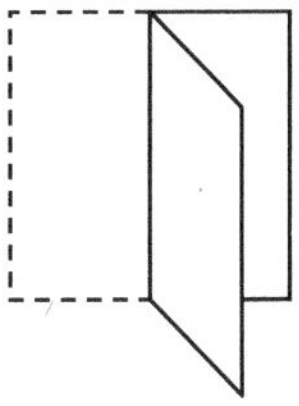

Fold the half in half

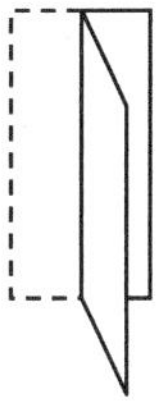

After the two folds, you have a piece 1/4 the size of the original piece. Without unfolding the final piece, fold it into thirds in the opposite direction, as illustrated.

Fold up one third

Fold again

Now color or shade the top only.

Unfold the paper. How many equivalent pieces are represented by the folds? What fraction of the original square is shaded? _______ Explain your answer.

You shaded 1/3 *of* 1/4 of the square, so 1/3 × 1/4 = _______

c. Use the paper-folding method on another square piece of paper to find 1/2 × 1/3. Draw a picture to show the unfolded square. This shows that 1/2 × 1/3 = _____

d. Use the paper-folding method for each of the following problems. In each case, you will need to partially unfold the paper before shading. Then, write a sentence explaining what you did and record a sketch of the unfolded square.

1/2 × 2/3 = _______ 2/3 × 4/5 = _______

2. Notice, in problem #1(d), when you opened the paper for 2/3 × 4/5, there are two rectangles, (1) the shaded rectangle, and (2) the entire rectangle (square), that have been divided into smaller equal-sized parts. In one direction the square has been divided into 3 lengths and in the other into 5 lengths.

   a. How many parts was the entire square divided into? _______ In the usual multiplication algorithm, where do you see 3 × 5?

   b. Of the unit square, the part you are considering has been shaded. In one dimension you shaded 2 (of the three parts), and in the other you shaded four (of the five parts) so the shaded region is 2 by 4 or contains 8 parts. In the usual algorithm, where do you see 2 × 4?

   c. Summarize how the usual fraction multiplication algorithm is modeled by folding paper.

OBJECTIVE: **Model fraction division**

You Will Need: **Centimeter strips from Materials Card 3.1.1 (eManipulative option: *Dividing Fractions*)**

By extending our number system to include fractions, we are able to think about division problems in a new way. Whereas you first thought of division problems as having a whole-number quotient and remainder, you will now see that the quotient and remainder can be a fraction. Despite the differences, whole-number division and fraction division have many similarities, as you will see in the next activity.

1. First, recall a model for division of whole numbers. Suppose you have 12 eggs, for example, and a cookie recipe that calls for 4 eggs. You want to figure out how many batches of cookies you can make. The question to ask is: "*How many* 4's are there in 12?"

   a. Using the measurement concept of division, we measure out groups of 4 eggs as shown in the diagram to the right. How do you obtain the answer to the division problem 12 ÷ 4, using this diagram?

   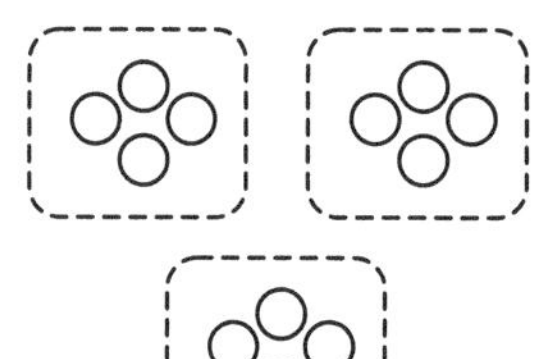

   b. Now take out your centimeter strips (W = White, R = Red, G = Green (light), P = Purple, Y = Yellow, D = Dark green, K = BlacK, N = BrowN, E = BluE, O = Orange) to model the following division problems. In each case, state the appropriate question to ask (such as, "How many reds are there in a purple?") and give the result. Record a sketch of your centimeter strips.

   E ÷ G = ________ N ÷ R = ________ O ÷ Y = ________

2. Next, suppose you only have 11 eggs to make the same cookies as in problem #1(a).

   a. How many batches of cookies can you make? _______ Draw a picture similar to the one in #1(a) to show your solution.

   b. In part (a), you had enough eggs to make 2 whole batches of cookies and had a remainder of 3 eggs. In this case, the unit is a group of 4 eggs, the number of eggs needed for one whole batch of cookies. What fraction of a whole recipe could you make using the extra 3 eggs? Explain your answer.

   This shows that 11 ÷ 4 = ________ (Write your answer as a mixed number.)

c. Use your centimeter strips to find the quotient for the following division problems. State remainders in fraction form. In each case, record a sketch of your centimeter strips.

K ÷ G = _______ O ÷ P = _______ (O + R) ÷ Y = _______

3. Now suppose you only have 3 eggs to use for the same cookie recipe.

   a. The question to ask now is "*How much* of a recipe could you make?" _______Draw a picture and use it to explain your answer.

   This example shows 3 ÷ 4 = 3/4. This is one of the ways of thinking of a fraction – as the result of dividing.

   b. For each of the following division problems, state the appropriate question to ask (such as, "How much of a purple is red?") and give the result. Record a sketch of your centimeter strips.

   R ÷ Y = _______ G ÷ D = _______ P ÷ (O + R) = _______

4. In some division problems, it is most appropriate to ask the question, "How many...?" In other situations, it is most appropriate to ask the question "How much...?" Explain the difference between these two types of division problems.

5. For each of the following division problems, let O + R be the unit. Find pieces that represent the lengths involved in the problems given. In each case, state the appropriate question to ask (such as "How many 1/2's are in 3/4?"), record a sketch of your centimeter strips, and then find the quotient. To determine the answer, it may be helpful to find equivalent fractions with a common denominator. Show the equivalent fractions with your centimeter strips.

   5/6 ÷ 1/4 = _______ 3/4 ÷ 1/6 = _______ 2/3 ÷ 5/6 = _______

   5/12 ÷ 2/3 = _______ 5/4 ÷ 1/3 = _______ 1/2 ÷ 7/4 =_______

EXERCISE

YOU CAN'T FIT A SQUARE PEG INTO A ROUND HOLE!

Directions: Use the fractions given on the left to fill in the missing holes on the right. The results should give true equations across and down.

1/12	11/12	7/12
1/4	1/2	5/6
3/4		

1/6	+	( )	+	( )	=	1
+		+		+		+
7/12	+	( )	+	5/6	=	5/3
+		−		−		−
( )	+	5/12	−	( )	=	( )
=		=		=		=
5/3	+	( )	−	5/12	=	11/6

3/4	1/3	3/2
3	1/2	6
1/4		

24	×	[ ]	×	[ ]	=	9
×		×		×		×
[ ]	×	24	×	1/12	=	[ ]
×		×		×		×
[ ]	×	[ ]	×	24	=	2
=		=		=		=
9	×	[ ]	×	1	=	54

CONNECTIONS TO THE CLASSROOM

1. A student says, "I know that if I multiply the numerator and denominator of a fraction by the same number, I get an equivalent fraction. So it must work the same with adding, like this: $\frac{3}{4} = \frac{3+2}{4+2} = \frac{5}{6}$." Does the student's procedure work? Why or why not?

2. Another student says, "I know when I want to simplify a fraction, I can cancel out common factors in the numerator and denominator. Like with 10/30, I have 1 × 10 upstairs and 3 × 10 downstairs so I can cancel out the 10's and I get 10/30 = 1/3. What if I have a fraction like $\frac{11}{13} = \frac{1+10}{3+10}$? Can I cancel out the 10's?

   a. How would you respond?

   b. The term "cancel out" does not really describe what we do when we simplify a fraction. What is another term you could use that would indicate the operation involved in this process? Explain your answer.

3. When comparing the fractions 1/2 and 1/3, a student draws the picture shown to the right and says, "this shows 1/3 is bigger than 1/2." What is wrong with the student's reasoning?

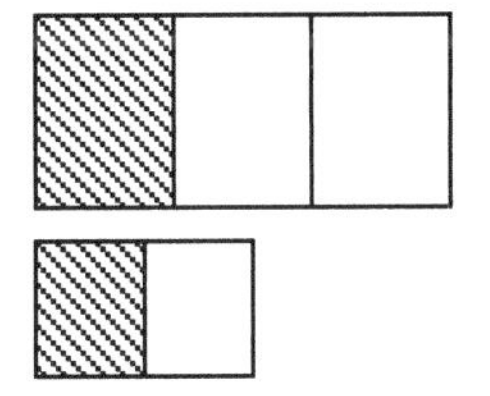

4. A student says, "I know 5 > 4 > 3 > 2, so 1/5 > 1/4 > 1/3 > 1/2." Is this reasoning correct? If not, what could you do to help this student?

5. When comparing the fractions 13/20 and 17/32 a student says, "13/20 is less than 17/32 since 13 is less than 17 and 20 is less than 32."

   a. Is this student's reasoning correct? Why or why not?

   b. How could the student use the fraction 1/2 to mentally check his answer?

6. When asked to give an example of a problem where they could use cross-multiplication to solve it, two students answered as follows.

Alex says, "I'd do the problem 1/5 × 2/3 like that. You cross-multiply 1 × 3 and 5 × 2. So the answer is 3/10."

Jamal says, "If I want to know whether 3/4 and 15/20 are equal, I would cross multiply. Since 3 × 20 = 60 and 4 × 15 = 60, I know the fractions are equal."

Are either or both of these students correct? If so, who? Explain why cross-multiplication works to solve problems of the type in his example. If one of the students is incorrect, how could you help him?

7. A student says, "I know when I multiply two fractions, I just multiply the two numerators and multiply the two denominators. I don't understand why I can't do the same thing when I add fractions, like 1/2 + 1/3 = 2/5. My picture seems right to me."

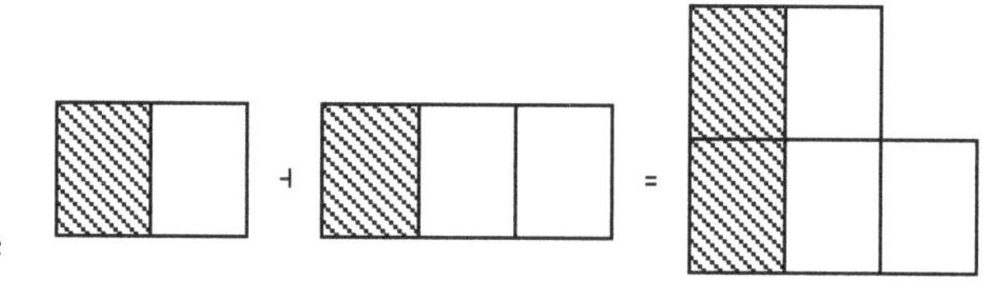

   a. Is this student's method correct? Why or why not?

   b. What question(s) can you ask the student to make sure that the student understands a correct method for adding fractions?

8. Another student shows the following work when asked to find the product of 2/5 and 3/4:

$$\frac{2}{5}\times\frac{3}{4}=\frac{2\times 4}{5\times 4}\times\frac{3\times 5}{4\times 5}=\frac{8}{20}\times\frac{15}{20}=\frac{120}{400}$$

   a. Is this student's answer correct? If so, is that a good method to use to multiply fractions? Why or why not?

   b. How can you use what you learned about equivalent fractions in this chapter to help this student simplify his process for finding the following product?

   3/6 × 10/15

MENTAL MATH

THE RANGE GAME

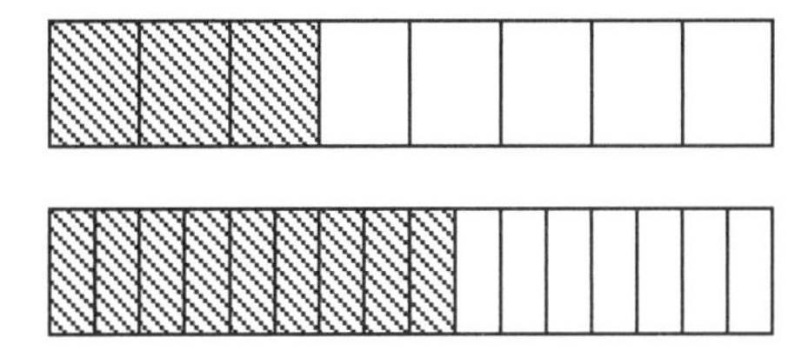

Use the digits 1 through 9 to see how many fractions you can find between 3/8 and 9/16.

Example: 3/8 < 1/2 < 9/16.

Write your fractions here:

DIRECTIONS IN EDUCATION
Teacher Beliefs/Student Beliefs
Would You Believe?

The beliefs about mathematics held by students and teachers can have a profound effect upon the outcome of learning activities. According to the NCTM Standards, the beliefs held by students "exert a powerful influence on students' evaluation of their own ability, on their willingness to engage in mathematical tasks, on their ultimate mathematical disposition."

Some of these beliefs are only myths and to guard against transmission of these myths to their students, teachers must examine their own beliefs and become aware of the myths. Teachers who understand the common myths about mathematics learning can be more effective in changing student beliefs.

What are some of the common math myths which negatively impact mathematics instruction?

- **MYTH: Students who are good in mathematics have a good memory.** Teachers who hold this belief tend to see math instruction as a series of rules and facts to be memorized. Students who are not good at memorization may begin to see themselves as unable to learn math.

- **MYTH: There is one best way to do an individual math problem.** Teachers reinforce this belief by requiring students to show their work and by counting the entire problem wrong if any step is omitted or done incompletely. During instruction the teacher demonstrates only one way to do the problems assigned, and does not encourage discussion of alternative approaches. Students learn not to tune in on their own insights or to value those insights.

- **MYTH: Counting on the fingers is unacceptable.** Teachers who do not value the use of concrete representations of math concepts view the use of fingers as a form of "cheating." Such teachers tend to view math only as an abstract thinking activity and present instruction symbolically. Students who need concrete examples to assist them in working with abstract concepts begin to see themselves as inferior to fellow students who can simply do the work in their heads or on paper.

- Other myths about math which can impact classroom instruction include:
  - Math requires the use of login, not intuition.
  - You must always know how you go the answer.
  - Math is done by working intensely until the problem is solved.
  - Men are better in math than women.
  - Mathematicians do problems quickly, in their heads.
  - Math is not creative.
  - There is a magic key to doing math.
  - The math I learned in school is good enough for today's students.
  - Increased reliance on standardized tests will improve student performance in mathematics.

What are some common math beliefs that can block progress?

- **Mathematics is computation.** Students tend to view math as the four basic operations. While study of other mathematical ideas such as geometry or statistical concepts may be interesting to students, they view their mathematical ability only in terms of facility with computation.

- **Math problems are quickly solvable in just a few steps.** Students who cannot quickly arrive at solutions to math problems tend to question their ability to solve the problem or to think that there is something wrong with the problem itself.

- **The purpose of doing math is to obtain the "right answer."** Students tend to view math in terms of absolutely right or absolutely wrong. They are unable to appreciate progress in the use of a difficult process if they do not obtain the right answer.

- **The role of the math student is to receive new mathematical knowledge and to be able to demonstrate it.** Students generally view math as something to be passively received and then duplicated on homework assignments and tests.

- **The role of the math teacher is to impart mathematical knowledge and to check student answers for correctness.** Students view the math teachers as the "Sage on the Stage" rather than the "Guide on the Side." As a result of this view, students limit their attempts to interact with the teacher in any discussion of mathematical ideas.

- **Math ability is related to innate ability more than to effort.** Students who see themselves as unable to learn math are seldom motivated to put much effort into the learning task. Since ability is not something within the student's control, math is also seen as outside the control of the student.

As you think about mathematical beliefs of yourself and your students, ask yourself:

1. Does my teaching reflect my acceptance of any math myths?
2. What can I do in my classroom to foster positive beliefs and attitudes about mathematics?
3. As a student of mathematics, do I maintain a positive attitude about my own learning?

7 Decimals, Ratio, Proportion, and Percent

THEME:

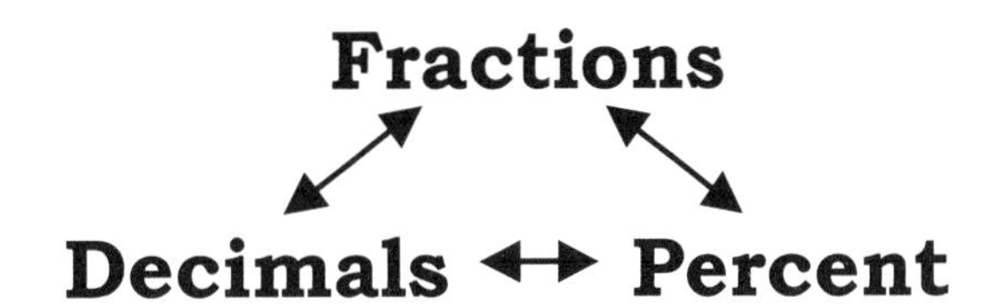

HANDS-ON ACTIVITIES

In this chapter, you will investigate several concepts that are related to fractions: decimals, ratio, proportion, and percent. Decimals offer an alternate way of writing fractions whose denominators may be expressed using powers of 10. In other words, decimals *are* fractions. A ratio is a comparison of two quantities that is commonly written in the form of a fraction. A proportion is a statement that two ratios are equal. Percent notation is used to represent a fraction whose denominator is 100. A deep understanding of these connections to fractions will help you answer questions such as:

Why do I need to line up the decimal point when I add these two decimals, but not when I multiply these two?

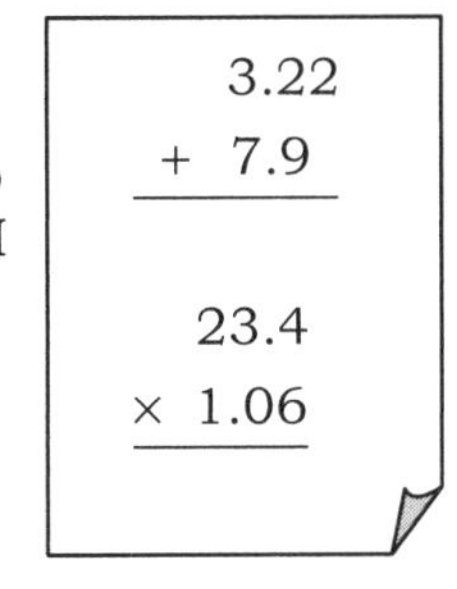

30% off
$24.00

Why can I multiply $24.00 by 0.70 to find the sale price of this item?

Many of the following activities involve the use of base ten pieces; however, in this chapter these have new names. A "flat", for example, may be called a decimal square or it may be thought of as a percent grid. Because you are already familiar with this model, you will be able to see the connections between the common whole number and decimal pencil-and-paper algorithms. Percent grids will help you visualize the concepts behind percent problems as well as develop mental methods to solve everyday percent problems. You will also work with a new manipulative, a set of geometric shapes, to understand the concepts of ratio and proportion.

OBJECTIVE: **Extend place value system to decimals and write decimal numerals and names**

You Will Need: **Materials Card 7.1.1 (Physical Manipulative option: *Base Ten Pieces*)**

Decimals provide a convenient way to represent fractions. The first activity introduces you to decimals and the numerals and word names we use to express them.

1. When you used base ten pieces to represent whole numbers, the unit cube represented the number 1. What numbers did the following pieces represent? Explain why.

   Block = Flat = Long =

   Complete the chart:

Piece:	B	F	L	U
Value:				

2. When representing numbers that have two decimal places, we use the flat for the unit amount of 1. Determine the value of each of the following pieces, expressed as a fraction, where the flat is the unit. Explain your answers.

   Block = Flat = Long =

   Complete the chart using fractions:

Piece:	B	F	L	U
Value:				

3. Review the charts you completed in #1 and 2.

   a. As you move from right to left, how are the values of the adjacent columns related?

   b. How are the values of the adjacent columns related as you move from left to right?

   c. Use your observations to label the missing place values on the top of this chip abacus.

		1			

   d. We can distinguish where the whole number part of a numeral ends and the fraction part begins by using a marker to represent a **decimal point**. Place the decimal point on the chip abacus in #3c.

4. Using chips and a chip abacus with decimal point marker, represent the following. Make a sketch of your results.

   a. 321.04 b. 3.2104 c. 32.104

   How are the representations of the previous numbers alike? How are they different?

5. Write the *expanded form* for the numbers from #4.

   321.04 = 3(100) +2(10) + 1(1) + 0(1/10) + 4(1/100)

   3.2104 =

   32.104 =

6. A chip abacus can be useful for explaining why we name decimal numbers the way we do.

   a. Write the numeral represented in the chip abacus to the right.________________

   a.

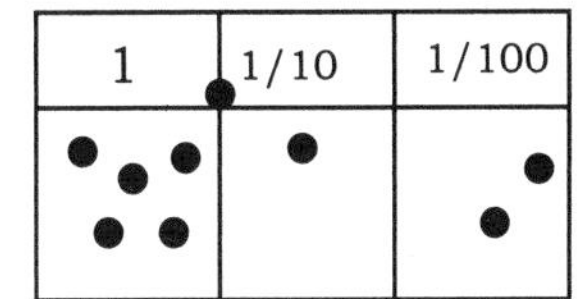

   b. Next, exchange the chip in the tenths column in the top chip abacus for hundredths chips. Show the result of these exchanges in the bottom chip abacus.

   b. 1 | 1/10 | 1/100

   Now how many hundredths do you have?

   _____________ hundredths

   c. This is the name given to the fraction part of the given number. We read numbers like this by first reading the whole number part as usual. Next, the decimal point is read as "and" and then the fraction part is read. How would you read the number in part (b)?

7. Write the word names for the numbers illustrated here:

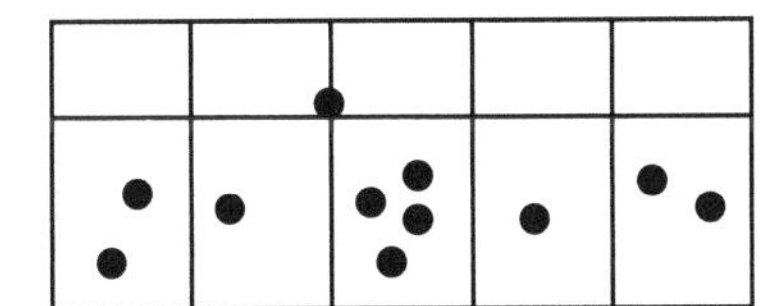

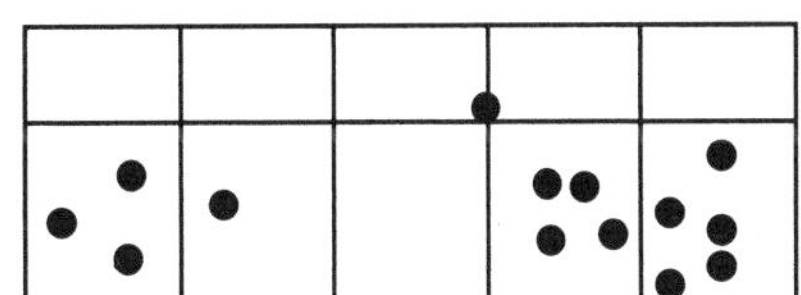

8. Write the decimal numerals to represent the following numbers.

   a. Thirty two and thirteen hundredths

   b. Five hundred and twenty one thousandths

   c. $19 \frac{10}{1000}$

   d. $34 \frac{11}{100}$

OBJECTIVE: **Represent and compare decimals using decimal squares**

In Chapter 6, fraction bars were used to represent fractions as equivalent parts of a whole. Similarly, decimal squares can be used to picture decimals and to illustrate relationships between them.

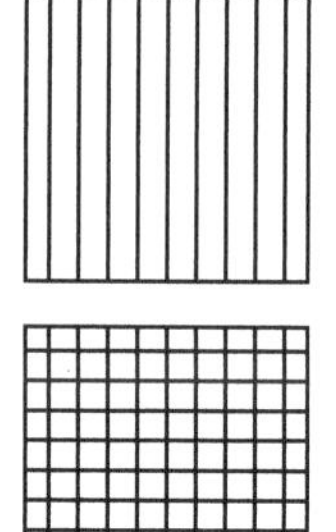

1. a. If a square represents our unit amount, you can divide it into parts in different ways to represent decimal fractions. For example, you can form ten columns. What decimal part does each column represent?

   b. If the square is divided into 100 equivalent parts as shown, what decimal part does each small square represent?

   c. If you were to divide each of these small squares into 10 equally sized parts, how many parts would be formed altogether? What decimal does each of these parts represent?

2. Write the decimal number represented by the shaded part in each of the following decimal squares.

   a. 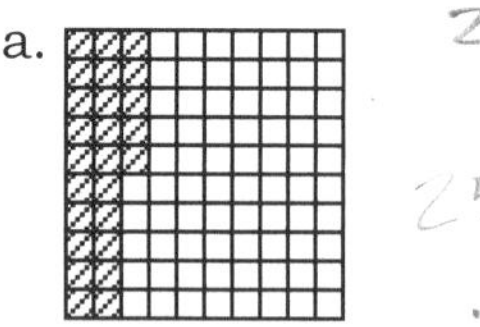 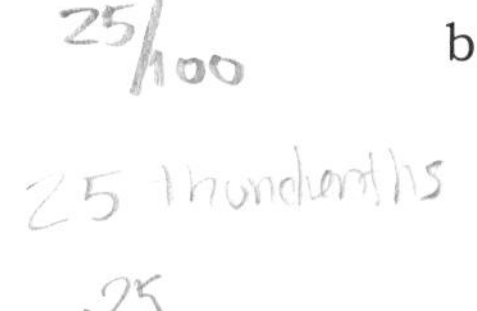 b. 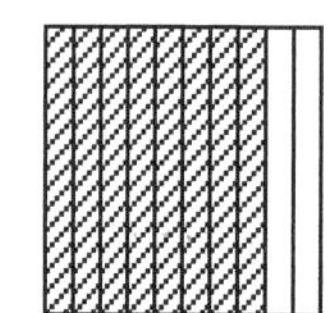 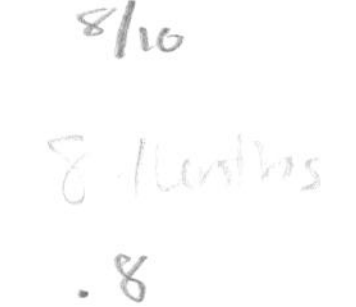 c. 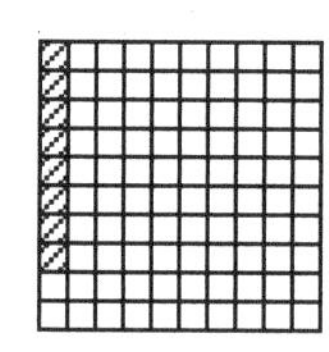

3. a. Shade the decimal squares as indicated and write the decimal number for each one:

   6 shaded parts out of 10 = _____ 60 shaded parts out of 100 = _____

   b. Compare the two decimal squares you shaded in part a. How are they alike? How are they different? What can you conclude about the decimals they represent?

4. In which of 0.36 or 0.4 is more area shaded? _____ Explain.

   Complete this statement with < or >: 0.36 ____0.4

5. Decimal squares provide a "parts-of-a-whole" way of comparing decimals. Another method involves place value. As an example, consider the decimals 0.709, 0.71 and 0.7.

   a. Write the *expanded form* for each of these numerals.

      0.709 = seven ~~tenths~~ hundred and nine thousanths $7(1/10) + 9(1/1000)$

      0.710 = seventy one hunduenths $7(1/10) + 1(1/100)$

      0.700 = seven tenths $7(1/10)$

   b. Explain how you can use the concept of place value, together with what you've written in part (a), to order these decimals. Write the numerals in order from smallest to largest.

      .710, .709, .700

   c. Compare the methods you used to order decimals in #4 and #5b. How are they alike? How are they different?

   d. Sketch decimals squares for 0.71 and 0.7 to show how the decimal-square method and the place-value method are equivalent.

   e. Express the decimals in part (a) so that they all have digits through their thousandths places – this is expressing all of the decimals so that they are all written to the same smallest place value. Arrange these three decimals from smallest to largest. Express your opinion about which of the three methods would be most effective for teaching students: decimal-square, expanded place value, or writing out place values to the smallest place value.

OBJECTIVE: **Multiply and divide decimals by 10**

You Will Need: **Chip abacus & decimal point marker from Materials Card 7.1.1**

Math students are often taught useful shortcuts for computing. One example is for multiplying whole numbers by powers of 10; when computing the product of 23 and 10, we think "put a zero on the right of 23 to get 230." In this activity, you will investigate shortcuts for multiplying and dividing decimal numbers by powers of 10, using a chip abacus to explain *why* these shortcuts work.

1. a. Recall the base ten pieces you have worked with. What can be done to simplify matters when you have 10 small cubes (units)? Answer the similar question for 10 longs and 10 flats.

   b. Suppose you have 2 small cubes. If you multiply the 2 by 10 by replacing each small cube with 10 small cubes, what will you have then? How could we express this using the fewest number of base ten pieces?

   c. Suppose you had 5 longs that are multiplied by 10 in the same way. What do you have as a result after making all possible exchanges?

2. Use chips to represent 3.57 on this chart:

B	F	L	U

   Multiply 3.57 by 10 and make all possible exchanges. Show the result on the second chart:

B	F	L	U

   In the first chart, there were 7 small cubes. What do these 7 cubes correspond to in the second chart? Explain your answer. Repeat for the 5 longs and 3 flats. How could you have easily obtained the second chart from the first?

3. Represent 1.72 on this chart:

B	F	L	U

   If you multiply this 1.72 by 100, predict what will happen.

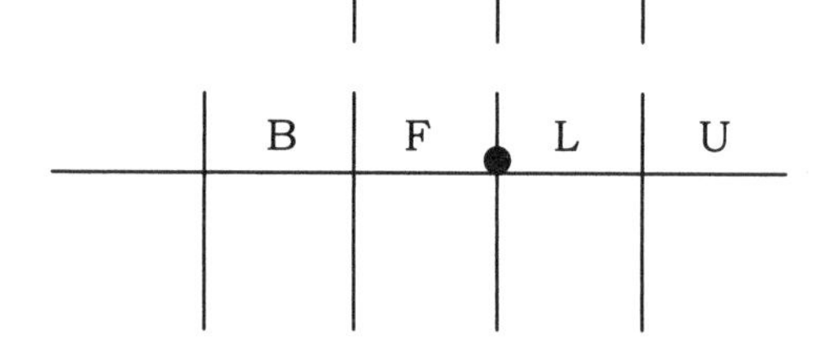

   Show this result on this chart.
   What do the 2 small cubes correspond to in the second chart? the 7 longs? the flat?

   How could you have easily obtained the second chart from the first?

Describe a shortcut to use when multiplying by 10, by 100, by 1000.

4. a. What do you have in each group if you divide a flat into 10 equal-sized groups?

   b. What do you have if you divide a long by 10?

5. a. Represent 26.3 on the chart:

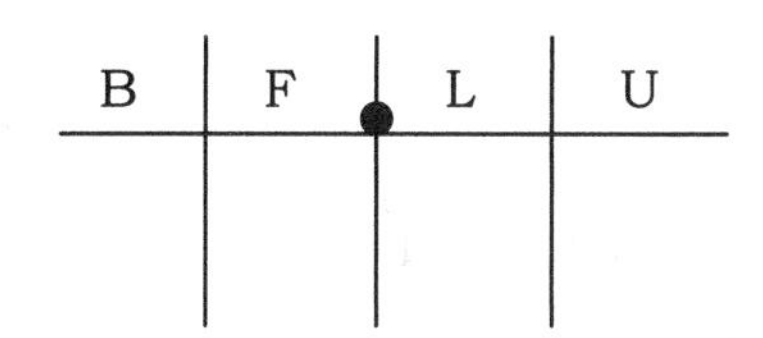

   If you divide this 26.3 by 10, what will the 2 blocks correspond to? The 6 flats ? the 3 longs?

   Show the result on this chart:

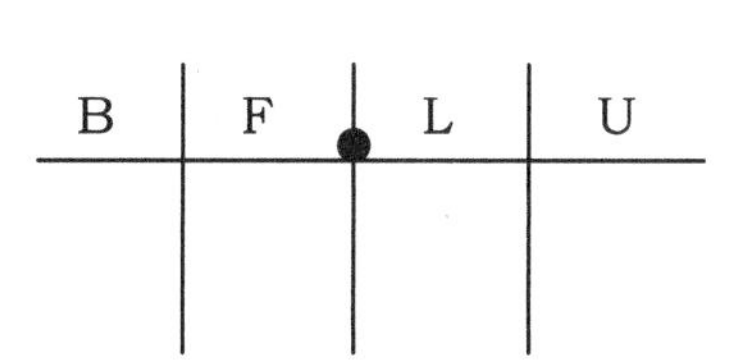

   How could you have obtained the second chart from the first?

   b. If you were to divide the original 26.3 by 100, how could you obtain the result?

Describe a shortcut to use when dividing by 10, by 100, by 1000.

6. What is the effect on a number when the decimal point is moved one column to the right?

   What about moving it two columns to the right?

   One column to the left?

   Two columns to the left?

7. Describe the decimal point moves that are equivalent to the following:

   Multiplying by 10:

   Dividing by 1000:

   Dividing by 10:

   Multiplying by 10,000:

OBJECTIVE: **Compare addition and subtraction of whole numbers and decimals**

You Will Need: **Chip abacus from Materials Card 7.1.1(eManipulative option: *Base Block Decimals*)**

The algorithms for adding and subtracting decimals are similar to those for whole numbers. In the next activity, you will use a chip abacus to model these operations and then make comparisons.

1. For each of the following problems:
   (i) Label the place values in the chip abacus and illustrate the computation with chips, showing any regrouping. Then record the sum or difference below the dotted line.
   (ii) Next to each chip abacus, write out the standard paper-and-pencil algorithm.

a. 3112 + 460

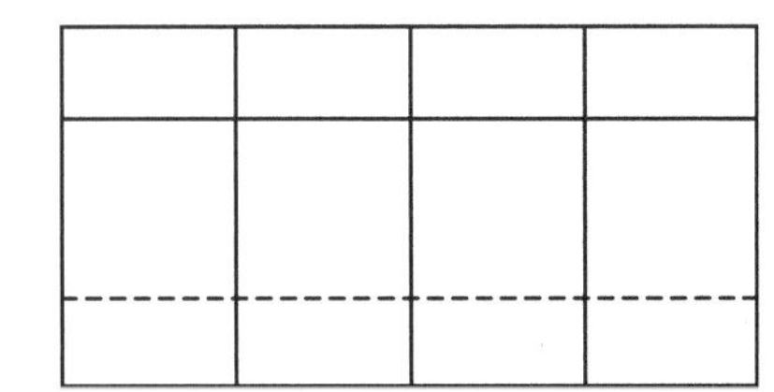

b. 5213 + 400

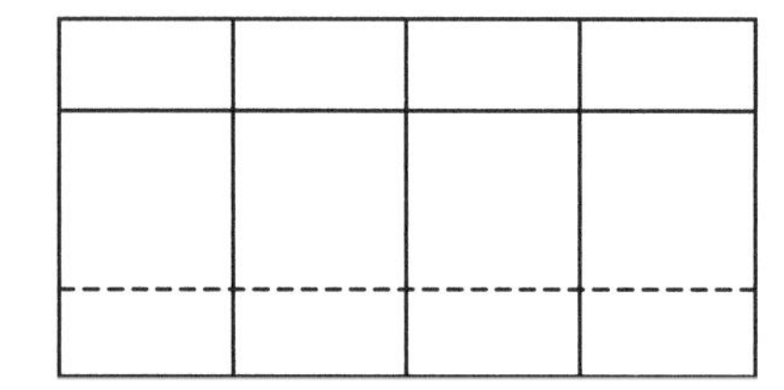

c. 312 – 150

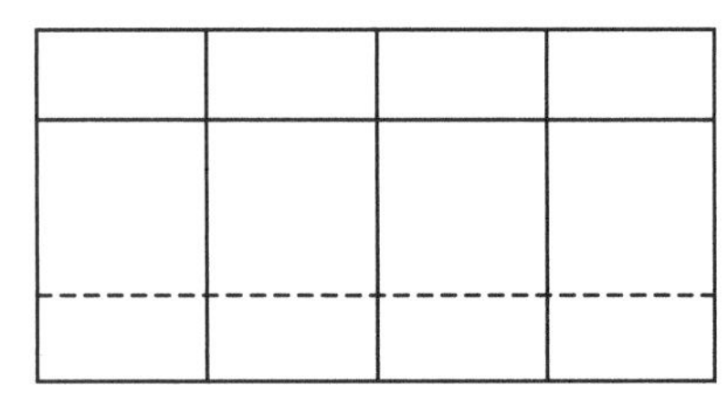

d. 13,213 – 7508

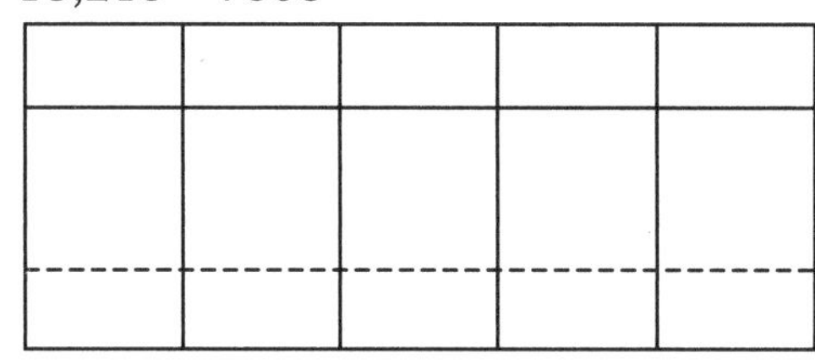

2. Now follow the directions in #1 for the following problems. Remember to indicate the decimal point for these on the chip abacus.

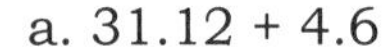

a. 31.12 + 4.6

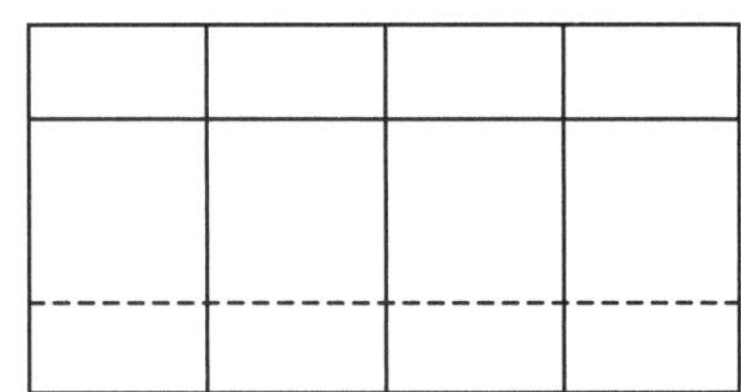

b. 5.213 + 0.4

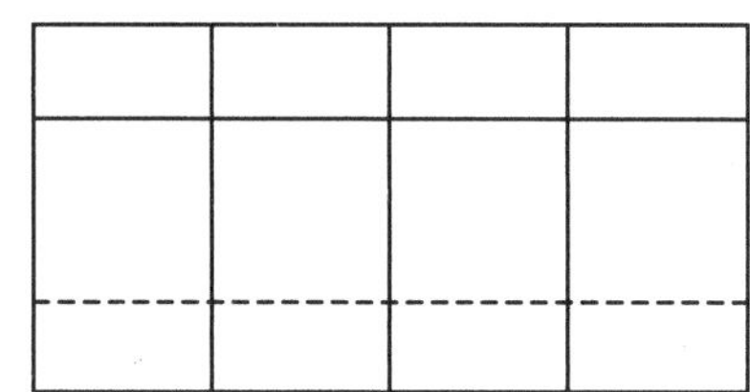

c. 3.12 – 1.5

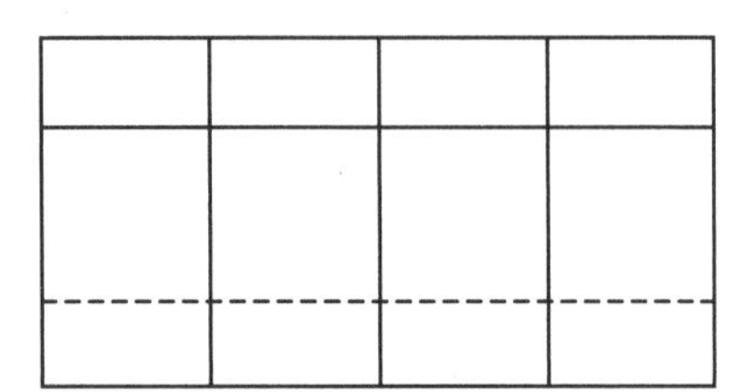

d. 13.213 – 7.508

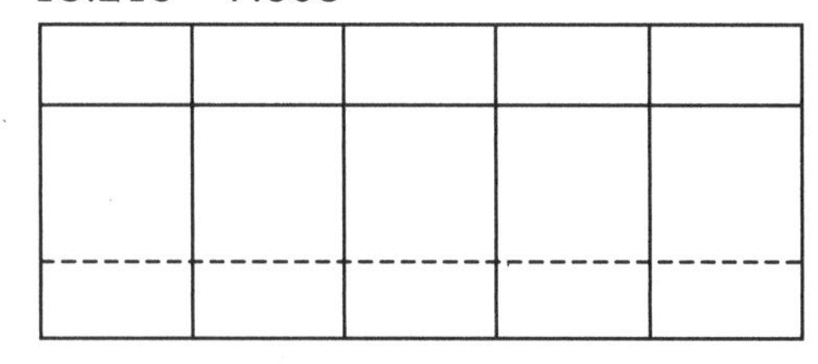

3. Compare the problems in #1 with their counterparts in #2. How are the algorithms for adding and subtracting decimals related to the usual algorithms for whole numbers? What is the difference?

OBJECTIVE: **Explore the standard multiplication algorithm for decimals**

You Will Need: **Base ten pieces from Materials Card 4.2.1 (Physical Manipulative option: *Base Ten Pieces*)**

Just as with whole number multiplication, we can illustrate multiplication of decimals with rectangles made up of base ten pieces. Viewing the product of two decimals in this way can help students discover the similarities to the whole number algorithm, as well as understand the unique aspects of decimal multiplication.

1. For this activity, you will use the flat as the whole. Thus, in this situation, the dimensions of the flat are considered to be 1 × 1, not 10 × 10 as with the case of whole numbers.

   In this case, the length of a long is 1. What is its width? Using this length and width, what is the area of a long (relative to the 1 × 1 flat)?

   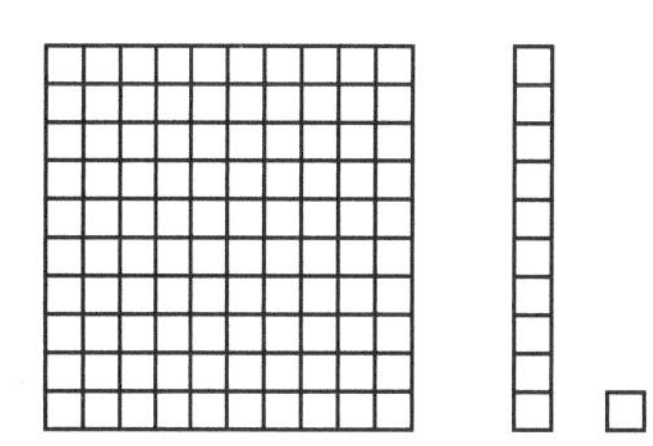

   What are the length and width of each small square? Using this length and width, what is the area of a small square (relative to the 1 × 1 flat)?

2. a. Record the dimensions and area of each of the sub-rectangles shown in this flat where the dimensions of the flat are 1 × 1 as in number 1 above:

   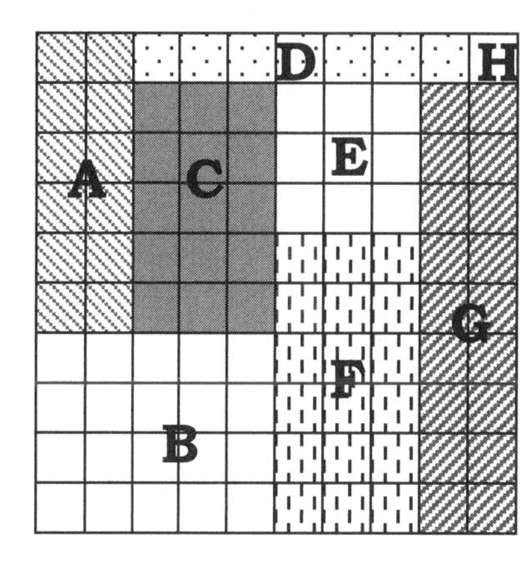

   Rectangle A _____ × _____ = _____

   Rectangle B _____ × _____ = _____

   Rectangle C _____ × _____ = _____

   Rectangle D _____ × _____ = _____

   Rectangle E _____ × _____ = _____

   Rectangle F _____ × _____ = _____

   Rectangle G _____ × _____ = _____

   Rectangle H _____ × _____ = _____

   b. Summarize how to multiply tenths by tenths.

3. Consider the following rectangle made up of base ten pieces, where the flat represents 1, each long represents 0.1 and each small square represents 0.01.

   a. Label the dimensions of this rectangle.

   b. What multiplication problem is modeled here?

   c. When the pieces representing the area are combined, what number does the area represent? Explain.

4. The following shows how to use expanded notation to find 1.2×2.3.

$$\begin{aligned} 1.2 \times 2.3 = (1 + 0.2)(2 + 0.3) &= (1 + 0.2)2 + (1 + 0.2)(0.3) \\ &= 1 \times 2 + 0.2 \times 2 + 1 \times 0.3 + 0.2 \times 0.3 \\ &= 2 + .4 + .3 + .06 \\ &= 2.76 \end{aligned}$$

   Explain how this method is modeled by what was done in problem #3.

5. In order to investigate the pencil-and-paper algorithm further, consider the multiplication problem 1.2×2.3 again.

   a. Write each of the decimals 1.2 and 2.3 in fraction form.

   b. Compute the product using these fractions.

   c. What does the denominator of the product tell you about the number of decimal places in the decimal form of this number?

   d. When multiplying decimals using the standard algorithm, how does one decide where to place the decimal point in the answer?

   e. How does the process in parts (a)-(c) help to explain the rule you wrote in part (d)?

6. Use fractions to compute the product 4.573×8.21.

   What does this tell you about the number of places to the right of the decimal point in the decimal form of this product?

OBJECTIVE: **Explore the standard division algorithm for decimals**

The next activity will help you make sense of the usual procedure for dividing decimals - the reasons for doing things the way we do, as well as why the algorithm works. Just as with decimal addition, subtraction and multiplication, we begin by comparing decimal division to whole number division.

1. Consider the division problem illustrated below.

   Using the measurement approach to division, you could ask:
   How many groups of size 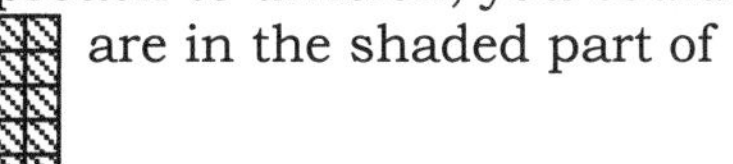 are in the shaded part of 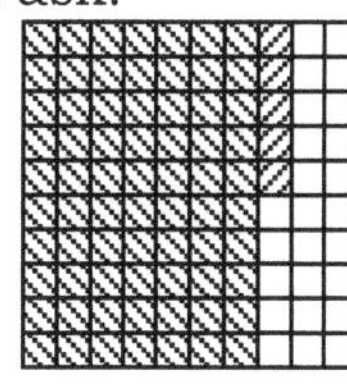

   a. If the 10 × 10 grid represents 100 (a flat) what division problem is modeled above?
   ______ ÷ ______

   b. If the 10 × 10 grid represents 1 (a decimal square) what division problem is modeled above? ______ ÷ ______

   c. Circle groups of size in the shaded part of

   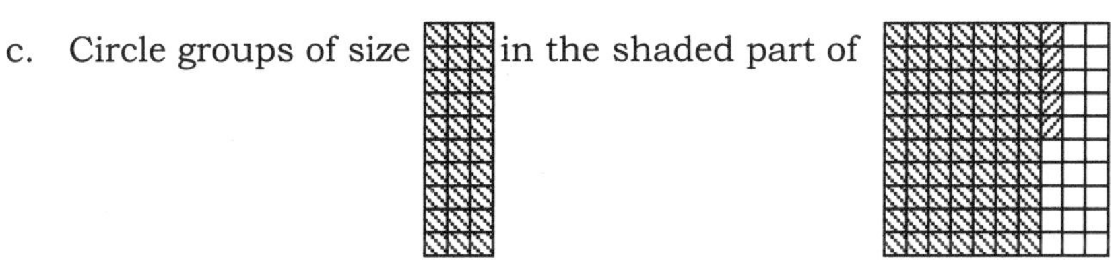

   How many whole groups did you circle? _____What fraction of the group of size  is left? _____

   d. Based on what you did in part (c), what is the answer to the division problems you wrote in parts (a) and (b)? _____

   e. In parts (a) – (d) above, you showed that 75 ÷ 30 and .75 ÷ .30 are equivalent division problems. How is this related to the usual algorithm for dividing decimals?

2. When we divide decimals, such as 5.184 ÷ 2.16, the first step in the standard algorithm is to move the decimal point in both the divisor (2.16, in this case) and the dividend (5.184, in this case) the same number of places until the divisor is a whole number.

$$\text{Thus, } 2.16\overline{)5.184} \text{ would be changed to } 216\overline{)518.4}.$$

   a. Why do you suppose we change the problem in this way?

   b. One way to make sense of moving decimal points in a division problem is to recall that moving a decimal point is equivalent to multiplying a decimal by a power of 10. In 5.184 ÷ 2.16, what power of 10 did we multiply the divisor and the dividend by in order to change it to 518.4 ÷ 216? ________

   This can be shown as follows:

$$5.184 \div 2.16 = \frac{5.184 \times 10^2}{2.16 \times 10^2} = \frac{518.4}{216} = \_\_\_\_\_\_ \div \_\_\_\_\_\_$$

   c. Write an equation, similar to the one in part (b), that shows how to change 203.415 ÷ 18.7789 to an equivalent division problem with a whole-number divisor.

OBJECTIVE: **Discover the meaning of ratios and find equivalent ratios**

You Will Need: **Materials Card 7.3.1**

Ratios allow us to compare the relative sizes of quantities. Using the concept of equivalent ratios can help you solve many day-to-day problems. If you wish to enlarge or reduce a recipe for example, the ratio of each ingredient to the whole amount must stay the same if the results are to be satisfactory. The next activity will help you discover the meaning of ratios as well as give you practice in determining when ratios are equivalent.

1. From the shapes on Materials Card 7.3.1, place the following in a pile in front of you:

   2 rhombuses 3 trapezoids 4 hexagons

   A **ratio** is an ordered pair of numbers, often indicating relative amounts. For example, the ratio 2 : 9 represents the ratio of the number of rhombuses to the number of all the shapes in your pile. Another way to show this ratio is in fraction notation:

   2/9 = number of rhombuses in pile /number of all shapes in pile

   Write the ratios that represent these relationships between the numbers of these shapes in your pile:

   a. rhombuses to trapezoids = ________ b. hexagons to all shapes = ________

   c. trapezoids to all shapes = ________ d. hexagons to non-hexagons = ________

2. A ratio can show the relationship of *part-to-part* or *part-to-whole*. For example, the ratio rhombuses : all shapes shows a relationship of part-to-whole.

   Look at the ratios you wrote in #1(a)-(d). Which examples represent a part-to-whole relationship?

   Which examples represent a part-to-part relationship?

3. Identify whether the following ratios compare part-to-part or part-to-whole.

   a. number of girls in class : number of students in class

   b. cups of lemonade concentrate : cups of water

   c. number of games a team won this season : number games the team lost this season

4. In problems #1 and #2, you used ratios to compare the relative sizes of the *numbers* of shapes in sets. We can also compare the relative sizes of *areas* of shapes.

   a. From the shapes on Materials Card 7.3.1, take a rhombus and cover it with triangles. How many triangles exactly cover the rhombus? ______ The area of the triangle is what fraction of the area of the rhombus? ________ Thus, the ratio of the area of a triangle to the area of a rhombus is _____ : _____.

b. In your ratio in part (a), the first number should have been smaller than the second since the area of the triangle is less than the area of the rhombus. Find the following ratios.

area of a triangle : area of a trapezoid = _____ : _____

area of a triangle : area of a hexagon = _____ : _____

5. Using only the rhombuses, trapezoids, and hexagons, find shapes whose areas have:

a. a ratio of 2 : 1 = ________________________ : ________________________

b. a ratio of 3 : 1 = ________________________ : ________________________

c. What is the ratio of the area of the triangle to the area of the hexagon? _____ : _____ Can you find another pair of shapes whose areas are in the same ratio? ________ Why or why not?

6. A triangle from the Materials Card has been cut in half to create two right triangles. Copies of this new right triangle exactly cover the rhombus as shown in the figures below.

a. What is the ratio of the area of the new right triangle to the area of the rhombus?

____

b. Show how to cover the trapezoid with this new triangle. What is the ratio of the area of the new triangle to the area of the trapezoid? ____________

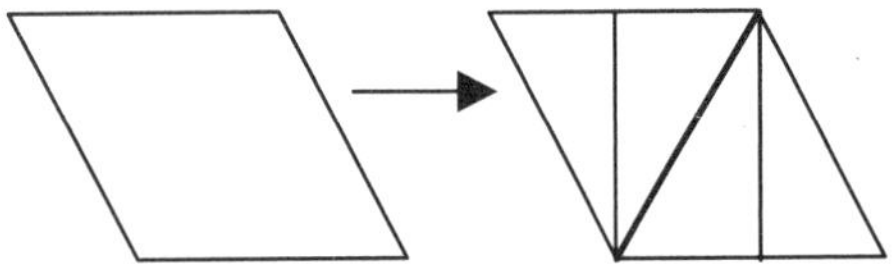

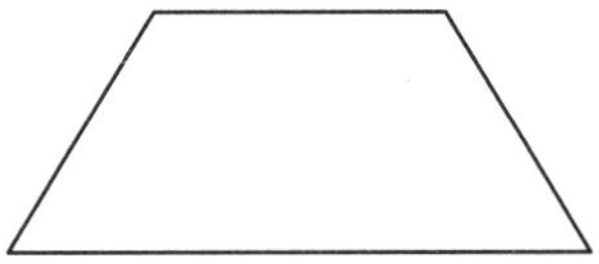

c. Now, use what you did in parts (a) and (b) to compare the area of the trapezoid to the area of the rhombus. Explain how you got your answer.

The ratio of the area of the trapezoid to the area of the rhombus is:________ : ________

d. You can find the ratio of the area of the trapezoid to the area of the rhombus in a different way. This time, use the *original* triangle and write an explanation of your answer.

area of trapezoid : area of rhombus = ________ : ________

e. Find the ratio of the area of the trapezoid to the area of the rhombus in yet another way. This time, compare the areas of the two shapes without using triangles. Explain your reasoning.

area of trapezoid : area of rhombus = ________ : ________

f. The ratios you wrote in parts (c), (d), and (e) are equivalent ratios. Show how to use cross multiplication to prove that the ratios from parts (c) and (d) are indeed equivalent.

OBJECTIVE: Use proportions to solve problems

When you know how to set up proportions, you are able to solve many problems more easily. Proportions are especially useful in solving percent problems or problems involving other ratios.

1. a. Compute your number of heart beats in one minute as follows: Find your pulse. Count the number of heart beats in 6 seconds. Write this number in the blank numerator below. The variable n is used to represent the unknown number of times your heart will beat in 60 seconds.

$$\frac{8}{6 \text{ sec.}} = \frac{n}{60 \text{ sec.}}$$

b. An equation between two ratios, such as the one above, is called a **proportion**. Use what you know about cross multiplication of fractions to solve this proportion for n.

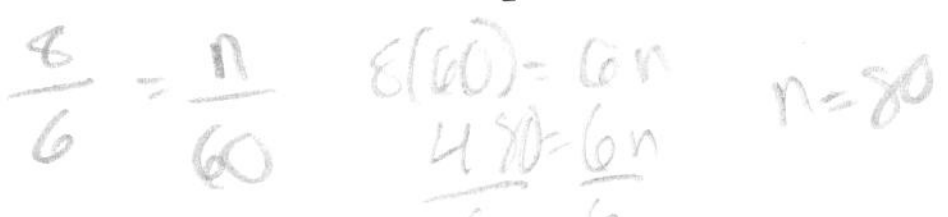

How many times did your heart beat in one minute? $n =$ 80

c. Use the information about your heart rate you found in part (b) to determine the following. Use proportions and show the work you do to solve them.

How many times does your heart beat in one hour? 4800

At this rate, how many times will your heart beat in one day? 115200

one week? 806400 one year? 38707200 80 years? 3096576600

2. Suppose you want to make a dessert to serve as many people as possible. The recipe you are using serves 4 people and calls for $1\frac{1}{2}$ teaspoons of vanilla. The recipe also calls for flour, but you only have enough flour to increase the recipe to serve 11 people. How many teaspoons of vanilla will you need?

a. There are two ways to correctly set up a proportion for this problem so that the unknown number of teaspoons of vanilla is in the numerator. Examine the four proportions shown next and decide which of these could be used.

(i) $$\frac{11 \textit{ people served by new recipe}}{4 \textit{ people served by original recipe}} = \frac{n \textit{ tsp of vanilla in new recipe}}{1.5 \textit{ tsp of vanilla in original recipe}}$$

(ii) $$\frac{1.5 \textit{ tsp of vanilla in original recipe}}{11 \textit{ people served by new recipe}} = \frac{n \textit{ tsp of vanilla in new recipe}}{4 \textit{ people served by original recipe}}$$

(iii) $\dfrac{4 \text{ people served by original recipe}}{1.5 \text{ tsp of vanilla in original recipe}} = \dfrac{n \text{ tsp of vanilla in new recipe}}{11 \text{ people served by new recipe}}$

(iv) $\dfrac{1.5 \text{ tsp of vanilla in original recipe}}{4 \text{ people served by original recipe}} = \dfrac{n \text{ tsp of vanilla in new recipe}}{11 \text{ people served by new recipe}}$

b. For each of the proportions you chose in part (a), examine the units in each ratio. What do you notice? Why does it make sense to set the two fractions equal to each other?

you have to put the common ones in the same position, you have to set it up so you can answer it so the two equations will be equal.

c. For those proportions in part (a) that were not set up correctly, explain why it does not make sense to set the two fractions equal to each other.

because the two equations were not set equal to one another.

d. Choose one of the proportions from part (a) and solve for the number of teaspoons of vanilla needed to serve 11 people.

$\frac{1.5}{4} = \frac{n}{11}$ $\quad \frac{16.5}{4} = \frac{4n}{4}$ $= 4.125$ tsp of vanilla

3. a. For a party, you want to make fruit punch. The recipe calls for 3 parts fruit juice concentrate and 8 parts lemon-lime soda. If you only have 2 cups of fruit juice concentrate, does it make sense to use 7 cups of soda? Why or why not?

$\frac{3}{8} = \frac{2}{7}$ $\quad 16 = 21$ $\quad$ no they are not equal

b. If you have 2 cups of fruit juice concentrate, how many cups of soda should you use?

$\frac{3}{8} = \frac{2}{x}$ $\quad \frac{16}{3} = \frac{3x}{3}$ $\quad x = 5.5$ $\quad$ 5.5 cups of soda

c. What are some other amounts of fruit juice concentrate and lemon-lime soda that could be mixed to make this fruit punch? Show how to answer this question using proportions.

Always maintain 3/8 as a constant ratio, then add a variable for either the concentrate or the soda you want to change, and leave the other as an open variable

$\frac{3}{8} = \frac{x}{20}$ $\quad \frac{60}{8} = \frac{8x}{8}$ $\quad x = 7.5$

for 20 cups of soda you need 7.5 cups of concentrate

OBJECTIVE: **Use a grid approach to model percent problems**

The concepts of percent problems can be visualized using a 10 × 10 grid. As you work through this activity, think about how this model is related to the ways in which you usually solve problems of this type.

1. The notation *n*% or *n* **percent** means *n* parts out of 100 equivalent parts. We will use a 10 × 10 grid to represent the **whole**, or 100%, and the shaded area to represent the **part**. For example, in the following grid 30 parts out of 100 are shaded. This represents 30% of the whole.

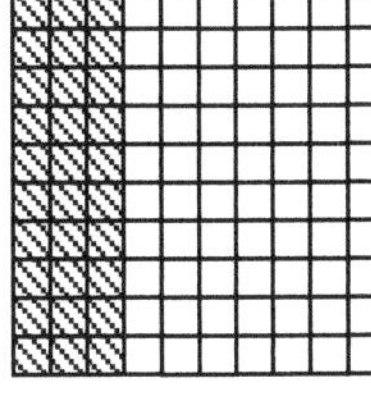

30 %

   a. For each of the following grids, shade the indicated percent of the whole.

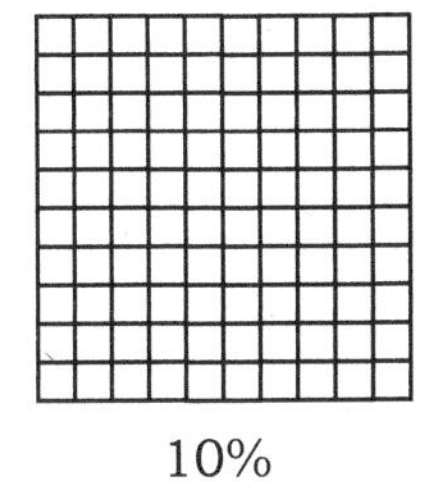

10%

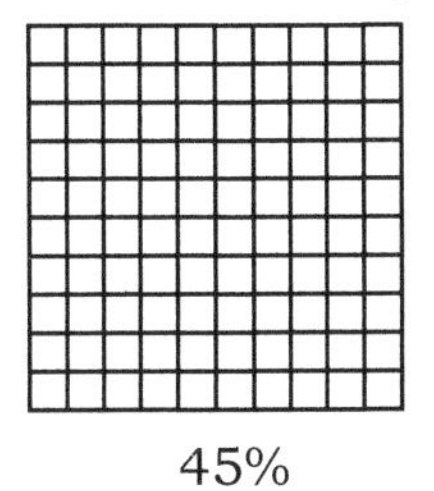

45%

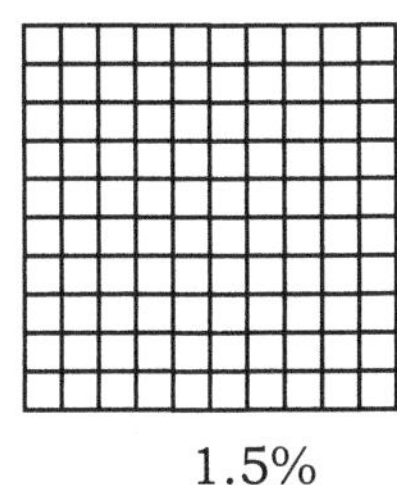

1.5%

   b. What would you have to shade to represent 250% of the whole? Make a rough sketch of this.

2. Recall that when we used base ten pieces to model whole numbers, a 10 × 10 grid (or flat) represented the number 100. In the case of decimals, we assigned a value of 1 to the 10 × 10 grid (or decimal square). When modeling percents with a 10 × 10 grid, the *value* of the grid varies from problem to problem, depending on the whole amount. However, the 10 × 10 grid *always* represents 100%.

   a. The value of the whole represented by this grid is 400.

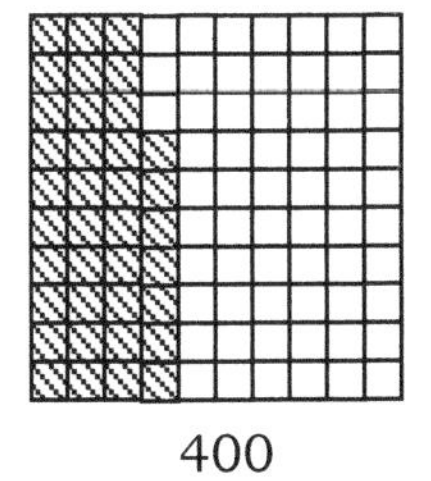

400

   When using a grid to solve a percent problem, it is often helpful to first find the value of one small square. What is the *value* of one small square in this grid? ________

   What percent of the whole is shaded? ________

   Find the *value* of the shaded part of the grid. Explain how to determine this, in terms of the grid.

b. Next, assign a value of 50 to the entire 10×10 grid. Use a process similar to the one you used in part (a) to find 37% of 50.

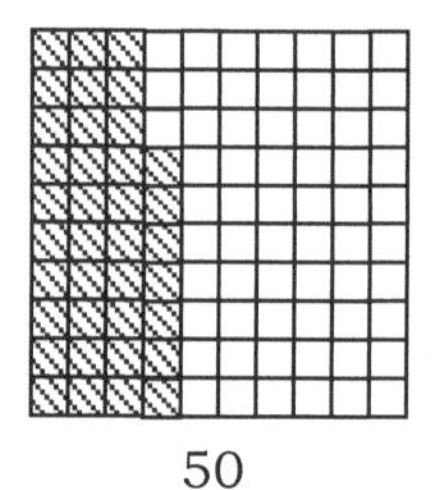

3. In problem #2, you were given the percent and the value of the whole, and you found the value of the part. That is, you answered the questions "What is 37% of 400?" and "What is 37% of 50?" Suppose, instead, you know the whole and the part and need to find the percent.

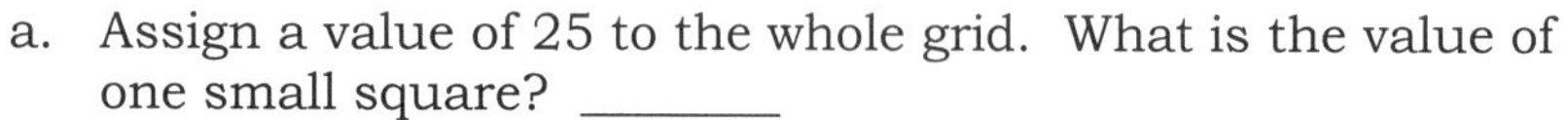

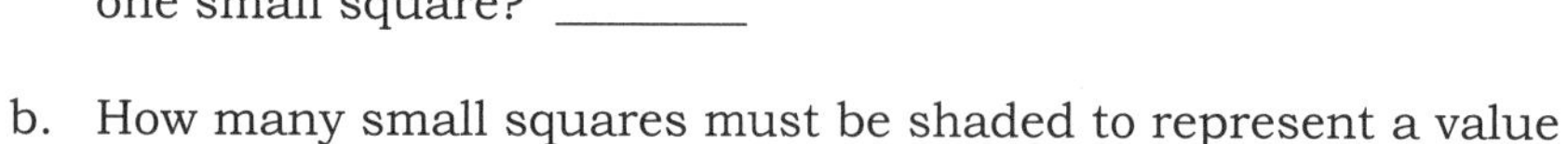

   a. Assign a value of 25 to the whole grid. What is the value of one small square? ________

   b. How many small squares must be shaded to represent a value of 15? That is, 15 is what percent of 25? Explain.

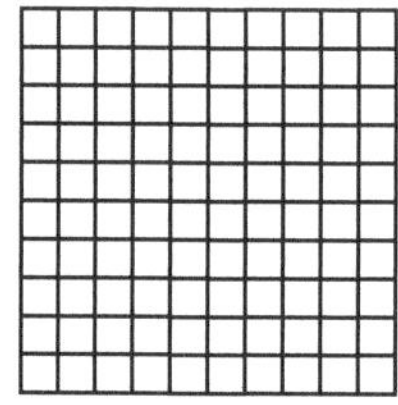

4. A third type of percent problem is when you know the value of the part and the percent and must find the value of the whole.

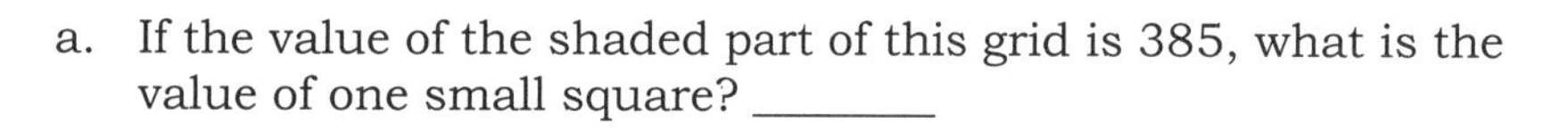

   a. If the value of the shaded part of this grid is 385, what is the value of one small square? ________

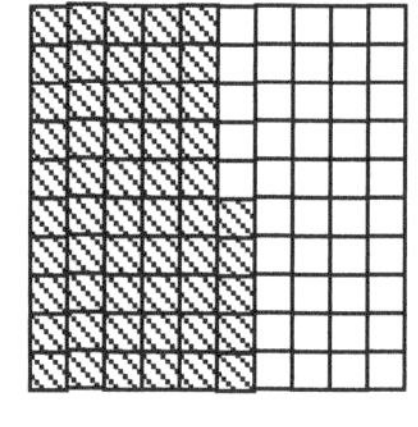

   b. Find the value of the whole grid, using your answer to part (a). That is, answer the question: 385 is 55% of what number?

5. Show how to use a grid approach to solve the following percent problems. In each case, first find the value of a 1% square. Clearly label the grids you use.

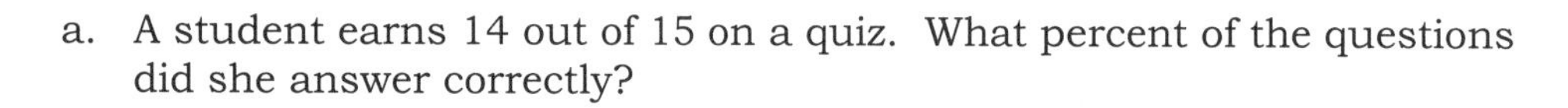

   a. A student earns 14 out of 15 on a quiz. What percent of the questions did she answer correctly?

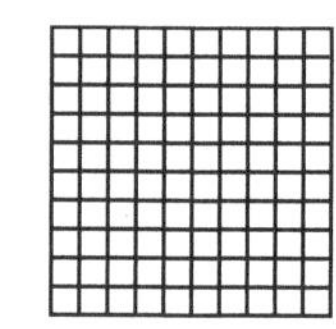

   b. A first-time homebuyer needs to have a 3% down payment for a mortgage loan. How much will she need for a down payment on a $97,000 home?

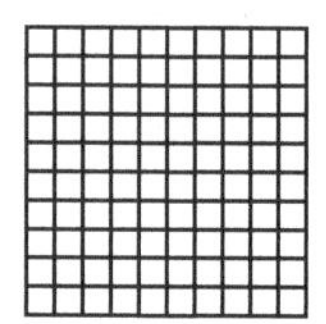

   c. A teacher's current salary is 115% of his original salary. If he now earns $38,000, what was his original salary?

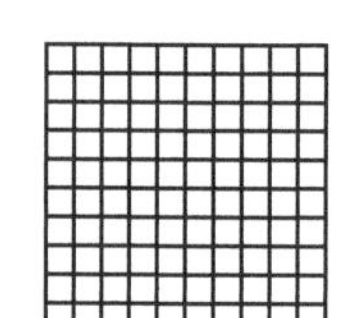

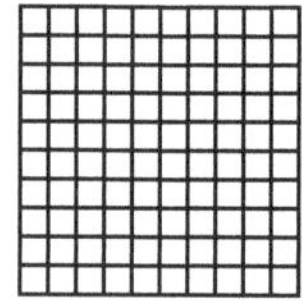

OBJECTIVE: Convert between percents, fractions and decimals

You may already be familiar with the common shortcuts for converting between percents and decimals. The next activity will help you to explain *why* these shortcuts work.

1. A shaded 10 × 10 grid can help you visualize equivalent fractions, decimals and percents.

   What fraction of this grid is shaded? ________

   What percent of the grid is shaded? ________

   What decimal does this shaded grid represent? ________

2. Converting a percent to a fraction is a matter of using the definition of percent. That is, n% means n parts out of 100 equivalent parts. Written as a fraction, n% = ________

   Write each of the following percents as a fraction.

   a. 87% = ________ b. 321% = ________ c. 1/2% = ________

3. In order to convert a percent to a decimal, you can use what you know about converting between fractions and decimals.

   a. In problem #2(a), you wrote 87% as 87/100. What decimal represents this fraction?

      Convert 19% to a decimal by first converting to a fraction.

   b. Based on your work in part (a) suggest a shortcut for converting from a percent to a decimal.

4. If you want to convert a decimal to a percent, you can simply reverse the process in #2.

   a. Write each of the following decimals as fractions with a denominator of 100. Then write the percent represented by the fraction.

Decimal	Fraction (over 100)	Percent
0.05		
0.7		
1.35		

   b. Based on your work in part (a), suggest a shortcut for converting from a decimal to a percent.

5. When converting a fraction to a percent, you must remember that a fraction can be interpreted as a division problem. That is, you first convert a fraction to a decimal, then to a percent.

   a. Convert each of the following common fractions to a decimal and then to a percent. Try to do this mentally, and then check with your calculator.

Fraction	Decimal	Percent
1/20		
1/10		
1/5		
1/4		
1/2		
3/4		

   b. The fractions in part (a) had terminating decimals. Each of the following fractions has a repeating decimal representation. Show how would you convert each of these to a percent.

Fraction	Decimal	Percent
1/3		
2/3		

OBJECTIVE: Solve percent problems mentally

Having a picture of a grid in your mind can be helpful when you need to solve percent problems *mentally*. Additionally, some common percents have convenient fraction equivalents that are sometimes easier to use when making mental computations.

1. Assign a value of 50 to the entire 10 × 10 grid.

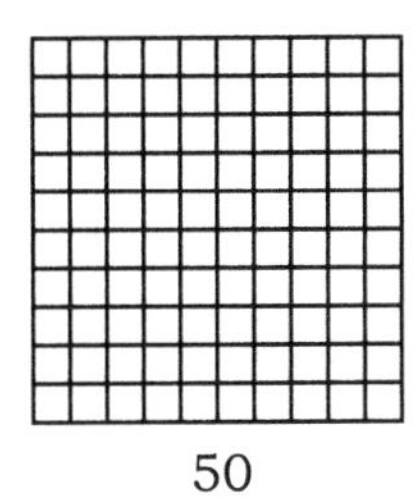

   a. Shade 10% of the grid.
   b. What is the value of the shaded part? ________
   c. What shortcut for computing 10% of a number does this example suggest?

   d. Next, find the value of 5% of the grid. ________

   e. How is finding 5% of a number related to finding 10% of that number?

   f. Show how to use what you did in parts (a) – (e) to find 15% of 50.

   g. Suppose you want to leave a tip of 15% of the bill when your dinner costs $24.60. How much is the tip? Compute your answer mentally. Then record your thought process.

2. For each of the following, make a mental computation. Record how you thought about each one.

   a. 50% of 1047 b. 20% of 25 c. 25% of 80 d. 75% of 60

3. In computing 50% of 1047, did you use a fraction equivalent? That is, did you think "1/2 times 1047" or "divide 1047 by 2"? Explain why these two ways to think of the problem are equivalent to the original percent problem. Sketch a shaded grid to accompany your explanation.

4. Explain how fractions could play a role in mentally computing each of the percent problems in #2(b)-(d).

OBJECTIVE: **Use an equation approach to solve percent problems (eManipulative option: *Percent Gauge*)**

As you saw in Activity 7.4.1, a 10 × 10 grid provides a concrete model for percent problems. In the next activity, you will see how this concrete model leads to an equation approach, a more abstract yet powerful and efficient method for solving problems involving percent.

1. In Activity 7.4.1, you used a 10 × 10 grid to answer the question, "What is 37% of 50?" Likely, your process went something like this:

   Value of the part = number of shaded squares × (value of a 1% square)

   $= 37 \times \left(\frac{1}{100} \times 50\right)$

   $= 37 \times 0.5$

   $= 18.5$

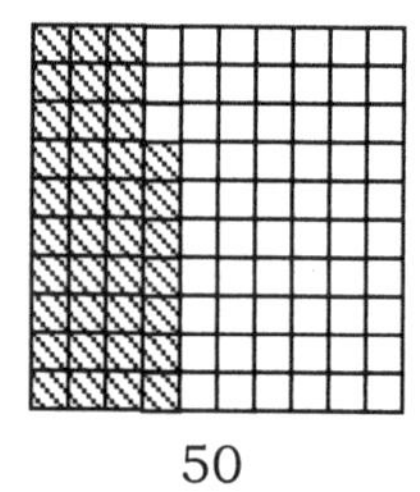

   a. Notice that the expression $37 \times \left(\frac{1}{100} \times 50\right)$ from above, could be rewritten as: $\left(37 \times \frac{1}{100}\right) \times 50$. What property of multiplication allows you to do this?________________

   This new expression above gives you $\frac{37}{100} \times 50$. That is, in order to find the part you multiply the percent (over 100) by the whole. Notice how there is a direct translation from the percent problem to an equation:

   What is 37% of 50?

   $$x = \frac{37}{100} \times 50$$

   You used x to represent the unknown (in this case, the *part*) and translated the word "is" to "=" and "of" to "×". The general form of this equation is:

   $$part = \frac{percent}{100} \times whole$$

   b. Use an equation with the above form to solve the following percent problems. In each case, two of the pieces of the above equation are given. Solve your equation to find the unknown piece.

   35 is what percent of 160?________

   150 is 18% of what number?________

2. Next, you will apply the equation approach to a percent decrease problem.

   During an end-of-the-season sale, a sporting goods store marks down all their ski equipment 25%. A pair of ski boots is on sale for $90. What was the original price of the boots? Complete parts (a) – (c) to solve this problem.

   a. The sale price is what percent of the original price?

   b. What is the unknown in this problem (the percent, the part or the whole)?

   c. We can reword the problem as "$90.00 is 75% of what number?" Translate this question into an equation and solve for the unknown.

   What was the original price of the boots? ________

3. Next, consider this percent increase problem:

   An elementary school has 315 students enrolled this year. If enrollment increased 5% over the enrollment for last year, how many students were enrolled last year?

   a. The enrollment this year is what percent of the enrollment for last year?

   b. What is the unknown in this problem?

   c. Restate the problem so that it asks a question similar to those in #1b and #2c.

   d. Translate your question in part (c) into an equation and solve.

4. Computer World is having a 30% off sale on all software. A customer brings in a coupon that allows him to take an additional 20% off the sale price. He exclaims "Wow! I just saved 50%!"

   a. If a software package had an original price of $125.00, what was the sale price with the 30% discount?

   b. How much did the customer pay if he used his 20%-off coupon to purchase the software package in part (a)?

   c. Is your answer in part (b) the same as a 50% savings? Why or why not?

   d. What percent of the original price did the customer pay for the software?

EXERCISE

☆ ☆ ☆ MAGIC STAR ☆ ☆ ☆

Solve the proportions and ratio problems below. Each problem is designated by a letter and that letter appears in one of the circles in the star. When you find the answer to each problem, put it in the corresponding circle.

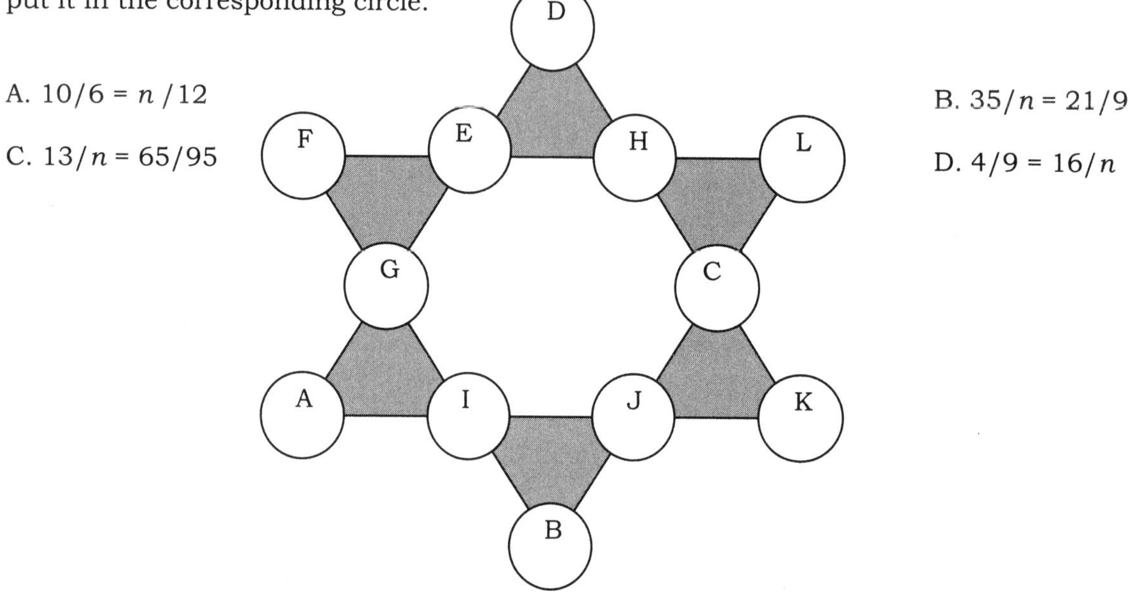

A. $10/6 = n/12$

B. $35/n = 21/9$

C. $13/n = 65/95$

D. $4/9 = 16/n$

E. A photo that is 3.5 cm by 6 cm is enlarged. Its new dimensions are n cm by 24 cm. What is n?

F. The scale on a map is given as 6 inches = 75 miles. If you take a 475 mile trip, what would it be on the map (in inches)?

G. The ratio of the weight of an object on Jupiter to its weight on Earth is 8 to 3. How much would an 11.25-pound rock on Earth weigh on Jupiter?

H. If 3.7 grams of salt will dissolve in 10 grams of water, how many grams of salt will dissolve in 100 grams of water?

I. The ratio of boys to girls in a class is 1:2. If there are 51 students in the class, how many boys are there?

J. If you can buy 24 pencils for 88 cents, how much will you pay for 15 pencils?

K. If 192 meters of pipe weighs 48 kg, how much pipe would weigh 2 kg?

L. Your car has traveled 93.5 km on 8.5 liters of gas. How many km/liter did your car average?

WHAT IS THE MAGIC OF THE STAR?

CONNECTIONS TO THE CLASSROOM

1. A teacher asked her students to write the decimal numerals to represent the following numbers:

   thirty two and thirteen hundredths

   five hundred and twenty one thousandths

   $19\,\frac{10}{1000}$

   $34\,\frac{11}{100}$

   a. One student wrote these answers: 32.013, 500.0021, 19.0010, and 34. 011. Is the student correct or incorrect? If incorrect, what is the student doing wrong? How might you help the student?

   b. Another student answered the second problem as 0.521. Is this correct? If not, why not? How would you name 0.521?

2. A student says "I know that when I put a zero at the end of a numeral, I get a bigger number, like 20 is greater than 2. I don't understand why it doesn't work the same with decimals, like 0.20 and 0.2. This is still 20 parts and 2 parts. Why isn't 0.20 greater than 0.2?" How could you help this student?

3. A student claims that $0.3 < 0.25$ because $3 < 25$. Is the student correct? Why or why not? Show how to use decimal squares to explain your answer.

4. When asked to compute 4.573×8.21, a student says, "If you count the number of decimal places to the left in the factors (2 in total), you could place the decimal two places from the left in the answer." Is this okay? Does this always work? Explain.

5. When asked to find these sums, a student showed the following work:

$$\begin{array}{r} .8 \\ +\ .5 \\ \hline .1\ 3 \end{array} \qquad \begin{array}{r} .9 \\ +\ .7 \\ \hline .1\ 6 \end{array} \qquad \begin{array}{r} .4 \\ +\ .2 \\ \hline .6 \end{array} \qquad \begin{array}{r} .6 \\ +\ .6 \\ \hline .1\ 2 \end{array}$$

   a. Explain how estimation could be used to check the reasonableness of each of this student's answers.

   b. Describe the error pattern this student is using. Why might she be using such a procedure?

   c. What does this student actually understand about adding decimals? What does she still not understand?

   d. What instructional procedures and/or hands-on materials might you use to help the student with this problem?

6. A teacher asked a student to solve the following problem: *A pair of shoes is on sale for 30% off the original price. If the sale price is $45.50, what is the original price?*

   The student claims that she must compute 30% of $45.50 and then add the result to $45.50 to find the original price. Is the student correct? Explain.

MENTAL MATH

Use ratios to solve these mental math problems. Which item is a better buy? Explain your thinking in each case.

DIRECTIONS IN EDUCATION
Mathematics as Communication
Mathematically Speaking

Young children develop language through verbal communication. When children use the language of the classroom to express their own ideas and thoughts, they invite the teacher "inside their head". They provide the teacher with a rich diagnostic tool through which the language of the classroom can be linked to what is already known by the student. Mathematical language allows students and teachers to express what is known and to use that knowledge base as the foundation for additional learning. The mathematics classroom can be as language rich as the reading/language arts classroom. Mathematical vocabulary can be placed on cards and displayed in the room as it is introduced. By keeping the language of mathematics visible and by encouraging the use of the new vocabulary in oral and written work in math, the teacher can set the stage for development of mathematical communication skills which empower students in the future study and application of mathematics.

Encouraging students to "speak mathematics" will:

- relate concrete materials, pictures, and graphic representations to mathematical ideas.
- allow students to evaluate and clarify their own thoughts about mathematics.
- help establish shared understandings of mathematics in a community of learners.
- connect their knowledge of the world around them to the mathematics of their classroom.
- extend their thinking about mathematics beyond pencil and paper skills
- encourage student to experiment with and expand on mathematical ideas.
- develop comfort with the everyday use of mathematical vocabulary.
- encourage students to listen and respond to the ideas of others in the classroom.
- enhance the teacher's ability to assess student needs and to make sound instructional decisions.

Mathematical communication can be encouraged by:

- **Talking and listening**. This can be encouraged in small groups as well as with the whole class. When done in small groups, some debriefing with the whole class is advisable so that the teacher can question the students and clarify any misconceptions.
- **Writing.** Students can explain the processes they used in problem solving, discuss individual feelings and attitudes toward mathematics, and create mathematical stories and problems to share with the class.

- **Representing.** When students use manipulatives or pictorial representations in conjunction with oral or written language to explain mathematical ideas, meaning is clarified.

- **Reading.** Children's literature is rich with books and stories containing mathematical problems and illustrating how other children have solved these problems. Children can use these stories as models to create their own mathematical stories for classmates to share. Interesting mathematics lessons can use children's literature as the starting point. This motivates student interest and involvement in the lesson.

As you think about mathematical communication, ask yourself:

1. What is the role of the classroom teacher in encouraging such communication?

2. Can I identify new vocabulary which should be developed in conjunction with a math concept?

3. Can I develop a math lesson based on a piece of children's literature such as *Alexander, Who Used To Be Rich Last Sunday,* by Judith Viorst, New York: Atheneum, 1977.

4. How is student learning enhanced by opportunities to communicate about mathematics either in writing or orally?

5. What is the relationship between mathematical communication and
   - cooperative learning?
   - higher order thinking?
   - forming connections with the real world?

6. How effectively do I speak mathematics?

8 Integers

THEME: Understanding Integers and Their Operations

HANDS-ON ACTIVITIES

The set of integers consists of the whole numbers, 0, 1, 2, 3,..., and their opposites, -1, -2, -3, ... The number -3 is said to be the opposite of 3 since it is the same distance as 3 from 0 on the number line, but on the opposite side of 0. Also, we say "3 is the opposite of -3."

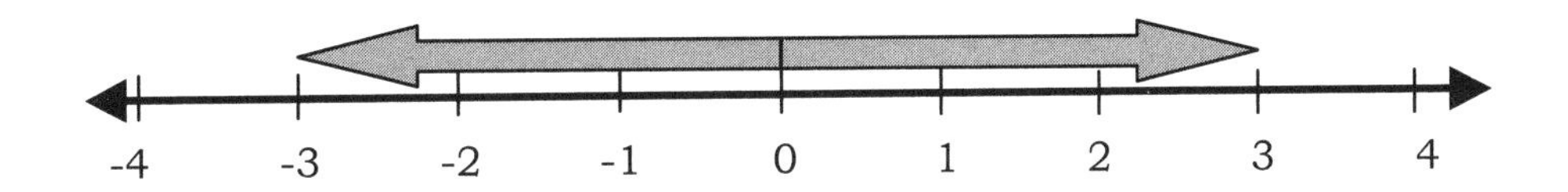

Extending the set of whole numbers in this way comes from the need to describe amounts that are less than or below a zero. When asked to give an example of a negative number, many adults will think of a negative balance in their checking account. Some examples that children may be able to relate to from their everyday lives are temperatures below 0° or elevations below sea level.

In Chapter 3, a set model was used to represent whole numbers and their operations. In this chapter, you will once again use sets of objects to model integers. However, in this case, you will use one color to represent positive integers and a different one to represent negative integers. In order to investigate operations with integers, you will revisit the concepts of set union for defining integer addition, the take-away approach for subtraction, repeated addition for multiplication, and partitive, measurement and missing factor approaches for division. You will also see a new way to think of subtraction – as adding the opposite.

OBJECTIVE: **Model integers**

You Will Need: **Materials Card 8.1.1**
(Physical Manipulative option: *Two-Color Counters*)

The set of integers consists of the whole numbers, together with their opposites. Integers, like whole numbers, can be represented by sets of objects. In this activity and throughout this chapter, you will use two different colored chips to model positive and negative integers and their operations.

1. The set of **integers** is the set **I** = {...,-4, -3, -2, -1, 0, 1, 2, 3, 4, ...} .

   a. The set of **positive integers** is {1, 2, 3, 4,...}. What is another name for this subset of **I**?

   b. The set of **non-negative integers** is {0, 1, 2, 3, 4, ...}. What is another name for this subset of **I**?

   The set of **negative integers** is {...,-4, -3, -2, -1}. This subset of **I** is sometimes called the set of "opposites of the counting numbers."

2. When representing integers using a set model, you will need objects of two different colors, one for positive integers and another for negative integers. A convention is to assign a black chip a value of 1 and a red chip a value of -1.

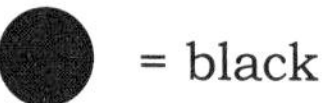

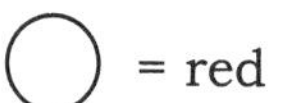

   What is the value of each of the following sets?

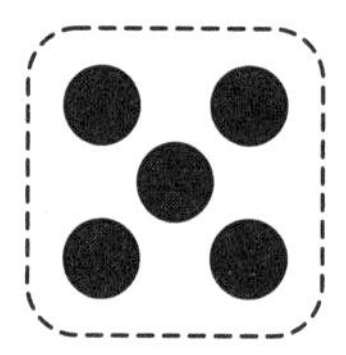

3. Consider the set consisting of black chips and red chips shown to the right. These chips can be matched in a 1-to-1 correspondence. Each pair of red and black chips "cancel" out. Thus, the entire set has a value of 0. -3 is called the **opposite** of 3 since together they represent 0. Also, 3 is called the opposite of -3.

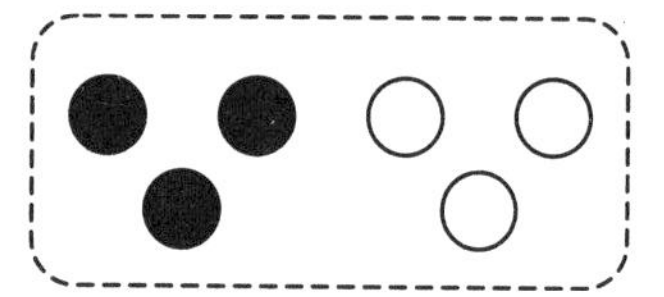

   Using black and red chips from Materials Card 8.1.1, show two different ways to represent 0 and record a sketch in each case. Name the pairs of opposites, as illustrated.

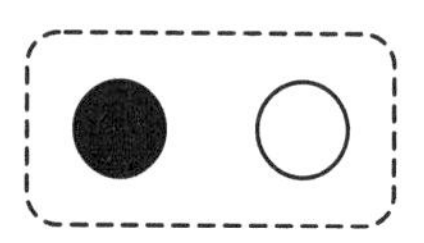

Opposites: 1, -1

4. As you found when representing zero, the representation of an integer with red and black chips is not unique; for example, the integer 5 could also be represented by the set shown to the right which differs from the representation of 5 in problem #1.

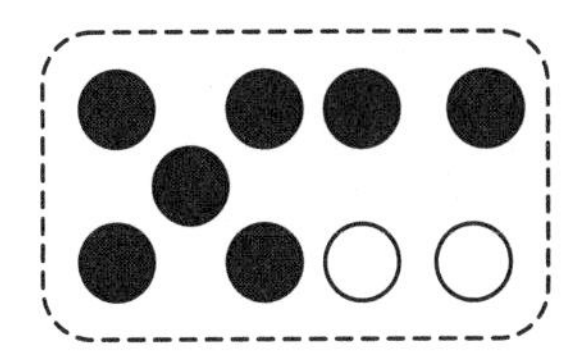

   a. Why does this set have a value of 5?

   b. Using your chips from the materials card, find two different ways to represent each of the following integers. Record sketches in each case.

   -2 3 -4

   c. Based on what you have learned, in how many ways can each integer be represented assuming that you have sufficiently many black chips and red chips?

OBJECTIVE: **Represent addition of integers**

You Will Need: **Red & black chips from Materials Card 8.1.1 (Physical Manipulative option: *Two-Color Counters*; eManipulative option: *Chips Plus*)**

Just as with whole-number addition, the sum of two integers can be represented as the union of two sets. Visualizing integer addition in this way can help students to discover and understand the usual rules we use for this operation.

Recall: 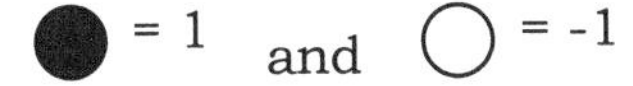

1. In the example to the right, two sets are shown being combined to form one new set.

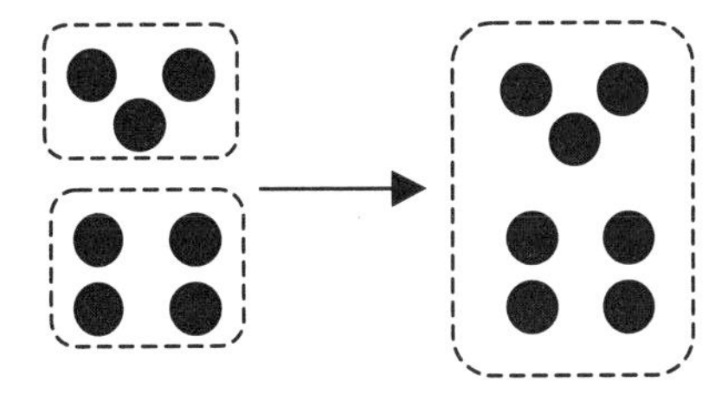

   a. This shows _____ + _____ = _____

   b. Using red and black chips from Materials Card 8.1.1, find the sum of -5 and -2. Record a sketch similar to the one in part (a) and label it.

   This shows -5 + (-2) = _____

c. Using red and black chips, find the sum of -5 and 2. Record a sketch and label it.

Explain how you obtain the sum -5 + 2 from your drawing. Then, show the process in your drawing above.

d. Using red and black chips, find the sum of 5 and -2. Record a sketch and label it.

Explain how you obtain the sum 5 + (-2) from your drawing. Then, show the process in your drawing above.

2. Refer to your work in problem #1(a)-(d).

   a. When will the sum of two integers be positive? Explain your answer.

   b. When will the sum of two integers be negative? Explain your answer.

   c. State a rule for finding the sum of two integers if both integers are positive or both are negative.

   d. State a rule for finding the sum of two integers if one integer is positive and one is negative.

3. Use your rules from problem #2(c) and (d) to compute the following sums without chips.

   a. 5 + (-4) = _____ b. -6 + (-8) = _____ c. -6 + 4 = _____

OBJECTIVE: **Represent subtraction of integers**

You Will Need: **Red & black chips from Materials Card 8.1.1 (Physical Manipulative option: *Two-Color Counters*; eManipulative option: *Chips Minus*)**

In Chapter 3, you examined subtraction of whole numbers with a set model and the take-away approach. This approach is also useful for modeling integer subtraction and leads to a new approach – adding the opposite. The following activity shows why we can change any subtraction problem to an addition problem.

1. The diagram at the right shows how to model the difference 5 – 3, using the take-away approach. We begin with 5 black chips and take out 3 black chips. The 2 remaining black chips show that 5 – 3 = 2.

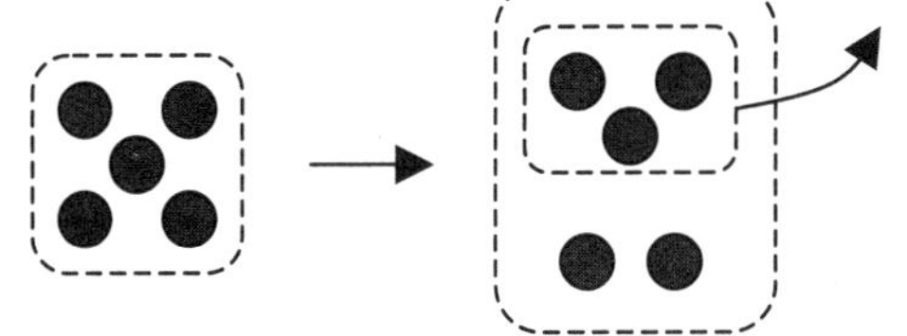

   a. Use the same approach as in the example above to find the difference -6 – (-2) using your chips from Materials Card 8.1.1. Record a sketch and the answer.

   -6 – (-2) = _____

   b. Now consider the subtraction problem 5 – 8. To model this problem, describe the chips you may start with (the number and color) ____________________ Describe the chips you would like to take away (the number and color). ____________________ Recall that 5 can be represented in many ways using black chips and red chips. Find a representation of the integer 5 that has enough black chips so that you are able to take 8 of them away. Show the take-away process in your drawing and record the difference.

   5 – 8 = _____

   c. You can also use the take-away approach to find a difference such as -3 – 5. Describe the chips you may start with (the number and color). ____________________ Describe the chips will you want to take away. ____________________

   Find a representation of the integer -3 that has enough black chips to be able to take 5 of them away. Show the take-away process in your drawing and record the difference.

   -3 – 5 = _____

2. Use your chips and the take-away approach to find each of the following differences. Record your pictures and label them.

   a. $2 - 6 =$ _____

   b. $-4 - (-2) =$ _____

   c. $5 - (-2) =$ _____

   d. $-3 - 4 =$ _____

3. Look back at your work for problem #2(d). You started with a set of 3 red chips and needed to take away 4 black chips.

   a. Since there were no black chips, you had to find another representation of -3 that had 4 black chips in it. Likely, you put 4 black chips and 4 red chips in with the original 3 red chips. What did you do with the 4 black chips after that?

   b. What happened with the 4 red chips you put in?

   c. What is a shortcut you could use to find $-3 - 4$?

4. Verify these equations using chips. Record your process with pictures.

   a. $-5 - 4 = -5 + (-4)$

   b. $3 - (-2) = 3 + 2$

   c. Summary: To subtract two integers, you can rewrite the difference as the first integer plus the ____________________ of the second integer and follow the rules for addition.

OBJECTIVE: **Model integer multiplication**

You Will Need: **Red & black chips from Materials Card 8.1.1 (Physical Manipulative option: *Two-Color Counters*; eManipulative option: *Chips Minus*)**

In this activity, you will see two different ways to explain the rules for multiplying integers. One way is to use red and black chips and the other way is to look for a pattern.

1. Recall from your work with whole numbers that multiplication may be viewed as repeated addition. So, 2 × 3 means 3 + 3. If you start with a set that represents 0 like this

 and then two times you put in three black chips, you get 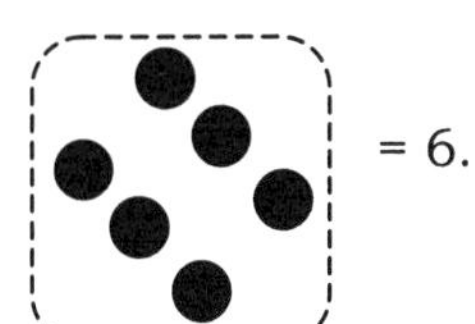 = 6.

Use red chips or black chips to compute the following products as in the example above. Record a drawing in each case.

a. 4 × 2 = 8 b. 3 × (-4) = -12 c. 2 × (-3) = -6

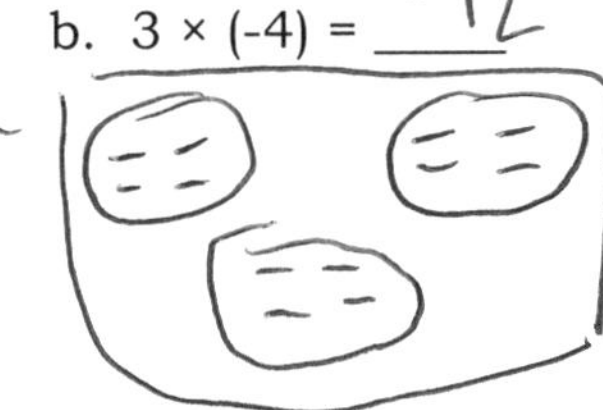

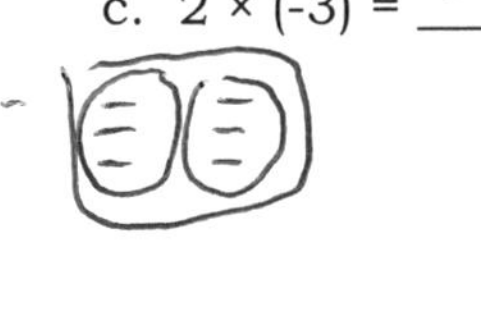

2. In problem #1, the first factor was positive in each of the products you found. Thus, you begin by *putting* chips *in* a set that represents 0. A chip model can also be used to find products where the first factor is negative. In this case, you *take out* chips from a set that represents 0, rather than put them in.

The problem 2 × (-3) means "*put in* two groups of three red chips to a set that represents 0." The product -2 × 3 means "*take out* two groups of three black chips from a set that represents zero." To find -2 × 3, you may begin with a representation of zero, such as the one shown to the right.

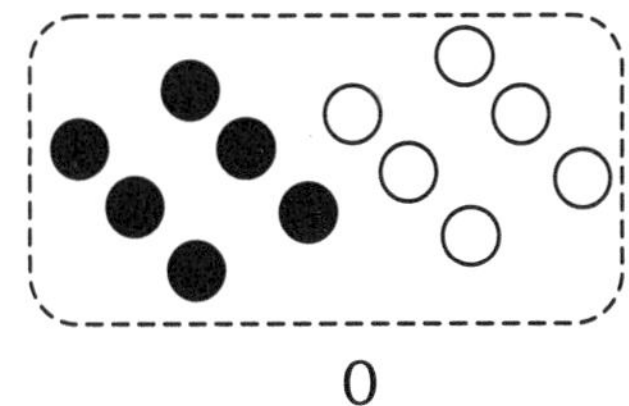
0

a. To find -2 × 3, take out 2 groups of 3 black chips. What chips remain?

b. What integer is represented by the remaining chips?

c. This shows that -2 × 3 = _____

3. Use your chips to find the following products. Record a sketch of your process in each case. It is important that you begin with an appropriate representation of zero.

a. -4 × 2 = -8 b. ~~-2 × (-2)~~ -3×-4 = 12 c. -4 × (-2) = 8

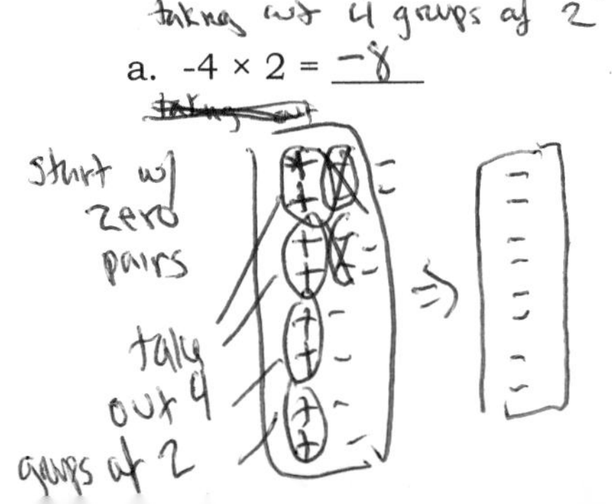

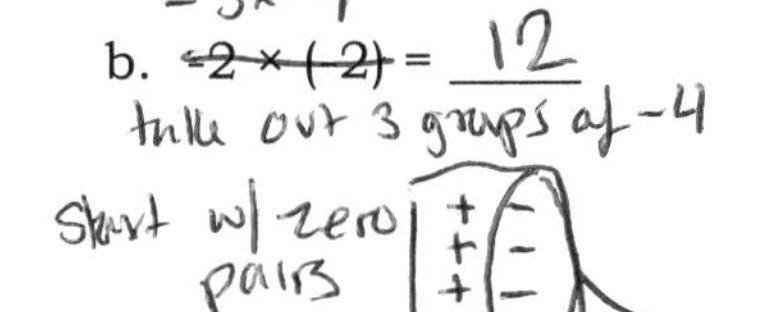

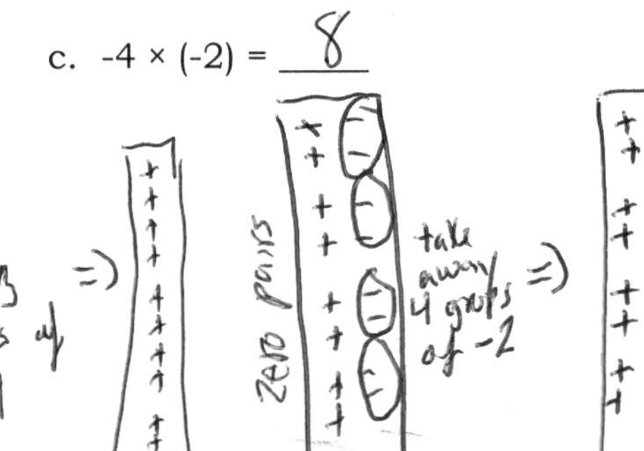

4. Refer to your work in problems #1-3 to complete the following.

   a. The product of a negative integer and a positive integer is ________________

   b. The product of a positive integer and a negative integer is ________________

   c. The product of two negative integers is ________________

5. Another way to justify the rules you wrote in problem #4 is to look at patterns.

   $5 \times 4 = 20$
   $5 \times 3 = 15$
   $5 \times 2 = 10$
   $5 \times 1 = 5$
   $5 \times 0 = 0$

   ____ × ____

   ____ × ____

   ____ × ____

   a. First, consider the beginning of a pattern shown to the right.

      What are the next three products in this list?

      Describe the pattern; that is, what must one do to get from one step to the next?

      Explain how you determine the answers to the next three problems you wrote, using your pattern.

      This pattern shows why the product of a positive integer and a negative integer ought to be negative.

   b. Next, consider the beginning of another pattern show to the right.

   $-5 \times 4 = -20$
   $-5 \times 3 = -15$
   $-5 \times 2 = -10$
   $-5 \times 1 = -5$
   $-5 \times 0 = 0$

   ____ × ____

   ____ × ____

   ____ × ____

      What are the next three products in this list?

      Describe the pattern; that is, what must one do to get from one step to the next?

      Explain how you determine the answers to the next three problems you wrote, using your pattern.

      This pattern shows why the product of two negative integers ought to be positive.

OBJECTIVE: **Model integer division**

You Will Need: **Red & black chips from Materials Card 8.1.1 (Physical Manipulative option: *Two-Color Counters*; eManipulative option: *Chips Minus*)**

When modeling whole number division in Chapter 3, you used several different approaches; these included partitive division, measurement division, and the missing factor approach. Each of these is useful for modeling division with integers, but only one of them applies in all cases, as you will discover in the next activity.

1. To model integer division using red chips and black chips, first consider the case where the divisor is positive, such as -6 ÷ 3. In this case, the partitive approach to division can be applied. Begin with six red chips and partition them into three groups.

   a. Show how to do this in the drawing at the right.

   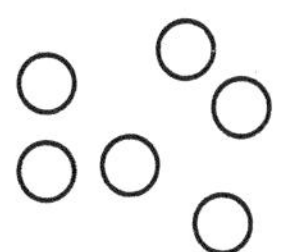

   Explain how to obtain the answer to the division problem -6 ÷ 3, using the drawing.

   b. Use red chips or black chips to compute the following quotients using the same approach in part (a). Record a drawing in each case.

   8 ÷ 2 = _____ -10 ÷ 5 = _____

2. If both the divisor and dividend are negative, as in -6 ÷ (-3), the measurement approach to division can be applied. In this problem, begin with six red chips and measure out groups of 3 red chips.

   a. Show how to do this in the drawing at the right.

   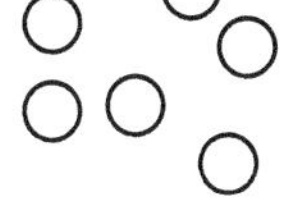

   Explain how to obtain the answer to the division problem -6 ÷ (-3), using the drawing.

   Could you use the partitive approach to model this division problem? Why or why not?

   b. Use chips to compute the following quotients using the same approach as in part (a). Record a drawing in each case.

   -8 ÷ (-2) = _____ -12 ÷ (-4) = _____

3. Now consider the case where the dividend is positive and the divisor is negative, as in 6 ÷ (-3).

   a. Explain why it is not possible to model this division problem using either the partitive or the measurement approaches.

   b. To explain how to find the answer to the problem 6 ÷ (-3), recall the missing factor approach to division.

      That is, find the missing factor n such that $6 = -3 \times n$.

      Use chips and the multiplication model from Activity 8.2.1 to check your answer. Record a sketch of this.

   c. The missing factor approach can be used to find the quotient, no matter what the sign of the divisor and dividend. Compute each of the following quotients using this approach. Use chips to check your answers and record a sketch of these.

      8 ÷ (-2) = _____ 4 ÷ (-4) = _____ -6 ÷ 2 = _____ -8 ÷ (-4) = _____

EXERCISE

Why did the bike racer go to the psychiatrist?

Solve each problem. Place the letter preceding the problem above the solution at the bottom of the page. Be sure to put the letter above the solution each time it appears.

U 27 – (-8) = _____
T -18 – 5 = _____
H 18 + (-12) = _____
A -45 + (-30) = _____
F 60 + (-40) = _____
E -11 – 6 = _____
V -15 + 41 = _____
L -3 + 15 = _____
S -47 – (-77) = _____

W 4 + (-22) = _____
D 7 + (-20) = _____
O 15 – 45 = _____
I -1 – (-20) = _____
G -8 + 4 – (-9) = _____
C -78 + (-37) = _____
N 22 + (-9) = _____
Y 1 – 12 = _____

___ ___   ___ ___ ___   ___ ___ ___ ___ ___ ___
6 -17   -18 -75 30   6 -75 26 19 13 5

___ ___ ___ ___ ___ - ___ ___ ___ ___ ___ ___ ___
-115 -11 -115 12 -17   12 -30 5 19 -115 -75 12

___ ___ ___ ___ ___ ___ ___   ___ ___ ___ ___ ___.
-13 19 20 20 19 -115 35   12 -23 19 -17 30

CONNECTIONS TO THE CLASSROOM

1. A student says, "I don't understand how there can be negative numbers. You can't have a negative number of things, right?" How would you respond?

2. A student has learned the following rule for adding one positive and one negative integer: *subtract the numbers and take the sign of the larger.* She says, "To do the problem -6 + 4, I'd do 6 – 4 and, since 4 is larger than -6, my answer is positive 2."

   a. Is the student correct? Why or why not?

   b. What part of this rule is the student confused about? Using models presented in this chapter, show how you could help this student state the rule correctly.

3. A student says, "You taught us that a negative and a negative make a positive, so why is -4 + (-6) = -10?" What is this student confused about?

4. a. A student says that $-x$ is negative because of the negative sign. Is the student correct? Explain.

   b. The same student says that $(-x)(-y) = xy$ because the product of two negatives is positive. Is the student correct? Explain.

MENTAL MATH

At 1:00 A.M. the temperature was -3°. By 5:00 A.M. it had dropped 13°. It then began to rise at the rate of 2° per hour. This rate continued until noon when the rate of increase changed to 3° per hour. At 4:00 P.M. it reached its peak and began dropping 2° per hour. What was the temperature at 5:00 P.M.?

DIRECTIONS IN EDUCATION
Learning Styles
I'll Learn It My Way

If students don't learn in the way I teach, then I must teach in the way they learn.

Individual learning styles are as diverse as individual faces or personalities. Although teachers accommodate many differences in their classrooms, they consistently expect students to adapt their learning styles to the teaching style of the classroom teacher. Some advocates of learning styles theory would suggest that the teacher assess and accommodate every student's learning style on an individual basis.

Classroom attributes which may require modification to address needs of individual students include temperature, lighting, noise level, movement patterns, grouping patterns and the ways in which information is transmitted. By considering the number of variables and the number of students in a single classroom, a teacher may find a significant number of modifications to take into account. It is, however, essential that the classroom teacher find a practical way to manage instruction which provides for the diverse learning styles within every classroom.

Students learn constantly at both the conscious and unconscious levels in our classrooms. If learning styles are not accommodated so that meaningful, comprehensive learning can occur, the brain simply engages in personally meaningful activity. However, the personally meaningful activity may not necessarily be related to the lesson or to what the teacher hopes to teach.

What are some of the traits which make up an individual's learning style?

- **Perceptual preferences:**
  - ❖ Auditory learners need to hear information.
  - ❖ Visual learners need to see and to use visual instructional resources.
  - ❖ Tactile learners need to touch and manipulate objects.
  - ❖ Kinesthetic learners need to involve the large muscles of the body as they learn.

- **Instructional environment preferences** (which may change as children develop) include:
  - ❖ The need for strong light versus soft light.
  - ❖ The need for quiet versus sound.
  - ❖ A preference for warm or cool temperatures.
  - ❖ The need for formal versus informal seating arrangements.

- **Sociological preferences** (which may also change as the child matures) include:
  - ❖ Students who are motivated by interaction with their peers.
  - ❖ Students who wish to learn directly from their teachers.
  - ❖ Students who learn best by themselves with appropriate resources.

- **Time-of-day preferences:**
  - ❖ Some students are morning people who can deal with complex ideas as soon as they arrive at school.
  - ❖ Some students are night owls who start to come alive around lunch time. (These students are often thought to be underachievers!)

- **Mobility needs:**
  - ❖ Some students flourish in an activity-based learning environment.
  - ❖ Some students are quite content to be seated passively for long periods of time.

- Brain research suggests a strong correlation between learning styles and **hemisphericity.** In addition, hemisphericity may determine a preference for:
  - ❖ A concrete, linear presentation of information in small incremental steps which lead to understanding.

❖ A holistic approach which begins with the broad concept and then focuses on the details.

How can learning styles be accommodated in the classroom?

- Remember that most children *can* master the required content, but their learning style will dictate *how* they master it.

- Present information more than once and in more than one way to increase the appropriateness of instruction for individuals.

- Provide an environment rich in instructional materials which address the needs of various learning styles.

- Recognize that there is not one best learning style.

- Realize that you will tend to teach in a style that matches your own learning style preferences. Know your own style and become aware of how often you design instruction to fit yourself.

As you think about learning styles, ask yourself:

1. How would I describe my own learning style?

2. How can I teach so that the needs of students with various learning styles are met?

9 Rational Numbers, Real Numbers, and Algebra

THEME: Extending Number Systems

HANDS-ON ACTIVITIES

Irrational numbers, algebraic equations, and graphs of functions are all part of the middle school and high school mathematics curriculum and are studied in depth in these grades. However, the ideas underlying each of these important concepts should be introduced in the early grades. As an elementary teacher, you will help your students lay the foundation they will need in order to be successful in their future studies.

Real numbers are classified as one of two types: rational or irrational. The set of rational numbers is an extension of both the set of fractions and the set of integers. A rational number can be written as the ratio of two integers, where the denominator is nonzero; for example 1/2, -3/5, 2/-3, -2, and 4 are all rational. Just like fractions, every rational number has a repeating decimal representation. On the other hand, irrational numbers cannot be written as the ratio of two integers and their decimal representations are non-terminating and non-repeating.

The basic ideas behind irrational numbers, as well as algebraic equations, are approachable for young children through the use of hands-on materials. One way for students to "see" irrational numbers is by solving problems that require the use of the Pythagorean Theorem. In the first activities of this chapter, you will use a manipulative, called a geoboard, to first develop the Pythagorean Theorem and then to model irrational lengths. A balance scale provides a useful model for visualizing the idea of equality and the procedures we use to solve simple equations. In the last activity, you will revisit functions, first introduced in Chapter 2, by looking at a visual representation called a graph.

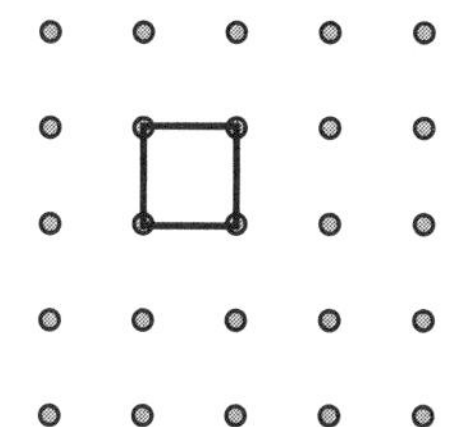

Using the Pythagorean Theorem, one could find the length of the diagonal of the square shown on the geoboard above – it is an irrational number.

Using a chip to represent one unit and a box to represent an unknown number of chips, the balance scale below shows the equation $2x + 3 = 9$.

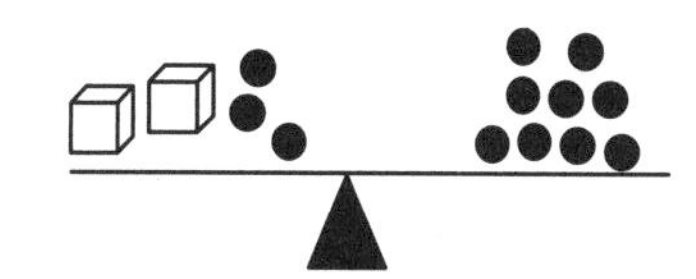

OBJECTIVE: **Use a geoboard and algebra to investigate the Pythagorean Theorem**

You Will Need: **Materials Card 9.2.1 (Physical Manipulative option: *Geoboard*; eManipulative option: *Pythagorean Theorem*)**

There are many ways to help students discover the special relationship among the three sides of any right triangle, known as the Pythagorean Theorem. One way to discover the Pythagorean Theorem geometrically is to use geoboard, a manipulative used in elementary classrooms for investigating concepts of geometry. You will also see the Pythagorean Theorem evolve algebraically.

1. A geoboard is a square array of pegs around which rubber bands can be stretched to form shapes. In this activity, the geoboard pegs will be represented by a square lattice and the rubber bands by line segments.

   The square shown on the lattice at the right is called a **square unit** since it is a square that measures one unit on each side. The **area** of a shape is determined by finding the number of unit squares it takes to cover the shape. What is the area of the square shown at the right? ________

   Find the area of each of the following shapes. Show how you determine your answers.

   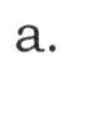

   a. 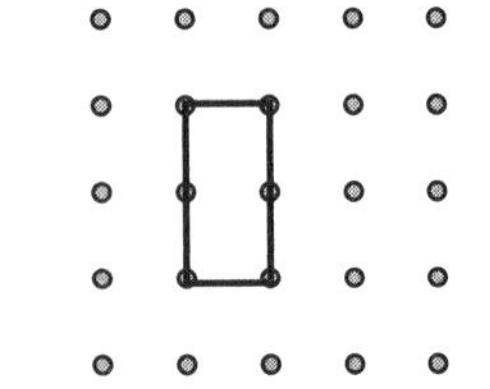

   Area = __________

   b. 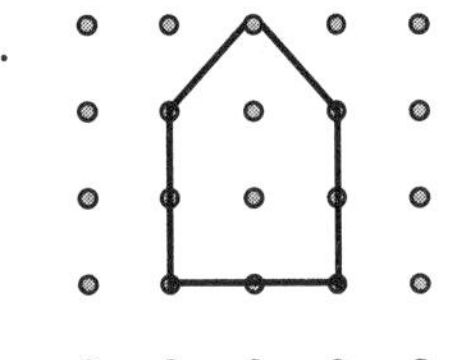

   Area = __________

   c. 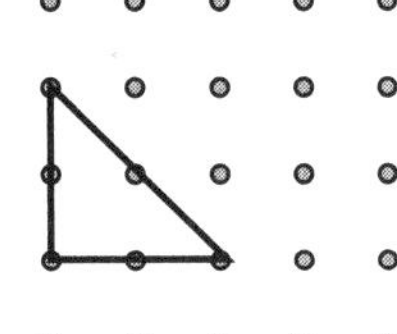

   Area = __________

2. On each lattice below, a right triangle (a triangle with a 90° angle) having a square shaped on each side is shown. Find the area of each of the squares.

   a. 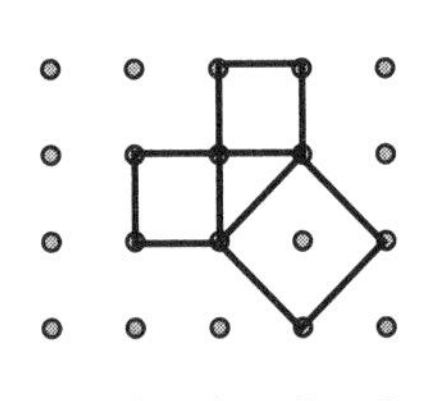

   b. 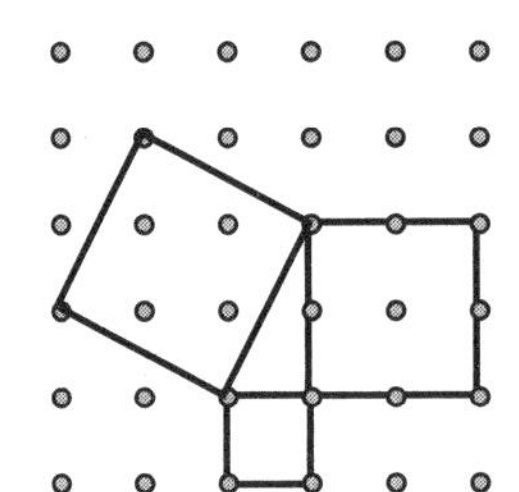

   c. In each of parts (a) and (b), what relationship do you notice between the area of the large square and the areas of the two smaller squares?

   d. One way to find the area of the large square in part (b) is to find the areas of the regions formed by the dashed lines as illustrated at the right. This shows the area of the large square = ________ Can the five shapes formed within the largest square be arranged to cover the other two squares? If yes, show how.

   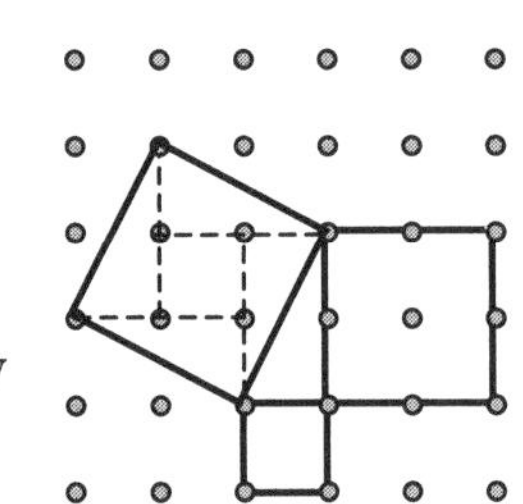

e. The **Pythagorean Theorem** is usually stated as follows: for any right triangle with legs a and b and hypotenuse c, $a^2 + b^2 = c^2$. Explain how your work in parts (a) – (d) is related to this familiar equation.

3. The geoboard methods you used in problem #2 give a way to visualize the relationship between the sides of a right triangle by thinking of the expression $a^2 + b^2 = c^2$ in terms of areas of squares. Notice though, that the Pythagorean Theorem is true for *any* right triangle, not only those that can be formed on a geoboard. To develop this theorem more generally, you will now use both algebra and geometry.

   Cut out the four copies of the right triangle with sides of length x and y, and hypotenuse of length z from Materials Card 9.1.1. Arrange them to form a square "donut" with a square hole as shown to the right. Answer these questions in terms of x, y, and z and label as appropriate in the drawing at the right.

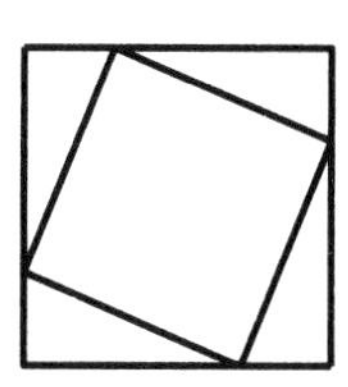

   a. The length of any outside edge of the large square is side ____ of the triangle plus side ____ of the triangle, or ____ + ____. What is the area of the large square? ____________ We will call this area A_1.

   b. What is the length of any side of the small square (the donut hole)? ____ What is the area of the hole? _____ We will call this area A_2.

   c. What is the area of each triangle? (Remember that the area of a triangle is 1/2 the base times the height. This is easy to see here since two of the triangles can be put together to form a rectangle with dimensions x and y.) _________

   d. What is the sum of the areas of the four right triangles? ______________ We will call this area S.

   e. Write an equation involving A_1, A_2, and S. ____________________________ Now substitute the x, y, and z values for A_1, A_2, and S and simplify your equation.

   f. Explain your simplified equation in geometric terms that apply to the right triangle.

4. Using your equation from problem #3(f), we know that 3, 4, and 5 could be the lengths of the legs and hypotenuse of a right triangle because $3^2 + 4^2 = 5^2$. Which of the following triples could be the lengths of the sides of a right triangle? Explain.

   a. 6, 8, 10 b. 5, 12, 13 c. 15, 112, 113 d. 7, 25, 26

OBJECTIVE: **Construct irrational lengths on a geoboard**

You Will Need: **(Physical Manipulative option: *Geoboard*)**

Numbers that have non-terminating, non-repeating decimal representations are called **irrational numbers**. In this activity, you will use techniques developed in Activity 9.2.1 to construct certain types of irrational number lengths on a geoboard. Seeing line segments of irrational number lengths can help demystify these numbers for students.

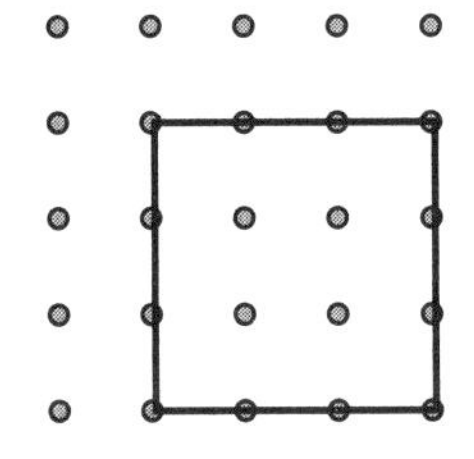

1. Consider the square shown on the lattice at the right.

   a. What is the area of the square? ___________

   b. What is the length of each side of the square? ___________

   Notice that $3^2 = 9$. That is, the side length squared equals the area of the square. Thus, we say "3 is a **square root** of 9."

   Every positive number actually has two square roots, one positive (called the **principal square root**) and one negative. We use $\sqrt{9} = 3$ to denote the principal square root. Since $(-3)^2 = 9$, -3 is the negative square root of 9, and is denoted $-\sqrt{9} = -3$. Since the square roots involved in this activity represent lengths of line segments, we will concern ourselves only with principal square roots.

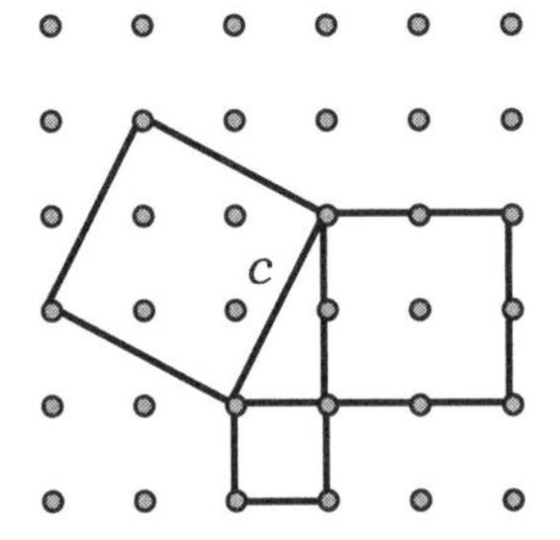

2. Consider the right triangle shown on the lattice to the right. A square is drawn on each leg and on the hypotenuse of the triangle. In Activity 9.2.1, you showed that the area of the large square equals the sum of the areas of the two smaller squares. In other words, if c is the length of the hypotenuse, then we can say $1^2 + 2^2 = c^2$ or $1 + 4 = 5 = c^2$. Since c is the number that can be squared to get 5, what is c? _____

   a. Find the length of the hypotenuse of each right triangle shown below. Give your answer in exact form as a square root.

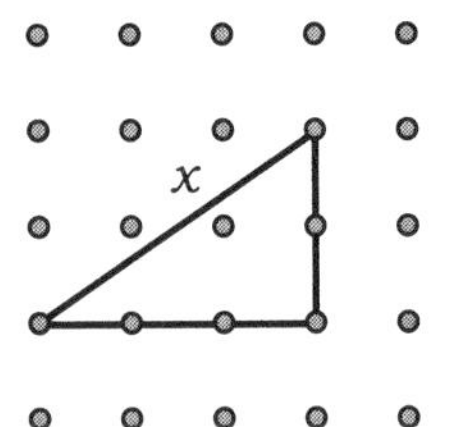

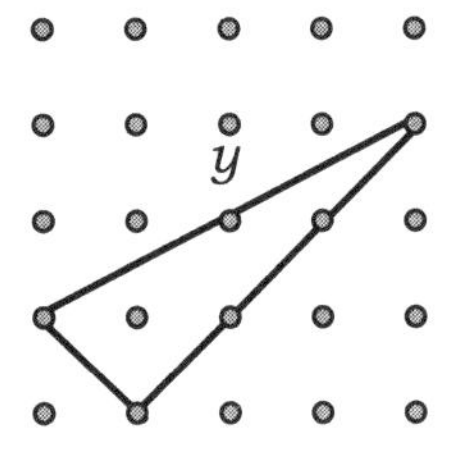

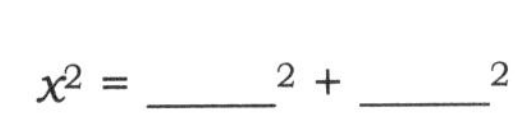

$x^2 =$ _____^{2} + _____^{2} $y^2 =$ _____^{2} + _____^{2}

$x =$ _____ $y =$ _____

Numbers such as $\sqrt{5}$, $\sqrt{13}$, and $\sqrt{20}$ are irrational numbers. In fact, $\sqrt{n}$ is irrational whenever n is not a perfect square. (For a proof of this, see your textbook.)

b. There are 14 different lengths that can be represented on a 5 by 5 geoboard; some are rational number lengths and some are irrational number lengths. Record all 14 of the possible lengths on the lattices below. You can fit several of these on each lattice. Be sure to do this systematically so you don't have any repeats or any missed lengths.

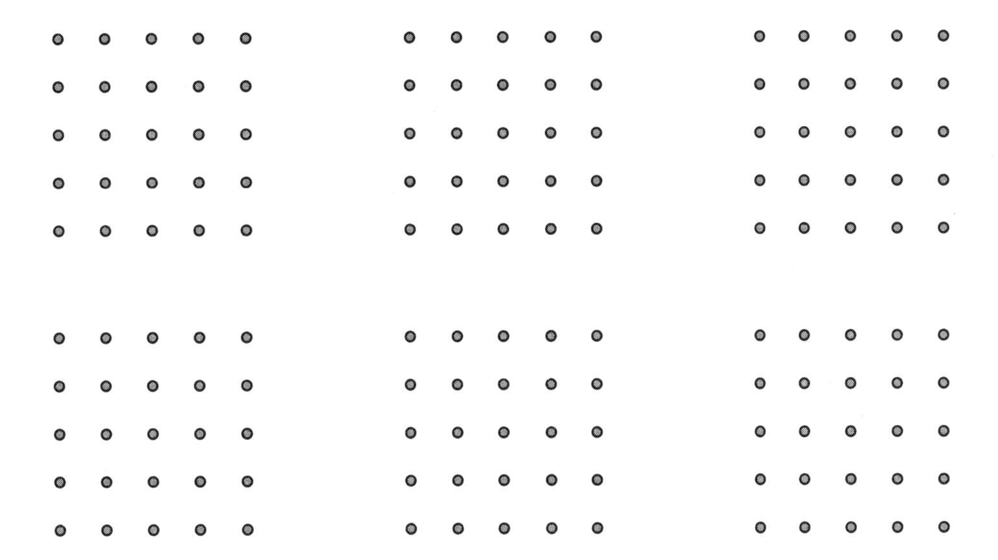

c. Show how to determine the length of each line segment you drew on the lattices in part (b). In some cases, you will need to draw a right triangle whose hypotenuse is the line segment under consideration.

d. Write all the lengths you found in part(c) from smallest to largest.

_____, _____, _____, _____, _____, _____, _____, _____, _____, _____, _____, _____, _____, _____

e. In your list in part (d), circle all the irrational numbers.

OBJECTIVE: **Use a balance scale model to solve equations**

You Will Need: **(eManipulative option: *Balance Beam Algebra*)**

The basic concepts of algebra can be introduced in the early grades. A balance scale is one model for introducing equations in the elementary classroom. The idea of balance is related to the concept of equality in that, for example, the same amount must be put on or removed from each side of the scale to maintain a balance. Similarly, with an equation, the same operation must be performed on each side to maintain equality.

1. An **equation** is a statement that two quantities are equal. To **solve an equation** means to find all the values for the variable that make the equation true. The usual approach is to replace an equation by simpler equations that all have the same solution as the original.

   In a balance scale representation of an equation, each chip represents one unit and each box on the scale may be thought of as taking the place of (or hiding) some unknown number of chips. As an example, consider the equation $3x + 5 = 14$. The balance scale method and the corresponding algebraic method for solving this equation are shown below.

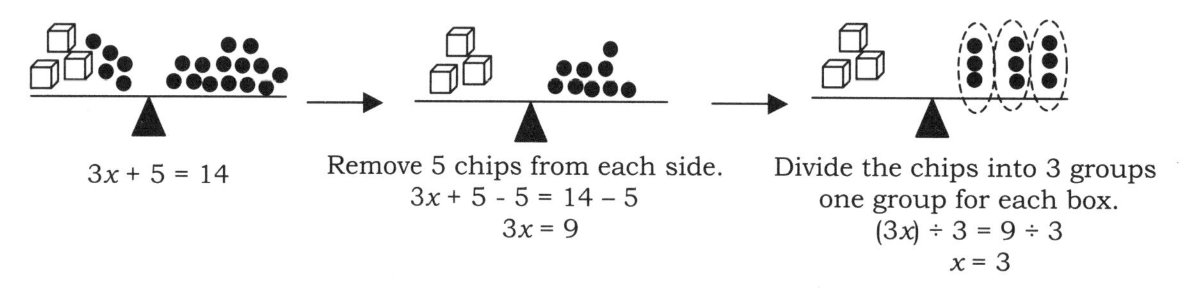

3x + 5 = 14
Remove 5 chips from each side.
3x + 5 - 5 = 14 – 5
3x = 9
Divide the chips into 3 groups
one group for each box.
(3x) ÷ 3 = 9 ÷ 3
x = 3

   Show how to check the solution to the equation $3x + 5 = 14$ in each of the following ways:

   a. Use the *original* balance scale.

   b. Use the *original* equation.

2. Use the balance scale model to solve the equation $5x - 9 = 2x + 15$ and show the corresponding algebra steps for each drawing. Be sure to label all drawings (including the intermediate steps) as shown in the example. (You will need to use 2 different colored chips for this one.)

3. The balance scale model can also be used to represent inequalities such as $3x + 5 > 14$.

   a. How would this be different from solving an equation using a balance scale?

   b. Use the balance scale model to solve the inequality $2x + 4 < x + 6$, making sure to show that one side is heavier than the other. Write out the corresponding algebra steps for each drawing. Label all of your drawings.

4. Another way to solve equations of the form $a\,x + b = c$ is to work backward through the operations performed with the variable. As an example, consider the equation $3x + 5 = 20$. The left side of the equation shows that x was first multiplied by 3 and then 5 was added to the product to get 20.

   a. Complete the diagram at the right by performing the opposite operations and in the reverse order.

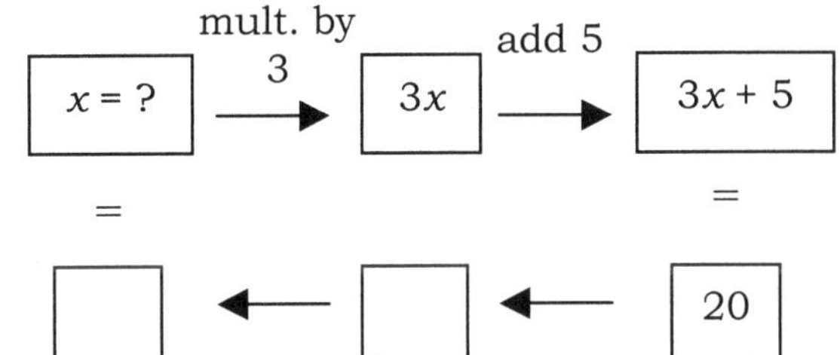

   b. Use the method of working backward to solve the equation $\frac{3}{4}x - 17 = -2$. Check your solution algebraically.

OBJECTIVE: Investigate representations of functions

In Activity 2.4.1, you represented functions using arrow diagrams. In Section 2.4 of your textbook, you saw two other ways that many functions can be represented: one was to choose several representatives from the domain set and construct a table; another was to write an equation that expresses the relationship between the sets. As you will see in this activity, it is often possible to represent functions in yet another way, with a graph in the coordinate plane.

1. Recall from Chapter 2 that a function is defined to be a correspondence between a first set of objects, the domain, and a second set of objects, the codomain, where the correspondence has a specific property. What is that property?

2. Laurel wants to make 2 cupcakes for each guest who attends her birthday party.

   a. The total number of cupcakes Laurel needs to make is related to the number of guests who will attend her party. Explain how this relationship meets the criteria for the definition of function that you wrote in problem #1. We'll call this function *f*.

   b. Shown at the right is a table for the function *f* from part (a), where several domain elements appear in the first column and their corresponding codomain elements appear in the same row of the second column. Complete two more rows of the table.

number of guests	number of cupcakes
2	4
3	6
7	14

   c. For many functions, it is possible to represent the function relationship using an algebraic equation. For the function *f* above, the number of cupcakes needed depends on the number of guests. So we will use x as the independent variable to represent the possible numbers of guests and $f(x)$ to represent the dependent variable, the number of cupcakes needed for x guests. Write an equation that expresses $f(x)$ in terms of x: $f(x)$ = _________

   If the domain of a function is other than the entire set of real numbers, it is important to list the domain set with its equation. Describe the values of x that make sense for this function.

   d. Many functions can be represented visually with a graph. A graph of *f* is shown in the coordinate plane at the right. Label the axes and explain how you decided the appropriate label for each axis.

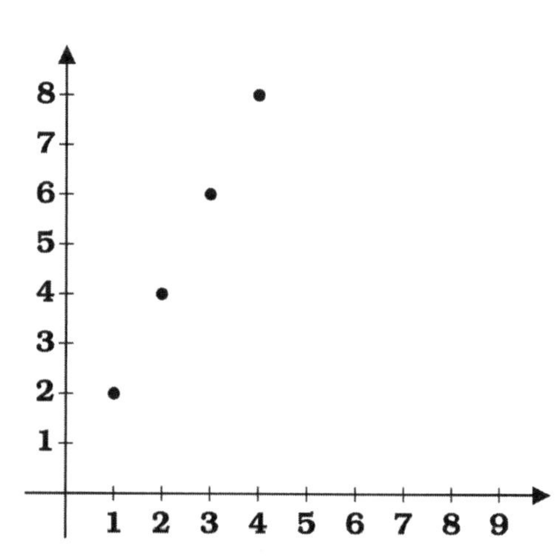

   Would it be appropriate to connect the points on this graph with a line? Why or why not?

3. A plumber charges $12.00 for every 15 minutes, or *any portion* of 15 minutes, that he spends working at a job.

   a. How much would the plumber charge if he worked 10 minutes? _______
   13 minutes and 45 seconds? _______ 20 minutes? _______ 35 minutes? _______
   2 hours and 40 minutes? _______

   b. The relationship between the amount of time the plumber spends working on a plumbing job and the amount of money he charges for that time is shown in a graph at the right. Explain why this graph makes sense for this situation.

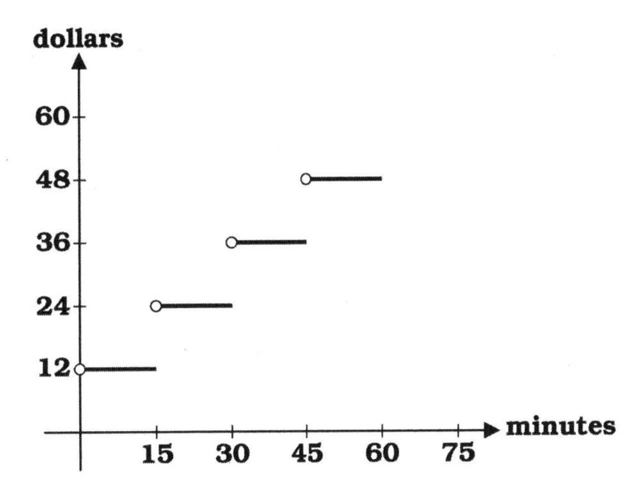

   c. As you can see, for example, the plumber will charge a customer $36.00 for all the different times greater than 30 minutes but not more than 45 minutes. Does this represent a function? Why or why not?

   d. The function graphed in part (b) is called a **step function**. Notice that there are open circles at one end of each step of the graph. Why do we need to do this? If we showed only the segments without the open circles, would the graph still represent a function? Explain your answer, referring to the definition of function.

4. A visual technique for determining whether a graph is the graph of a function is the **vertical line test** which states: if any vertical line crosses a graph *more than once*, the graph does *not* represent a function. If every vertical line crosses a graph in *at most* one point, the graph *does* show a function.

   a. Move a vertical pencil over each of the following graphs to determine whether they represent the graph of a function. Then explain your answers.

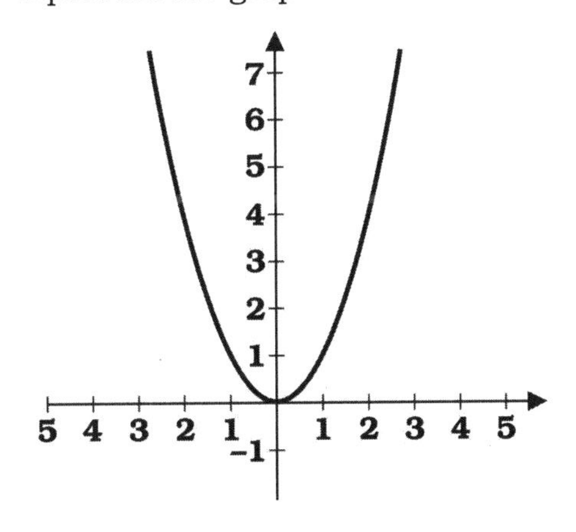

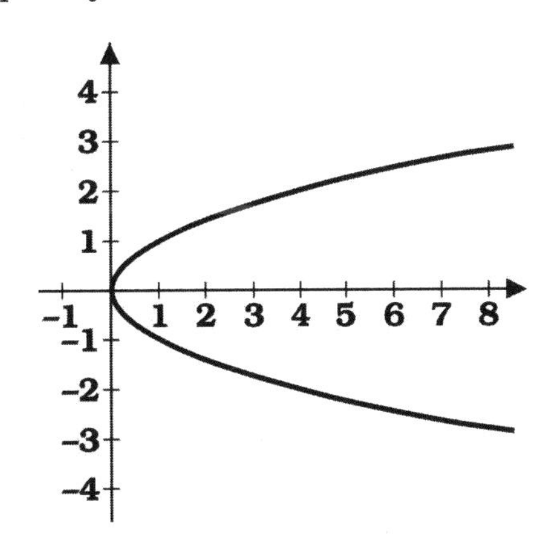

b. Reasoning in terms of the vertical line test, explain why the graph in problem #3(b) represents a function.

c. Reasoning in terms of the vertical line test, explain why the graph in problem #3(b) would not represent a function if we showed only the segments without the open circles.

d. Refer to the definition of function to explain why the vertical line test works.

5. A phone company offers two long-distance calling plans. Plan A includes a $4.95 charge per month plus $0.05 per minute. Plan B is a flat rate of $20.00 per month with unlimited long-distance calls.

a. If you make only a few long-distance calls each month, which plan is cheaper? If you make a lot of long-distance calls, which plan is better? Explain.

If you only make a few calls Plan A is better, if you call alot plan A will be even more expencive so plan B is better.

b. Create two graphs, one for Plan A and one for Plan B. Show the number of minutes on the horizontal axis and the total cost on the vertical axis.

min	$
0	4.95
30	6.45
50	7.45
100	9.95
200	14.95
300	19.95
500	29.95

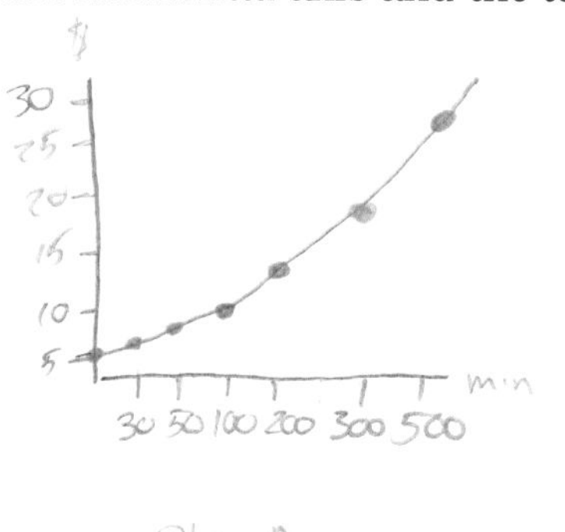

Plan A

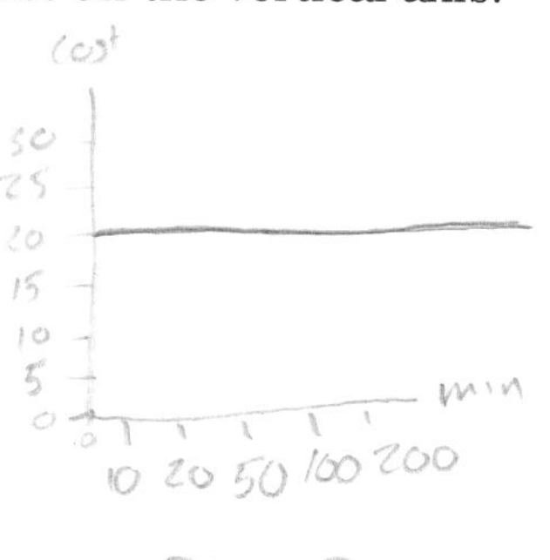

Plan B

c. At what point will the two plans cost the same? Explain how you determine your answer. How could you show your solution in a graph?

if you talk for 301 min. on plan A it will equal the exprence of plan B.

20

301

d. Is either of these plans a function? How does your graph help you answer this question?

Plan A is

$.5x + 4.95 = \$$

x = number of min used

EXERCISE

What happens to frogs who double park?

Decide whether each number is rational or irrational. Circle the letter in the appropriate column at the right. When you are finished, print the circled letters above the appropriate blanks at the bottom of the page to solve the riddle.

	Rational	Irrational
1. 22/7	E	I
2. π	W	T
3. 6.125	G	A
4. 0.571428571428...	W	A
5. $\sqrt{225}$	Y	S
6. $\sqrt{49}$	Y	S
7. $\sqrt{2}$	B	T
8. -4/3	A	R
9. 3.1415925385...	C	A
10. -.060060006...	A	H
11. 0.135135135...	T	E
12. -0.999199911999...	R	A
13. 3.40764076...	D	T
14. -9.7071212212221	O	A
15. $\sqrt{13}$	R	E

Letter Answer	__ __ __ __	__ __ __	__ __ __ __	__ __ __ __ !
Question Number	7 10 1 5	3 15 11	2 14 8 13	9 4 12 6

CONNECTIONS TO THE CLASSROOM

1. A student says "0.19119111911119... is a rational number because it repeats the same pattern over and over." Is the student correct? Why or why not?

2. When asked to solve the equation $3x + 4 = 5x$, a student says, "My first step is to get all of the x's on one side of the equation. To do this, I will take the $3x$ from the left side and move it over to the right side."

   a. Use the balance scale model to represent the equation $3x + 4 = 5x$. Then show in your drawing the student's first step he described above.

   b. Is this a correct first step? Why or why not? Use your drawing in part (a) to explain.

3. A student, Maria, says "To have a graph of a function, you can't have two different points on the graph that have the same x value." Josiah says, "No, for the graph of a function you can't have two different points with the same y value." Determine which of these statements is true and which is false. Explain your answer, in each case.

MENTAL MATH

A magician says she can tell you what your age is. She asks you to think of your age, but not tell her. Then she instructs you to add 12 to your age and multiply that result by 3. Then you are to add 9 to that result, divide by 3, and finally subtract 15. The magician asks what your result is and announces that is your age.

Is the magician correct? How does she know?

Make up some similar magic of your own.

DIRECTIONS IN EDUCATION
Cooperative Learning
Together We Can

Cooperative learning is a teaching model which provides for heterogeneous groupings of students within the classroom. By using this instructional strategy, the classroom teacher can combine the teaching of social skills with the teaching of content area goals and objectives. It is a strategy which must be carefully planned by the teacher so that the students need only concern themselves with being totally involved in the accomplishment of the task at hand. The lesson will usually begin with instruction required to accomplish both the content objective and the social skill. The students then participate in the group activity. The lesson ends with a discussion of both the content objective and the social skill.

Reasons to use cooperative learning:

- Cooperative learning is supported by research as an effective teaching strategy.
- Student involvement with the learning is enhanced.
- It reduces the tracking of students into ability groups.
- Cooperative learning promotes higher level reasoning and communication skills.
- Skills are used in a real world setting.
- It enhances self esteem by promoting acceptance of and by peers.
- It values individual contributions to the group task.
- Cooperative learning is fun!

Cooperative learning means more than just learning and working together in groups – it has 5 essential elements:

- **Positive Interdependence** – activities are structured in such a way that individual success is enhanced by group success. The objective is a feeling of "We're in this thing together" rather than a feeling of "We win, you lose" or of "Everyone for themselves."

- **Individual Accountability** – each person is responsible for his/her own learning within the cooperative group. It is the task of the teacher to clarify the criteria for success so that learners can monitor their progress toward the goal.

- **Face to Face Interaction** – students must be seated in such a way that they are literally face to face as they participate in the cooperative activity. Materials should be provided in group sets rather than individual sets to promote this interaction.

- **Collaborative Skills** – in addition to an academic goal, each lesson has a collaborative goal which provides for use of a previously learned skill or for instruction about the use of a new skill. It is generally wise to pair difficult academic goals with previously learned social skills and to pair difficult new social skills with previously introduced academic goals.

- **Group Processing** – the success of any cooperative learning classroom is closely linked to the quality of the debriefing which occurs after each lesson. It is during the debriefing that academic understandings are clarified and that social skills are refined. The debriefing should deal first with the academic content of the lesson and then with the social skill being practiced.

Cooperative skills which enhance the classroom environment and enrich the teaching/learning process:

- Communication skills such as active listening, asking good questions, clarifying, giving constructive criticism.

- On-task skills such as staying with the group, encouraging everyone, and using manipulatives in helpful ways.

- Skills which check and redirect the group's activities such as knowing when to ask for help, summarizing the group's progress, and checking the solution to a problem.

As you think about cooperative learning, ask yourself:

1. What is my role as a teacher in planning and presenting a successful cooperative lesson?

2. What kinds of learning objectives work best in cooperative groups?

3. Are there some kinds of learning objectives which may not be appropriate for cooperative groups?

4. How often should students be engaged in cooperative activities in an average week?

5. Which types of students will benefit the most from cooperative grouping in the classroom?

10 Statistics

THEME: **Organizing, Picturing, and Analyzing Data**

HANDS-ON ACTIVITIES

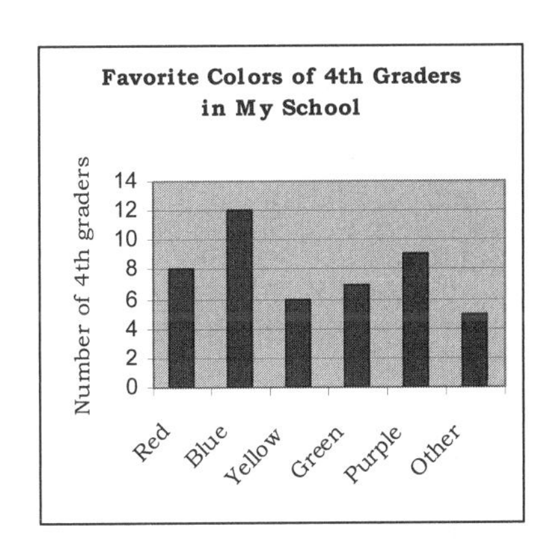

The science of statistics involves collecting, organizing, representing, and interpreting data and analyzing the results. The analysis process may include creating a graph, which gives a quick overall picture of the data. This can help the reader make comparisons or see trends, such as changes over time. Graphical representations of data are often shown in newspapers, magazines, scientific journals, websites, brochures, or any place people look for information. It is important to remember that the people who cite statistics and create graphs of data may be doing so to sway opinions or decisions people will make about the products they buy, how they will vote, etc. An understanding of the ways in which graphs may be constructed to emphasize certain aspects of a data set will help you decide when you are being deceived or misdirected. Consider the two graphs shown at the right; they were created using the same data, but each one gives the reader a different impression of the data. Activities in this chapter provide realistic experiences for students to draw on when thcy cncounter graphs in the future.

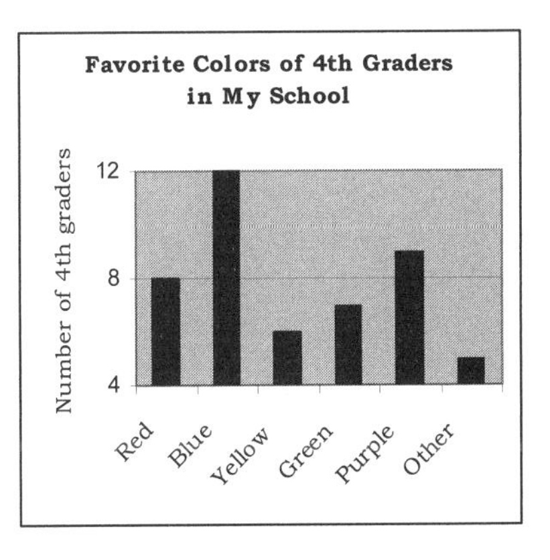

Data analysis also includes computing numbers that give us a sense of the "center" of a data set, such as mean and median, as well as numbers that help us understand the overall "spread" in a data set, such as range and standard deviation. Our interpretation of such statistics for a set of data may help us make decisions or in our attempts to predict future outcomes. One of the activities in this chapter will help you gain a deeper understanding of the concept of average, using a hands-on model. Finally, box and whisker plots, which provide one way to gain an understanding of dispersion in a set of data, will be studied.

OBJECTIVE: **Picture data using graphs**

You Will Need: **Materials Card 10.1.1**

Bar graphs and pie graphs are two common types of graphs. In this activity, you will collect and organize a set of data pertaining to the students in your class, then use a bar graph and pie graph to make visual summaries of your data.

1. Formulate a survey question you would like to ask each of the students in your class. You may want to ask about such topics as the number of siblings, the distance from your college to each person's hometown, the daily activity that requires the most amount of time, the grade level they want to teach, etc. Be sure there are no ambiguities in your question; for example, if your question is about the number of siblings, do you want to include step-brothers and step-sisters?

   a. Write your question, then conduct a survey of your classmates and record the results below.

      Look at your list of data above. What are your initial impressions?

   b. After collecting data, the next step is to organize it. Organize the information by grouping the data. Choose the number of categories that seems most appropriate for the data you have. Between 3 and 5 categories usually make a visually pleasing graph.

   c. Create a bar graph that summarizes your data. Clearly label each axis and give your graph an informative title.

d. Does your graph in part (c) change your impressions of the data from those you initially wrote after seeing only the data? If so, explain how.

e. Write at least two observations you can make about the data in your graph in part (c). Do you find any surprising results?

2. In a pie graph, a disk is used to represent the whole and the pieces of the pie represent the parts out of the whole. Follow the steps in parts (a) through (c) below to display the information from the survey of your classmates in a pie graph.

   a. First determine the percentage of the total responses represented by each category (round each percentage to the nearest hundredth of a percent). Show your computations below.

   What is the sum of the percentages above?_______ If you did not obtain a sum of exactly 100%, why do you suppose this is the case?

   b. Determine the central angles for the sectors that will be used to show each category. (The central angle in the sector is equal to the given percentage of _______degrees.) *Show* how you found each angle. What is the sum of the angles? ____________

   c. Use the protractor from Materials Card 10.1.1 to carefully measure out each sector in the circle at the right. Then, label each sector with its category name and percentage (not the angle.) Lastly, give your graph an informative title.

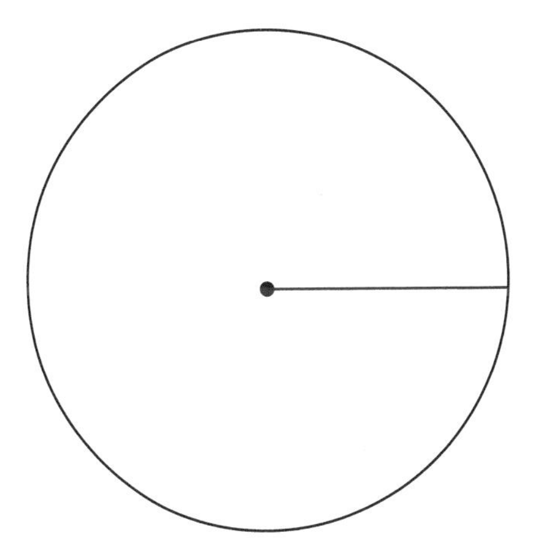

   d. What feature of the data set is lost when we use a pie graph rather than a bar graph? Explain.

OBJECTIVE: **Use scatterplots to make predictions**

You Will Need: **A straightedge**
(eManipulative option: *Scatterplot*)

A scatterplot is a type of graph useful for investigating the possible relationship between two variables. There are times when points representing ordered pairs of data lie approximately in a straight line; this line may be used to make predictions. There are precise methods for finding the line that best fits a set of data. However, in this activity, you will sketch such a trend line by "eyeballing" it.

1. The scatterplot below shows points representing data that was collected as part of an elk survey for a fish and game department study. To collect the data, observers in a helicopter drew an imaginary circle of radius of 30 feet around the elk group they were observing, and approximated the percentage of the ground covered by snow.

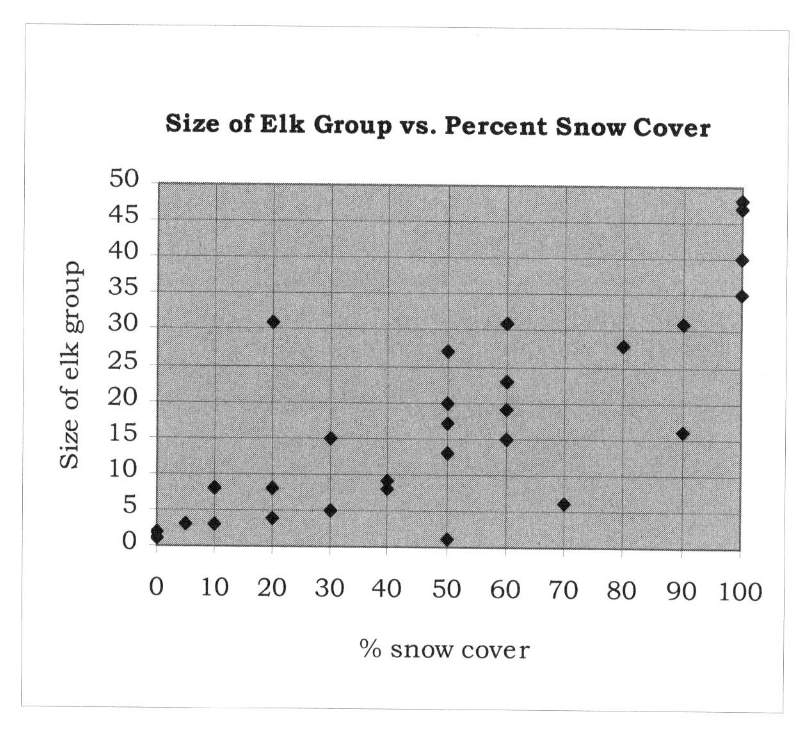

a. What does the point (60, 15) mean?

b. Use your straightedge to draw a straight line, through the data points so that it "fits" as well as possible – this line is called a regression line. One way to do this is to draw a line that contains as many of the data points as possible with about the same number of data points above and below the line.

Use your line to estimate the elk group size when there is a 75% snow cover. ______

c. What relationship do you notice between the percent of snow cover and the elk group size?

NOTE: We say, "There is a *correlation* between the percent of snow cover and the size of elk herd." Although the line represents the data, there is no indication that a greater percentage of snow cover *causes* a larger elk herd size. There are many other variables that may affect the size of an elk herd. Nevertheless, the scatterplot can give you a clue that two things might be related.

d. Which data point or points seem to be outliers? ____________________ Explain why you think these points are outliers.

If you removed these outliers from the data set, how would your regression line change?

2. For the next problem, you will collect data from your classmates and create a scatterplot to represent it.

   a. Think of two pieces of data to collect from 20 of your classmates that you believe may be related. Remember that the data you collect must be numerical (one example is height vs. shoe size). Record the name of each variable and the pairs of data you collect in the chart below.

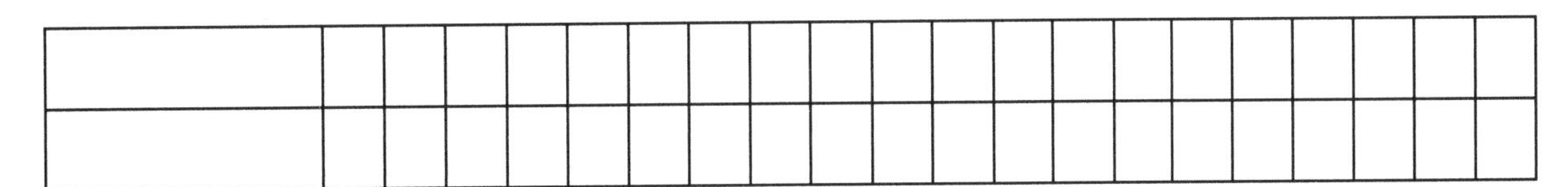

   b. Now create a scatterplot of the data.

   c. Does the data seem to fall approximately on a line? _______ If so, sketch the regression line.

   d. Use your regression line to make a prediction and record it below.

OBJECTIVE: **Discover the meaning of mean**

You Will Need: **Squares from Materials Card 3.1.2**

Most students' understanding of mean, or average, is a procedural one. That is, they can compute the mean of a set of numbers, but they do not have a conceptual understanding of what this number tells them about a data set. The next activity is designed to help you discover the meaning of this common statistic.

1. With the squares from Materials Card 3.2.1, make two stacks as pictured to the right. The goal is to rearrange the squares so that you have an equal number of squares in each stack, while leaving the number of stacks the same. Explain how you could do this.

2. Repeat problem #1 with stacks of squares as follows. Show your process in each case.

   a. 2 stacks: 3 squares and 11 squares

   b. 2 stacks: 4 squares and 8 squares

   c. 3 stacks: 3 squares, 10 squares, 5 squares

3. The "leveling off" process you used in problems #1 and #2 is a hands-on method for finding the **mean**, also known as the **average**, of a set of numbers. How is this process related to the usual way we compute mean?

4. For each of the following problems, use stacks of squares to find your answer. Show your process in each case.

   a. On four quizzes, Pham received scores of 8, 5, 6, and 9. What must he earn on the fifth quiz to have an average of 8 on all five quizzes?

   b. Leigh Ann has an average score of 7 on five quizzes. Give three different possibilities for her five scores.

OBJECTIVE: **Use statistics to analyze data**

In the last activity, you investigated one measure of central tendency, the mean. In this activity, you will use two other statistics that measure the "center" of a data set: mode and median. Also, you will consider one method for measuring the spread in a data set by computing the range.

1. The bar graph below shows the names of 20 students, arranged in order according to the length of their name. The bars are shaded to represent the number of letters in each student's name.

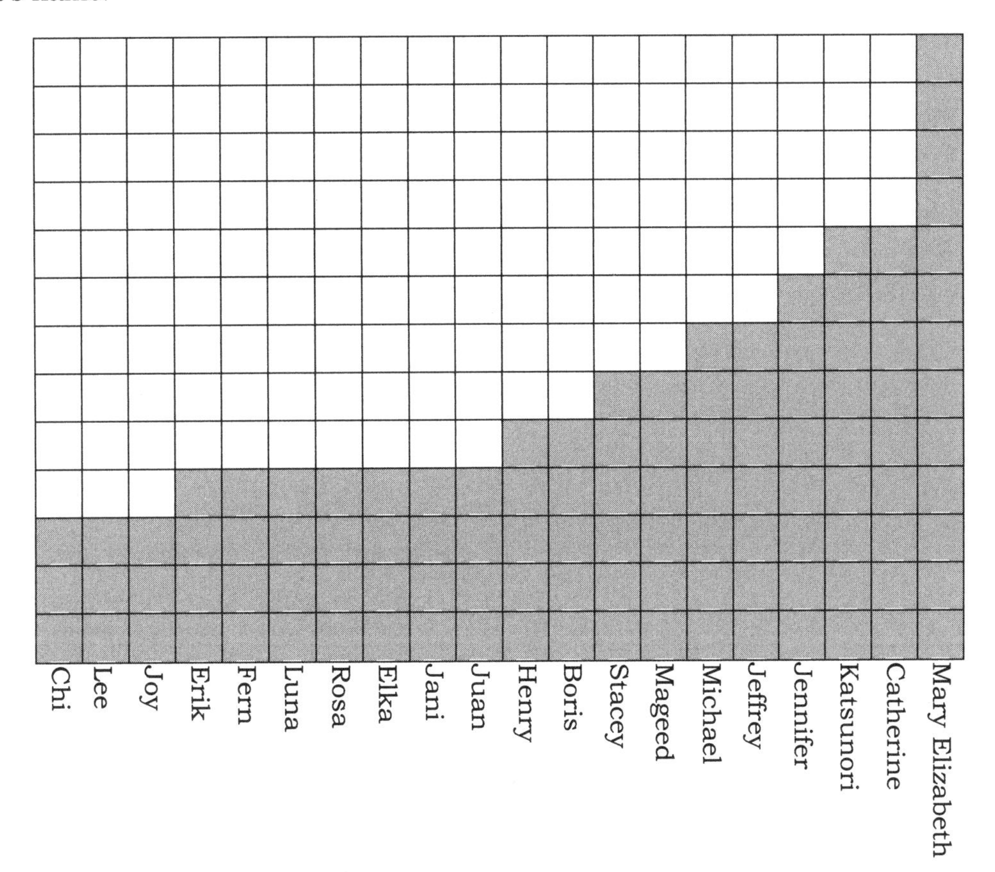

a. What is the mean name length for this group of students? ______ Draw a horizontal line through the graph to represent the mean length.

b. The number that occurs most frequently in a data set is called the **mode**. What name length is the mode for this set? ______

c. The **median** of a set of data that are *listed in order* is the middle number, if there is an odd number of data values. If there is an even number of data values, the median is the mean of the middle two numbers. What is the median name length? ______

d. Which of the mean, mode, and median do you think best represents a "typical" name length for these 20 students? ______________ Explain.

e. The lengths range from the shortest of ______ to the longest of ______. The range of a set of data is the largest number minus the smallest number. The range of the name lengths is ______.

2. a. How many people have a name longer than the mean? ______
   Shorter than the mean? ______

   b. How many people have a name longer than the mode? ______
   Shorter than the mode? ______

   c. How many people have a name longer than the median? ______
   Shorter than the median? ______

   d. Which of the three - mean, mode, or median –will divide your graph so that the same number of people are above and below that value? Why?

   e. What factors might cause the mean to be off-center?

   f. What factors might cause the mode to be off-center?

   g. Can the mean, median and mode of a set *all* be on-center? Explain.

3. a. If Mary Elizabeth decides to go by the name Elizabeth, will any of the mean, median, or mode change? If so, which one or ones? Explain.

   b. If she decides to use the name Mary, which will change? Explain.

   c. Under what circumstances will changing one element of a data set change the mean? the mode? the median? the range? Explain in each case.

OBJECTIVE: **Use box and whisker plots to analyze data**

You Will Need: **(eManipulative option: *Box Plot*)**

Measures of central tendency are numbers that represent the "center" of a data set. We describe the spread, or dispersion, of a set using numbers such as range and standard deviation. In this activity, you will investigate a graph that can you visualize both the center and the spread of a data set, without the need for standard deviation.

1. A relatively easy way to illustrate the spread in a data set is with a **box and whisker plot**, a visual summary of a data set that includes 5 numbers; the lowest data value, the highest data value, the median, and two other numbers called the quartiles.

   Suppose the quiz scores for 14 students on their first math test of the semester are:

   58, 60, 65, 70, 73, 74, 78, 82, 82, 84, 86, 87, 90, 94

   a. The lowest value for this data set is _____ and the highest value is _____.

   b. The median for this set is _____. Show how you determined your answer.

   c. Two other useful numbers indicated in a box and whisker plot are the **quartiles**. For this set of 14 students' test scores, we find the quartiles in the following way:

   58, 60, 65, 70, 73, 74, 78, 82, 82, 84, 86, 87, 90, 94

   The **lower quartile** is the median of the 7 smallest data values

   The **upper quartile** is the median of the 7 largest data values

   The lower quartile for this data set is _____ and the upper quartile is _____.

   Why do you suppose we call these numbers "quartiles"?

2. A box and whisker plot for the students' test scores in problem #1 is shown below.

   Scores on Test 1 for 14 students

   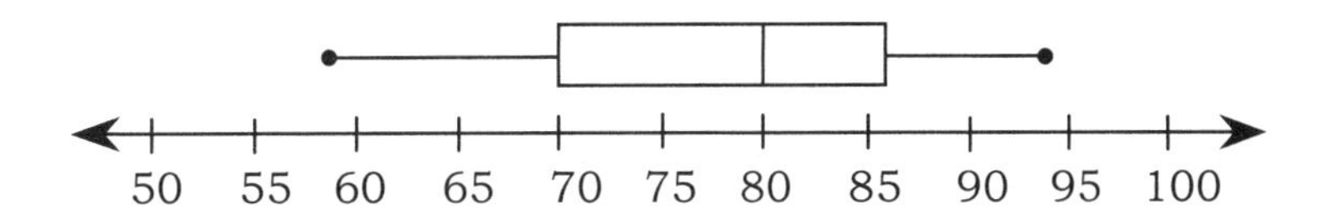

   a. The name of the plot comes from the fact that it is made up of a box, with "whiskers" that extend to the lowest and highest values of the data set. Label these in the plot as *L* and *H*, respectively.
   b. The median of the data set is shown by the vertical line drawn in the box. Label the median in the plot as *M*.
   c. The ends of the box represent the lower quartile and the upper quartile. Label these as in the plot as Q_1 and Q_3, respectively.

3. As you can see, a box and whisker plot shows a data set divided into four parts with approximately the same amount of data in each part. The box represents approximately 50% of the data and each whisker represents approximately 25% of the data. Notice that, depending on the data, the whiskers may be of different lengths. Also, the two parts of the box may be different lengths.

   Use the box and whisker plot in problem #2 to answer the following questions:

   a. How many scores were greater than or equal to 80? _____

   b. Among the scores represented by the box, where is there more spread - from the median to the lower quartile, or from the median to the upper quartile? Explain your answer.

   c. Among the scores represented by the whiskers is there more spread among the numbers represented by the left whisker or the right one? Explain

4. By the time the students in problem #2 took their second math test of the semester, three more students had added the course. These 17 students had the following scores on their second test: {38, 65, 65, 70, 70, 72, 80, 84, 85, 85, 86, 87, 90, 90, 90, 91, 94}.

   a. Record the values for the lowest test score, the median score, and the highest score.

   $L =$ _____ $M =$ _____ $H =$ _____

   b. For this set of 17 students' test scores, we find the quartiles in the following way: Note the process is different from that in problem #1(c). When the data set has an odd number of numbers (as here), we use the numbers below (and above) the median to determine the quartiles whereas when there is an even number of data, we use the lower half and upper half to determine the quartiles

   38, 65, 65, 70, 70, 72, 80, 84, (85), 85, 86, 87, 90, 90, 90, 91, 94

   The **lower quartile** is the median of the 8 smallest data values

   The **upper quartile** is the median of the 8 largest data values

   Record the values of the lower quartile and upper quartile: $Q_1 =$ _____ $Q_3 =$ _____

   c. Use the line below to make a number line and create a box and whisker plot for the students' scores on Test 2.

   d. Why is the whisker at the left end so long in relation to the rest of the plot?

e. A data value that is very large or very small when compared with the other numbers in the set is called an **outlier**. Which quiz score(s) do you suspect is an outlier? _____

   One way to define the term "outlier" is in terms of the **interquartile range**, or IQR. The IQR is the range between the upper quartile and the lower quartile; in other words, it is the length of the box. What is the IQR for the students' Test 2 scores?____

f. If a data value is more than 1.5 times the IQR above the upper quartile or below the lower quartile, the value is considered to be an outlier. Put another way, if a data value is farther away from either end of the box than 1.5 times the length of the box, the number is an outlier. Determine the outliers, if any, for Test 2 scores. Explain your answer.

5. An outlier is indicated in a box and whisker plot by an asterisk (*) above the data value. In the case that a data set does have outliers, the whisker ends at the data value farthest away from the box that is not an outlier.

The box and whisker plot for Test 2 is shown, with the outlier indicated above the number line, at the right.

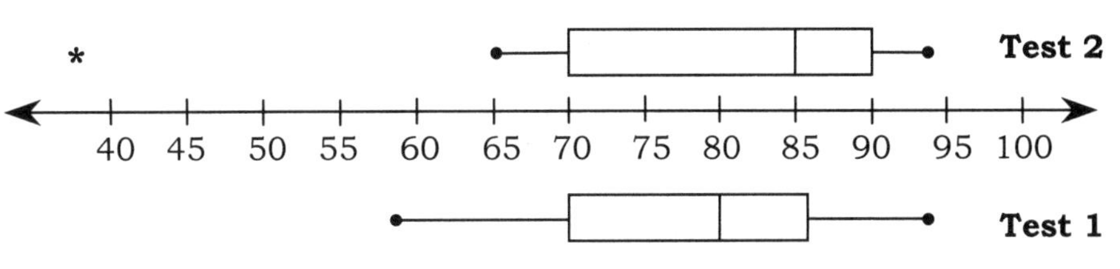

In order to compare the performance of the students on the two math tests, the box and whisker plot for Test 1is shown below the number line.

a. Refer to the box and whisker plots to decide which test the students had a better performance on. Write at least three reasons to defend your conclusion; include the ideas of spread and center in your explanation.

b. How did the one score of 38 affect your decision in part (a)? If you had compared the plot you created in problem #4c with the one for Test 1, would you have made the same decision? Why or why not?

OBJECTIVE: Investigate misleading graphs

People create graphs to help their audience get a visual picture of data, but it is important to remember that politicians, companies, advertisers, and others use graphical presentations of data in order to make a point. When choosing how to construct a graph, people often make choices that best help them to make their point. In this activity, you will examine ways in which elements of a graph can be manipulated to create different impressions of the same data.

1. Examine the double bar graph shown at the right.

   a. In your own words, describe what is being shown in this graph.

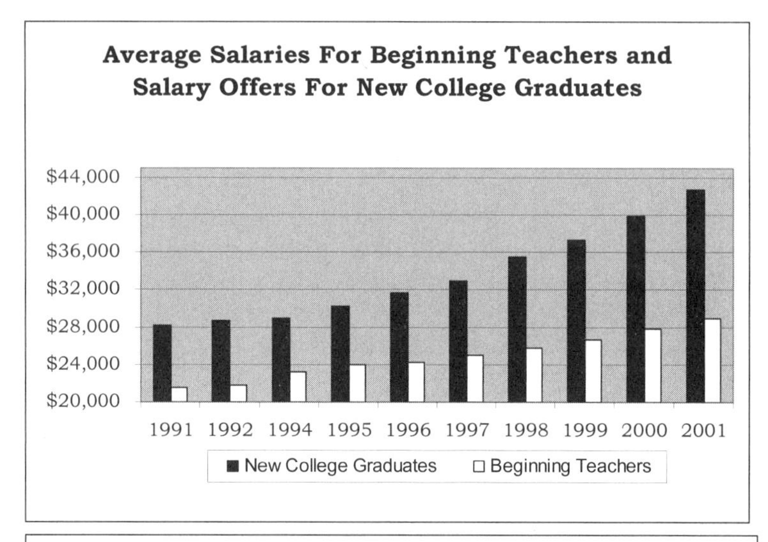

   b. Now compare the graph in part (a) with the one pictured at the right. Do these graphs appear to represent the same data? Explain.

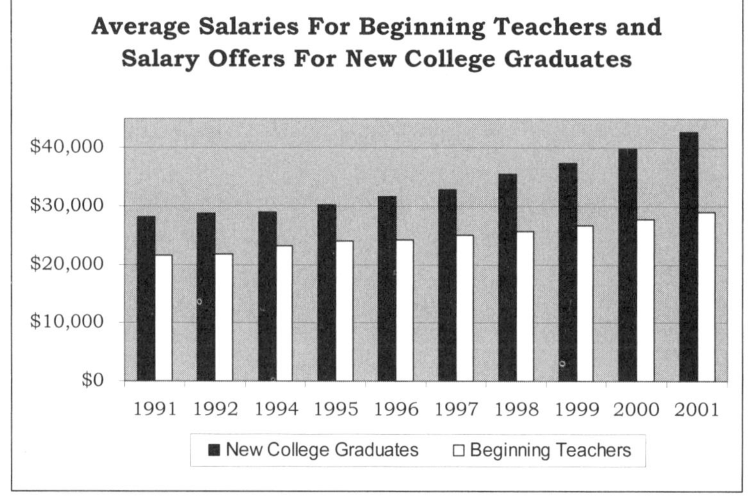

Source of data: http://www.aft.org/research/survey01/tables/tableIII-2.html

   c. What are the differences between these two graphs?

   d. How do these differences affect your impression of the data?

2. Refer to the two line graphs shown below to answer the questions in parts (a) – (f).

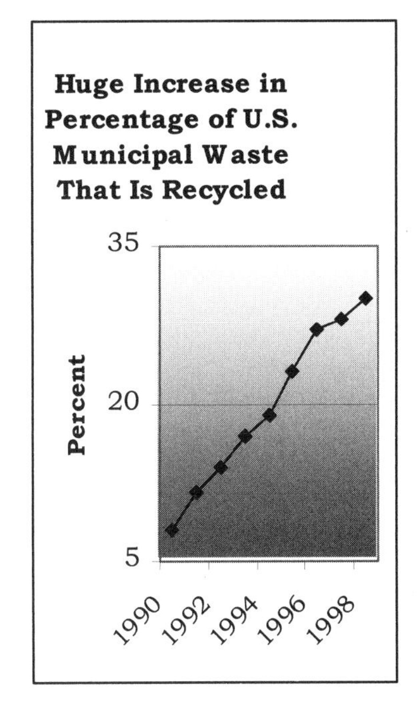

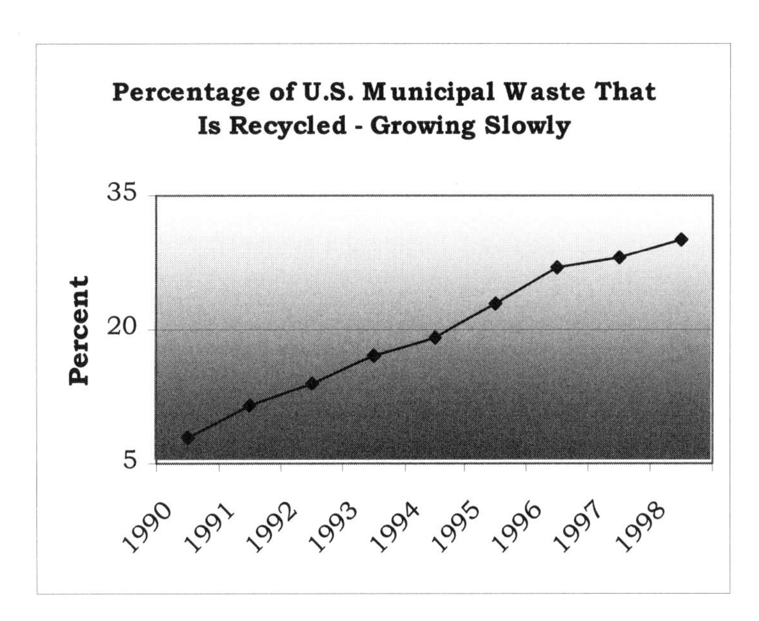

Source of data: http://www.zerowasteamerica.org/MunicipalWasteManagementReport1998.htm

a. Does the title of each graph seem reasonable? Why or why not?

b. Do these graphs appear to represent the same data? How can you tell?

c. Describe what has been done to give such different impressions of the data.

d. Are both of these valid ways to present this data? Why or why not?

e. Why might someone want to present this data as in the graph on the left?

f. Why might someone want to present this data as in the graph on the right?

3. Consider the line graph shown at the right.

   a. Explain two different ways you could alter the vertical axis on this graph so as to create a different impression of this data.

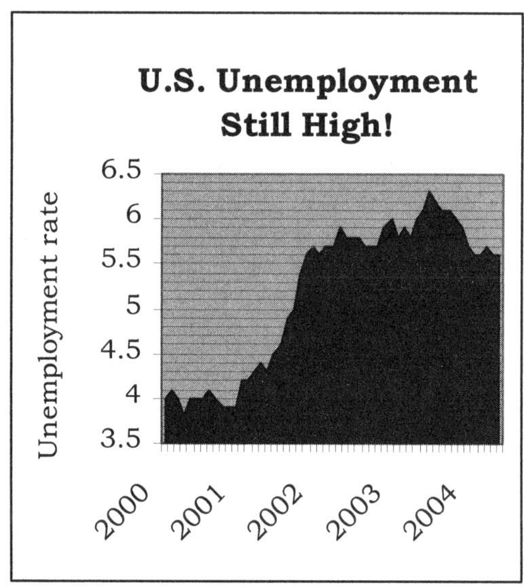

Source of data: http://data.bls.gov/servlet/SurveyOutputServlet?data_tool=latest_numbers&series_id=LNS14000000

   b. How could you change the horizontal axis to create a different impression of the data than that shown in the original graph?

   c. Choose one or more of the techniques you described in parts (a) and (b) and redraw the graph of unemployment rates, using the space at the right.

      Give your graph a title that seems appropriate for this new presentation of the data.

   d. Who might benefit from having the unemployment data represented as in the graph in part (a)?

EXERCISE

How is a sick bird like robbing a bank?

A bowling team bowled 3 games. Use their scores to answer the following questions. Put the letter of the correct answer on the line with the same number at the bottom of the page.

	Game 1	Game 2	Game 3
Dolly	142	136	157
Molly	173	131	209
Pauly	169	171	180
Wally	216	154	172
Holly	153	167	161

1. Who had the greatest range of scores?	S. Wally	O. Molly
2. Who had the lowest mean score?	R. Holly	H. Dolly
3. What was the overall mean score for the team (for all 15 games)?	L. 166	P. 171
4. Who bowled the median game scores (for all 15 games)?	J. Pauly	R. Holly
5. What was the team mean for game 1?	Y. 171	G. 181
6. Who had the highest mean score?	C. Molly	B. Pauly
7. Whose mean was above the overall team mean?	I. Wally	K. Dolly
8. Who was nearest the team mean in game 3?	V. Pauly	S. Wally
9. For which game was the team mean lowest?	A. #2	U. #1
10. Is the overall team mean above or below the median game?	E. Below	F. Above
11. Who had the least range of scores?	L. Holly	G. Pauly
12. What was the team's total score?	T. 2491	W. 2563

___ ___ ___ ___   ___ ___ ___   ___ ___ ___ ___   ___ ___ ___ - ___ ___ ___ ___ ___ ___!

12 2 10 5   9 4 10   6 1 12 2   7 3 3 10 9 11 3 10 8

CONNECTIONS TO THE CLASSROOM

1. When asked to find the median of the set {1, 3, 3, 5, 5, 7, 25, 0, 5, 3, 1}, a student said, "The middle number is 7, so the median is 7." Is the student correct? Why or why not?

2. A student asked each of his classmates how many siblings they have. He found the following statistics: the fewest number of siblings is 0 and the greatest number of siblings is 3. The median number of siblings is 3 and the mean number of siblings is 2.1.

   a. The student claims that most people in the class must have around 2 siblings. Has he correctly interpreted his data? Explain.

   b. How could you help this student understand what his statistics tell him about his data?

3. A student created the bar graph shown at the right to display the results of a survey she conducted in her school.

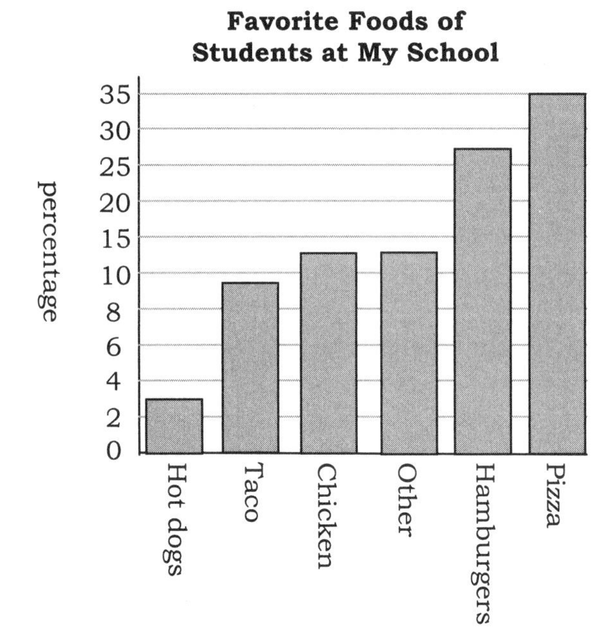

   a. Is there anything inappropriate about this graph? Explain.

   b. If the student had used a unit of 2% for the entire vertical axis, how do you think your impression of the data would change?

   c. If the student had used a unit of 5% for the entire vertical axis, how do you think your impression of the data would change?

   d. Why do you suppose the student constructed the graph the way she did?

4. For a presentation on the development of the downtown area in his city, a student created the pictograph shown at the right.

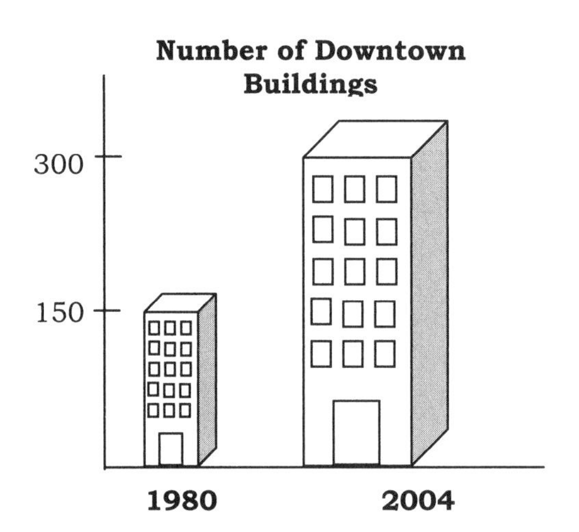

   a. From the scale of the graph, we see that the number of buildings downtown in this city in 1980 was________ and by 2004 had grown to________. So the number of buildings in 2004 is about _______________as much as that in 1980.

   b. Looking at the picture of the two buildings, we get the impression that the volume of the taller one is *more* than twice the volume of the other. Approximate the number of small buildings it would take to fill the large building. Explain how you got your answer.

   c. Part (b) shows that if one focuses on the picture and not the vertical scale, the comparison between the two buildings may be misinterpreted. Use a separate sheet of paper to show how a 3-D pictograph could be constructed in an attempt to avoid any possible misinterpretation.

MENTAL MATH

The histograms at the right picture three different data sets, all with the same range of values. Indicate the set with (1) the highest median; (2) the lowest median; (3) the highest standard deviation; and (4) the lowest standard deviation.

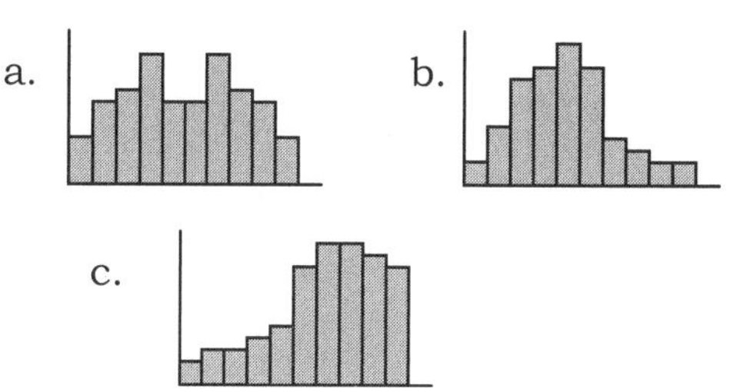

DIRECTIONS IN EDUCATION
Matching Strategies to Outcomes
The Teacher as Decision Maker

Learner outcomes may vary from lesson to lesson. Some lessons have as their goal the mastery of basic skills or the recall of factual information. Other lessons are designed to assist students in the development of conceptual understanding. Still other lessons may provide opportunities for students to apply learning in new or diverse ways. Some lessons may even have as their underlying objective social skills which enhance the learning environment of the classroom. Academic goals in such lessons may be of secondary importance. Teachers must make key decisions about the type of learner outcome expected for each lesson. In conjunction with the consideration of these outcomes, the teacher must select appropriate instructional delivery models with which outcomes may be achieved. While instructional models exist in great variety, each model is better matched to certain types of learner outcomes than to others. It is the ability to appropriately match teaching models to expected learner outcomes which distinguishes the professional teacher.

Teaching models which support various learner outcomes:

- **Mastery strategies.** Direct instruction, mastery learning, programmed instruction, and teacher-directed instruction are some of the terms used to describe models of teaching in which the desired outcome of the lesson is skill acquisition or factual recall. These models are designed to be structured and directed by the teacher or by a carefully sequenced set of instructional materials to lead every student in the class to the same destination, but perhaps at different rates.

- **Conceptual strategies.** Concept attainment, concept formation, and inquiry and discovery learning are some of the models which may be used when the desired outcome of the lesson is conceptual processing or the formation of links with prior knowledge to expand conceptual understanding. While the lesson may be structured in advance by the teacher, the learner assumes greater responsibility for the learning with the teacher serving in the role of mediator or facilitator.

- **Divergence strategies.** Problem solving, inventive strategies, synectics (or metaphorical thinking), and open-ended discussions are instructional models which may be used when the goal of the lesson is application of knowledge in divergent, self-directive ways. In these lessons, the teacher provides the catalyst for the learning by posing interesting problems or questions – providing the "cognitive itch". The learner, once engaged in the activity, has sole responsibility for the outcome of the process. The role of the teacher is to serve as a resource when called upon and to pose additional problems or questions to keep the process in motion once begun.

- **Involvement strategies**. Some strategies have as their primary objective the involvement of students in the learning process. Cooperative learning, team games or tournaments, quality circles, and peer partnerships are some of the strategies which increase student participation in the lesson. These strategies can be used to teach the social skills necessary for the classroom as well as to facilitate the acquisition of skills and concepts or the application of prior learning. These strategies may also be used in conjunction with the previous strategies to enrich the classroom environment.

Thinking skills are enhanced by the use of varied teaching strategies:

- Students can be helped to understand that they move through various levels of knowledge acquisition and use by being told the desired outcome of each lesson. They can learn to match what they are doing to a taxonomy of knowledge levels such as that of Benjamin Bloom. In this way, they can be taught to engage in metacognition – thinking about their own thinking process.

- By engaging in divergent and involvement strategies, students can begin to see themselves as being in charge of their own learning and as being able to serve as a resource for their fellow students. In this way, students begin to sense the power of learning and to see themselves as lifelong learners.

- Teachers who can move comfortably from the directive role to the mediative role to the supportive role and back again are able to provide for learning at all levels of the taxonomy.

- Teachers who are able to engage in divergent thinking and decision making project this open-minded, flexible attitude in their classrooms. Students are given a clear message that thinking is valued.

- A classroom environment rich in varied teaching and learning models is interesting to students and promotes active participation in the learning. While some routine is desirable for the handling of everyday tasks such as roll taking and paper collection, daily repetition of a single lesson model over time can create detachment of learner interest and limitation of learner outcomes.

As you think about matching strategies to outcomes, ask yourself:

1. Which of the teaching models have I observed in classrooms?

2. How comfortable do I feel about the directive role? mediative role? supportive role?

3. In what ways can I promote higher-order thinking skills in my students?

11 Probability

THEME: Using Mathematics to Predict

HANDS-ON ACTIVITIES

The concepts of probability allow us to describe numerically the likelihood of uncertain occurrences. The many applications of probability include weather forecasting and determining insurance premium rates. When a weather forecaster says, "There is a 75% chance of snow tomorrow," she is indicating the degree to which she believes it will snow tomorrow. In isolation, the weather tomorrow is unpredictable. However, by considering tomorrow's weather as part of a pattern, the forecaster can give some idea of how likely it is that it will snow. Her claim is based on the regular weather patterns that have been observed over a long period of time. Insurance companies use probability when they decide how much to charge their policy holders; for example, the probability that a 16-year old male will have a car accident is greater than the probability that a 45-year old male will have an accident. Thus, insurance premium rates for young male drivers are higher than rates for older male drivers.

The origins of the science of probability were based in games of chance, such as those involving playing cards, dice, or wheels of fortune. In the elementary classroom, students are often introduced to probability concepts through experiments that involve rolling numbered cubes, spinning spinners, and flipping coins because experiments such as these are easy to conduct and have readily-observed outcomes. In addition, students are better able to focus on and develop an understanding of probability concepts when they are related to these types of simple, hands-on activities, rather than more complicated situations such as weather forecasting.

When you flip two coins, what is the probability that both coins will land on heads?

What is the probability of rolling "snake eyes" when you roll two dice?

The activities in this chapter will help you investigate the meanings of experimental versus theoretical probability, develop properties of probability, and apply counting techniques useful for computing probabilities. You will also conduct a simulation to solve a complex probability problem. Additionally, you will see how Pascal's Triangle is related to some probability experiments.

OBJECTIVE: **Compute experimental and theoretical probability**

You Will Need: **Materials Card 11.1.1 (eManipulative option: *Simulation*)**

Young students are first introduced to the ideas of probability through simple experiments involving hands-on materials such as coins, dice, spinners, or blocks. In the first activity, you will make use of one such experiment in order to discover the meanings of experimental probability and theoretical probability, as well as investigate properties of probability.

1. The **experimental probability** of an event is found by performing an experiment several times and then comparing the number of times the event occurs to the total number of times you conduct the experiment. Using symbols, the experimental probability [denoted $Pr(E)$] is:

$$Pr(E) = \frac{\text{number of times } E \text{ occurs}}{\text{total number of times the experiment is performed}}$$

As an example, consider drawing one block from a set of three blocks of different colors. Cut out one block of each color from Materials Card 11.1.1. Place them face down on the table and mix them up. Draw one block, record its color, and return it to the table face down. Repeat 14 more times, tallying your results in the chart below.

blue	red	yellow
▢	▢	▢

a. Find the following.

$$Pr(\text{blue}) = \frac{\text{number of times blue was drawn}}{\text{total number of draws}} = \_\_\_\_\_$$

$Pr(\text{red}) = \_\_\_\_\_$ $Pr(\text{yellow}) = \_\_\_\_\_$

b. Repeat the process above and compute the experimental probabilities again.

$Pr(\text{blue}) =$ $Pr(\text{red}) = \_\_\_\_\_$

$Pr(\text{yellow}) = \_\_\_\_\_$

Did you get the same experimental probabilities as in part (a)?

blue	red	yellow
▢	▢	▢

2. The **theoretical probability** of an event, if all the outcomes are equally likely, is found by considering an ideal experiment and comparing the total possible ways an event can occur to the total number of possible outcomes. The set of all the possible outcomes in an experiment is called the **sample space**.

Consider drawing a block from the blue, red, and yellow blocks as you did in problem #1. The sample space in this experiment is the set {blue, red, yellow}, all the possible colors of blocks. Thus, the size of the sample space is 3. Find the theoretical probability, or just "probability" (denoted $P(E)$), for each of the following events.

$$P(\text{blue}) = \frac{\text{number of blue blocks}}{\text{total number of blocks}} = \_\_\_\_\_$$

$P(\text{red}) = \_\_\_\_\_$ $P(\text{yellow}) = \_\_\_\_\_$

3. Cut out all the blocks from Materials Card 11.1.1 and again consider drawing one block from this set of 8 blocks. Note that even though there are only three colors, now there are *four* blue blocks, *three* red blocks, and *one* yellow block.

   a. Compute each of the following theoretical probabilities.

   P(blue) = _____ *P*(red) = _____ *P*(yellow) = _____

   b. If you drew a block and returned it to the pile, repeating this process for a total of twenty-four draws, about how many times would you expect to get a blue block? _____ Explain.

4. Place all 8 of the blocks face down on the table. Draw one block, record its color and replace the block on the table. Do this 10 times, tallying your results in the table at the right.

blue	**red**	**yellow**

   a. Compute each of the following.

   Pr(blue) = _____ *Pr*(red) = _____

   Pr(yellow) = _____

   b. Continue drawing one block at a time, recording your results in the chart above. Do this 10 more times. Then, compute the experimental probabilities again, based on 20 trials.

   Pr(blue) = _____ *Pr*(red) = _____ *Pr*(yellow) = _____

   c. Continue drawing blocks and recording your results in the chart above. Do this 20 more times. Then, compute the experimental probabilities again, based on 40 trials.

   Pr(blue) = _____ *Pr*(red) = _____ *Pr*(yellow) = _____

   d. Look back at the experimental probabilities you computed in parts (a) – (c). Which of these most closely resembles the theoretical probability you computed in problem #3?

5. Suppose that you drew one block from this set of 8 blocks, repeating this process for a total of 10 times.

   a. What is the fewest number of times that a blue block could have been drawn? _____

   What is the smallest possible value for *Pr*(blue)? _____

   b. What is the greatest number of times that a blue block could have been drawn?_____

   What is the largest possible value for *Pr*(blue)? _____

   c. Fill in the blanks: _____ $\leq$ *Pr*(blue) $\leq$ _____.

6. Suppose that a box contains only three red blocks. Compute the following probabilities.

   a. *P*(green) = _____ b. *P*(red) = _____

   The event of selecting a green block is an **impossible** event. The event of drawing a red block is **certain** to occur.

7. Suppose a box contains 2 blue blocks, 4 red blocks, and 6 yellow blocks. Compute the following probabilities.

   a. *P*(blue) = _____ b. *P*(red) = _____ c. *P*(yellow) = _____

   d. *P*(blue) + *P*(red) + *P*(yellow) = _____

   e. In how many ways can you select a blue block *or* a red block? _____
   What is *P*(blue *or* red)? _____

   In this case, the event of drawing a blue block is disjoint from the event of drawing a red block. Is the following true?

   P(blue or red) = *P*(blue) + *P*(red)

   f. The event of "not selecting a blue block" is called the **complement** of the event of selecting a blue block." Compute the following probabilities.

   P(not blue) = _____ *P*(not red) = _____ *P*(not yellow) = _____

   How is the probability of an event related to the probability of the complement of the event?

8. Summarize the properties of probability.

   a. For any event E, _____ $\leq P(E) \leq$ _____.

   b. The probability of an impossible event = _____

   c. The probability of a certain event = _____

   d. If A and B are disjoint events, then $P(A \text{ or } B)$ = _____________

   e. If $\overline{A}$ is the complement of an event A, then $P\left(\overline{A}\right)$ = _____________

OBJECTIVE: **Compute probabilities geometrically**

You Will Need: **Materials Card 11.1.2, thumbtacks, pennies, and paper (Physical Manipulative Option: *Spinners*)**

In the next activity, you will discover connections between geometry and probability. As in the first activity, you will also compare experimental and theoretical probabilities of experiments of the type often used in elementary classrooms.

1. The spinner shown at the right has four sectors labeled R(red), Y(yellow), B(blue), and G(green). An experiment consists of spinning the pointer once, and recording the color.

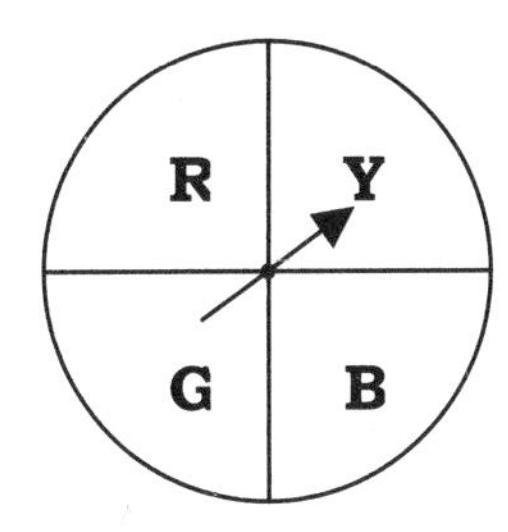

   a. What is the sample space for this experiment?

   b. Since the blue sector is one out of four equally –sized sectors, the probability that the pointer will land on blue is _____.

      Find the theoretical probability for the pointer landing the given color(s) for each of the following:

      P(green) = _____ *P*(yellow or green) = _____ *P*(not red) = _____

2. The spinner shown at the right has three sectors labeled B(blue), Y(yellow), and G(green). An experiment consists of spinning the pointer once, and recording the color.

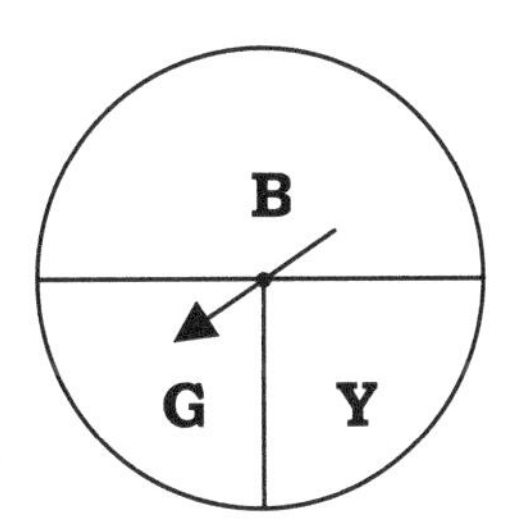

   a. Are the events "blue" and "green" equally likely on this spinner? Explain your answer.

   b. Imagine dividing this spinner into four sectors, as shown at the right. Explain how this can help you to compute each of the following probabilities.

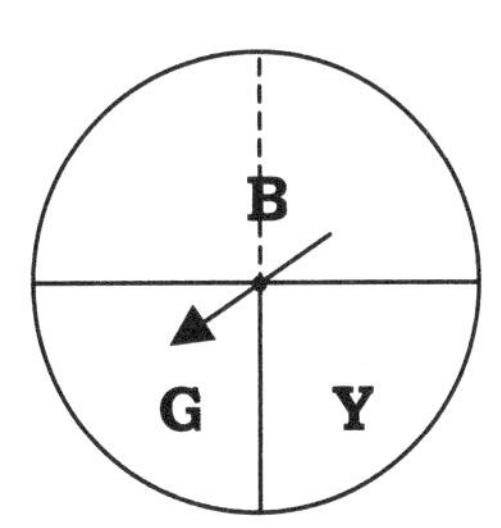

      P(blue) = _____ *P*(blue or green) = _____

   c. Use a technique similar to that in part (b) to find the following probabilities for the spinner shown at the right.

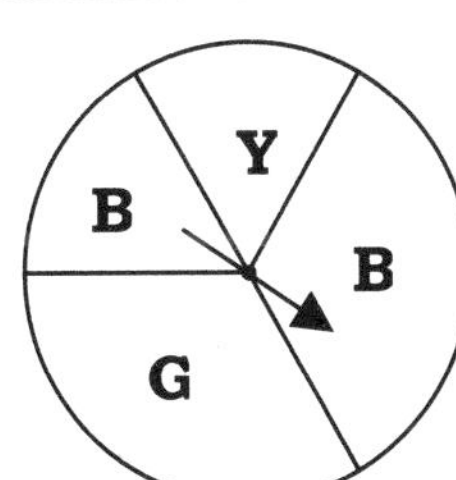

      P(blue) = _____ *P*(yellow) = _____

      P(yellow or blue) = _____ *P*(not green) = _____

3. An airplane drops a box of relief supplies onto a drought-stricken savannah. Because of wind currents and variations in release elevation and location, it is expected that the box will land in a random location on the field. The field is divided by fences as shown at the right.

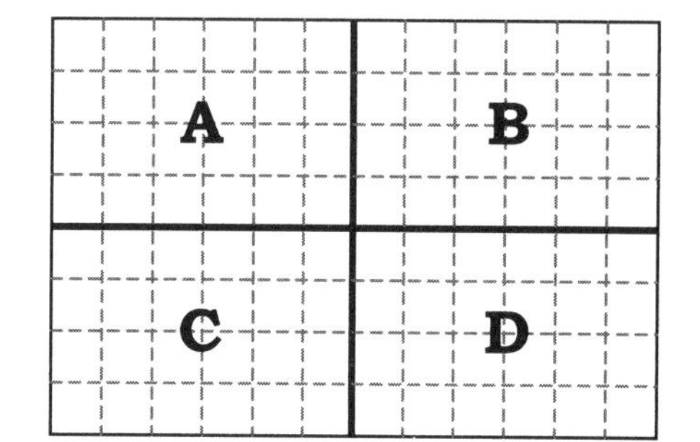

a. On Materials Card 11.1.2, use a pencil to divide the grid into regions as indicated. Model the situation by closing your eyes and randomly dropping thumbtacks from a cup onto the grid. The region in which the point lands determines the outcome. If a thumbtack bounces off the grid, do not record the toss. Tally your results until at least 50 drops have been recorded. In order to speed up the process, you can drop several tacks at the same time or, if working in a group, divide the 50 drops among your group members and share data at the end.

Plot A	**Plot B**	**Plot C**	**Plot D**	**Total**

The experimental probability of landing in region A = _____

The theoretical probability of landing in region A = _____

Does the box have an equal chance to land on all parts of the field? Why or why not?

b. Now divide the grid on Materials Card 11.1.2 as shown at the right and repeat the process in part (a) with these regions.

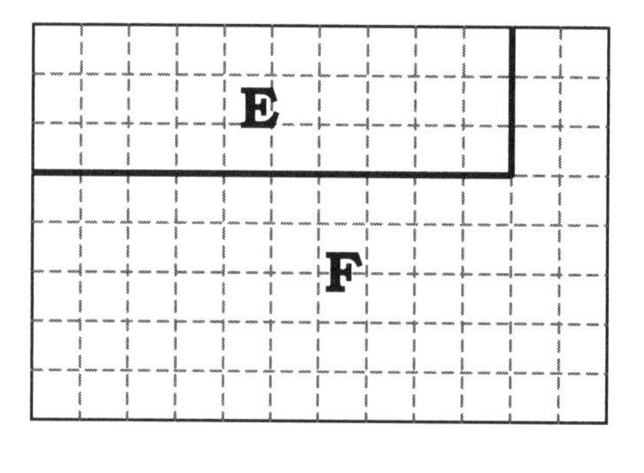

Plot E	**Plot F**	**Total**

The experimental probability of landing in region E = _____

The experimental probability of landing in region F = _____

What is the area of region E? _________ region F? _________ the total area? _________

What is the theoretical probability of landing in region E? _____ region F? _____

This problem illustrates how we can view probability in terms of geometry; for example, the probability that the box will land in region E is 5/16, since region E accounts for 5/16 of the total area of the field.

4. Next, take out a blank, unlined sheet of paper, draw parallel lines 1 inch apart as shown at the right, and lay it on the table.

   a. An experiment consists of flipping a penny randomly onto your sheet of paper. Do you think it is likely that a penny will land so that it is not touching a line? Why or why not?

   b. Flip a penny onto your paper, 50 times. Coins that land off the paper should be re-tossed. Tally your results in the chart below.

Number of coins on a line	**Number of coins not on a line**

Compute each of the following:

Pr(line) = _____ *Pr*(not a line) = _____

   c. The diameter of a penny is 3/4 inch. In what region must the *center* of the penny fall so that the penny is not touching a line? Shade this region on the lined paper shown in part (a).

   How does the area of these shaded regions compare with the total area of the paper?

   What is the theoretical probability that a coin will not land on a line? _____

   d. If the lines were drawn in both directions to form a 1-inch grid on the paper, then what is the probability that a coin will not land on a line? Show how you got your answer.

5. The following famous problem was first presented by Count Buffon, a French naturalist, in the 1700's. Buffon's Needle Problem is a way of approximating the irrational number π, the Greek letter 'pi' (pronounced 'pie') using probability methods.

   a. As a "needle", use a short needle, pin, or toothpick. Find the length, l, of the object you are using.

   b. Using $d = 2l$, draw parallel lines d units apart on a sheet of paper.

   c. Drop the "needle" 50 times randomly onto the sheet of paper. Count the number of the times the needle lands on a line. What is the experimental probability that a "needle" dropped on this piece of paper will touch one of the lines? _________

   d. Compute $1/P$, where P is the probability you computed in part (c). Using calculus, it can be proved theoretically that $1/P = \pi$. How close was your experimental value?

OBJECTIVE: **Compute probabilities when tossing dice**

You Will Need: **Two dice**

Dice are readily available and outcomes of dice-tossing experiments are easily observed, making them a popular hands-on model for teaching probability concepts in the elementary classroom. In the next activity, you will use dice to investigate an experiment in which the outcomes are not equally likely. Also, you will see one way to systematically list all the outcomes in the sample space of an experiment.

1. Toss a pair of dice 30 times (or a single die twice each time), and tally the sum each time in the table below.

2	3	4	5	6	7	8	9	10	11	12

   a. Compute the experimental probabilities of getting the following sums.

   $Pr(7)$ = _____ $Pr(3)$ = _____ $Pr(12)$ = _____ Pr(less than 5) = _____

   Pr(greater than 6) = _____

   b. Why are some sums more likely to occur than others?

2. To determine the theoretical probabilities of each sum when rolling two dice, first fill in the sums in the chart below.

die #1

die #2

+	1	2	3	4	5	6
1						
2						
3						
4						
5						
6						

Summarize the number of ways of getting each possible outcome in the table below.

Sum	2	3	4	5	6	7	8	9	10	11	12
Number of ways											

What is the total number of possible outcomes when tossing two dice?_____

3. Compute the following probabilities, using the information you recorded in problem #2.

$P(7)$ = _____ $P(3)$ = _____ $P(12)$ = _____ P(less than 5) = _____

P(greater than 6) = _____

OBJECTIVE: **Use tree diagrams for multi-stage experiments**

You Will Need: **Colored blocks from Materials Card 11.1.1 and a coin (eManipulative option: *Coin Toss*)**

For many experiments involving multiple stages, listing all the outcomes in the sample space can be made easier by the use of a tree diagram. Tree diagrams, in turn, can be helpful in developing a property used for computing the total number of possible outcomes in the sample space of an experiment.

1. Consider the experiment of flipping a coin two times (or two coins at the same time), and recording heads or tails on each of the two flips.

   a. In the space below, list all the possible outcomes for this experiment. To represent the outcomes, it can be helpful to use abbreviations; for example, the outcome of a head on the first coin and a tail on the second coin may be written as "HT."

   Sample space for flipping two coins = ______________________________

   b. A tree diagram for this experiment is shown below.

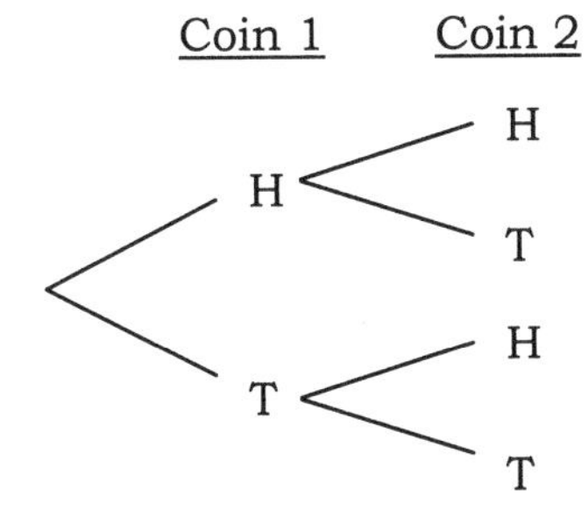

   Explain how the tree diagram was made.

   What information does the diagram show?

   Why do you suppose this is called a tree diagram?

   c. Compute the following probabilities. Show how you got your answer in each case.

   P(heads on both coins) = _____ P(tails on at least one coin) = _____

   d. In the space at the right, draw a tree diagram for the experiment of flipping a coin *three* times (or three coins all at once).

   e. Use the tree diagram from part (d) to compute the following probabilities. Show your process for each problem.

   P(heads on exactly two coins) = _____

   P(heads on at least two coins) = _____

   P(exactly two heads in a row) = _____

2. Now take out these colored blocks from Materials Card 11.1.1: 2 blue, 1 red, and 1 yellow. An experiment consists of drawing 2 blocks from this set of 4 blocks (which are placed face down on the table) in the following way: draw the first block, note its color, and return it to the pile before drawing the second block.

   a. The first set of branches for the tree diagram in this experiment (representing the possible outcomes on the first draw) is shown at the right. Complete the tree diagram by drawing the branches for the second draw.

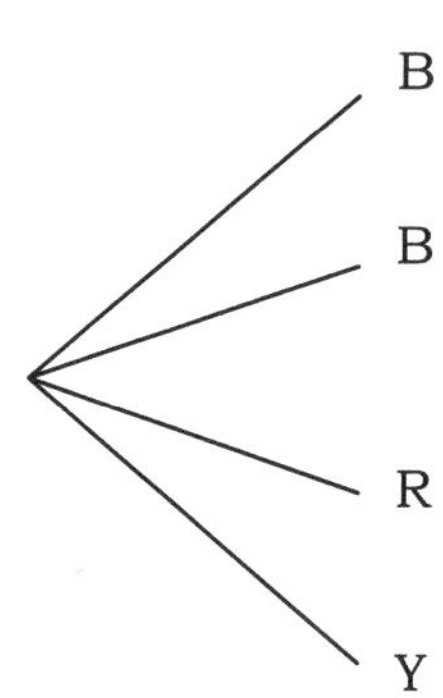

   b. List the outcomes in the sample space:

   c. Show how to use the tree diagram to compute the following probabilities.

      P(drawing at least one blue block) = _____

      P(drawing exactly 2 blocks of the same color) = _____

      P(drawing exactly one blue and one yellow block) = _____

3. The experiment in problem #2 is called **drawing with replacement** because the first block was replaced in the pile before the second block was drawn. Now consider an experiment of **drawing without replacement**, where two blocks are drawn from a set with 2 blue, 1 red and 1 yellow block, without replacing the first block before the second one is drawn.

   a. The first set of branches for this experiment is the same as for the experiment in problem #2. However, for example, if we draw a yellow block on the first draw, there are only 3 remaining blocks to select from for the second draw. This is indicated in the tree diagram as shown at the right. Complete the tree diagram for the experiment of drawing two blocks without replacement.

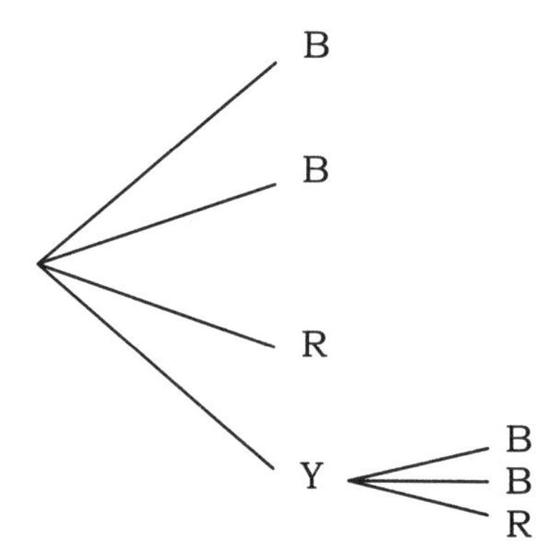

   b. Show how to use the tree diagram to compute the following probabilities.

      P(drawing at least one blue block) = _____

      P(drawing exactly 2 blocks of the same color) = _____

      P(drawing exactly one blue and one yellow block) = _____

4. When computing the probability of an event, you need to know the number of outcomes in the event and the number outcomes in the sample space. By observing the way tree diagrams are constructed, you can discover a way to compute the size of the sample space, without having to count all the possible outcomes.

   a. Look again at the tree diagram for the experiment of flipping two coins as shown at the right. Notice that there are two branches representing the first flip, *and for each of these branches*, there are two branches representing the second flip. Thus, there are 2 × 2 = 4 total possible outcomes.

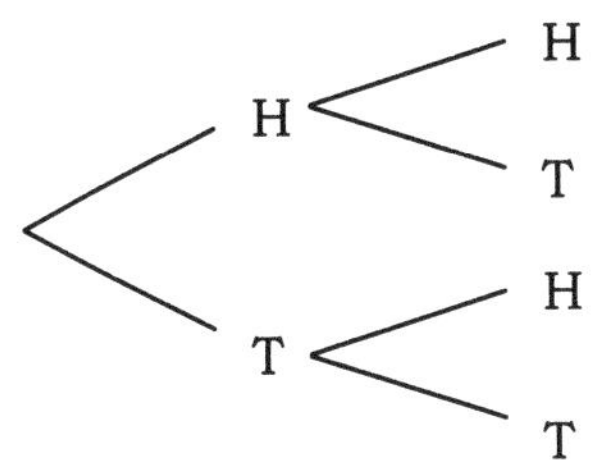

   Using similar reasoning, explain how you could find the total number of possible outcomes for flipping a coin *four* times coins (or four coins at the same time).

   b. Consider the experiment of drawing two blocks without replacement in problem #3. We see that there are 4 possibilities for the first draw. For each of these 4 possibilities, there are 3 possibilities for the second draw. Thus, there are ____ × ____ = ____ total possible outcomes.

   c. These examples illustrate the **Fundamental Counting Property** which states, "If an event *A* can occur in *r* ways, and for each of these *r* ways, an event *B* can occur in *s* ways, then events *A* and *B* can occur, in succession, in $r \times s$ ways."

   Use the Fundamental Counting Property to find the number of outcomes in the sample space for each of the following experiments.

   Roll two dice.

   Roll a die, and then flip a coin.

   Spin the spinner shown at the right, and then roll a die once.

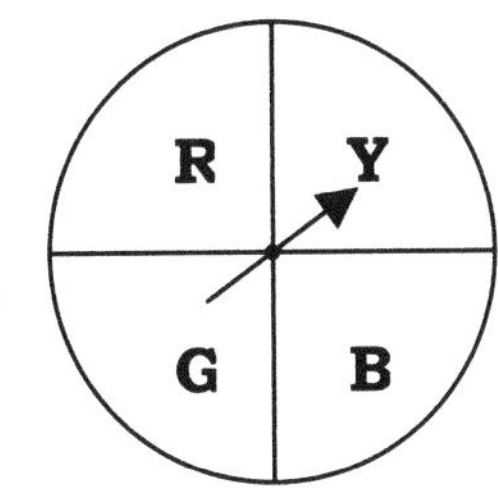

   Draw one block from a set with 2 red, 2 yellow, and 1 green block, and then spin the spinner shown at the right one time.

OBJECTIVE: Use probability tree diagrams

In Activity 11.2.1, you saw the usefulness of tree diagrams (also called outcome trees) for listing the outcomes in the sample space for an experiment. In this activity, you will see another type of tree diagram that can help you compute probabilities, without the need to list the entire sample space.

1. Consider an experiment that consists of randomly drawing one block from a set of 2 blue blocks, 1 red block, and 1 yellow block and noting its color. The outcome tree diagram for this experiment is shown below, on the left.

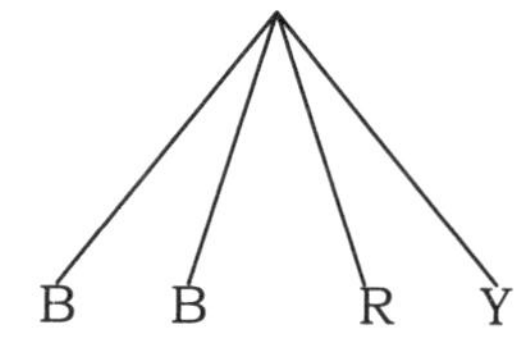

We can form a **probability tree diagram** from this outcome tree by labeling the probability of each outcome on the branches of the tree, as shown at the right.

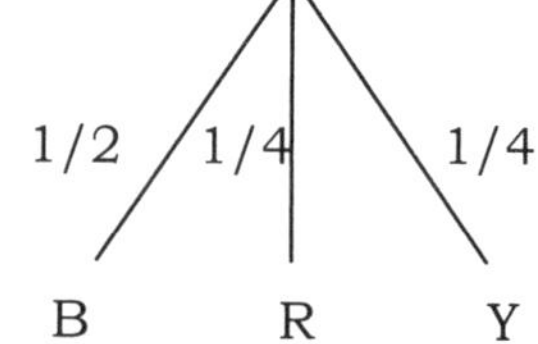

   a. Explain how these two diagrams are similar and how they are different.

   b. Now, consider the experiment of randomly drawing two blocks, *with* replacement, from the set of 2 blue blocks, 1 red block, and 1 yellow block. Recall the outcome tree diagram for this experiment, shown at the right.

      Shown at the right is the first set of branches of the probability tree for this experiment. Show how to extend the probability tree to represent the second draw, and then label each branch with the appropriate probability.

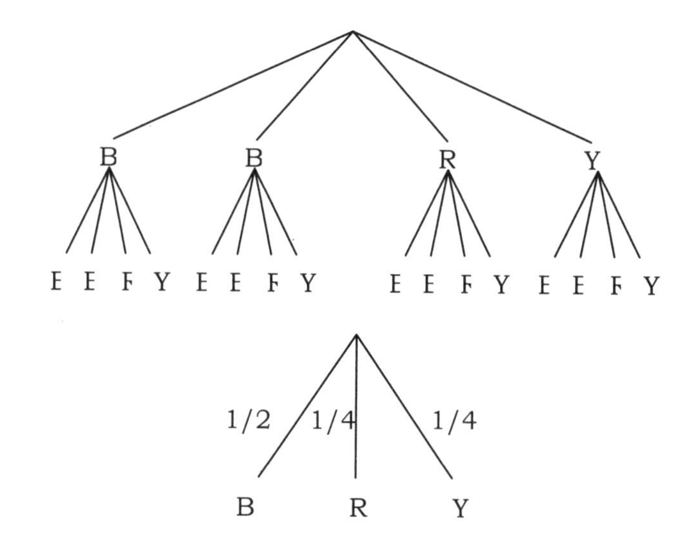

   c. Using the outcome tree diagram in part (b), you can see that "RR" is one outcome, out of 16 equally-likely outcomes, so P(RR) = 1/16.

      Now look at your probability tree diagram in part (b), and the branches leading to the outcome RR. The probability of drawing a red block on the first draw is 1/4 and the probability of drawing a red block on the second draw is 1/4. What inference can you make about how the probabilities of these two draws relate to P(RR) above?

d. As you saw in part (c), in a probability tree, you can find the probability of an outcome by multiplying all the probabilities on the branches for that outcome. This is the **multiplicative property of probability**. Using this property, label the end of each branch in the *probability* tree you drew in part (b) with the probability for that outcome. Show how to check these answers using the *outcome* tree given in part (b).

e. Add the probabilities in the bottom row of your probability tree. What do you notice? Explain why this makes sense.

2. Another property of probability can be useful when you want to compute the probability of an event with more than one outcome. To discover this property, consider the outcome tree diagram from problem #1(b).

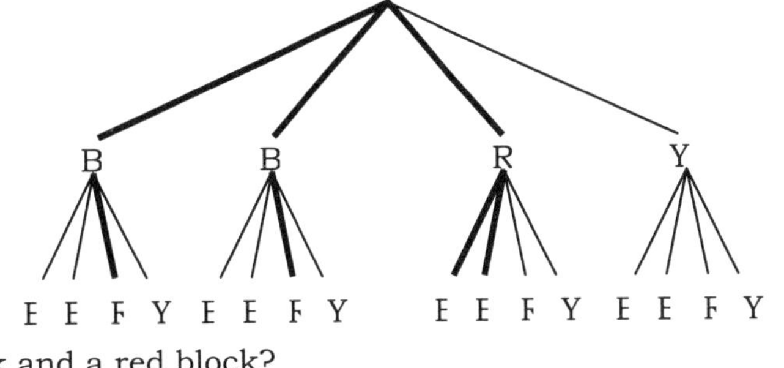

a. If you want to compute the probability of drawing a blue and a red block *in any order*, you must consider the four paths highlighted in the outcome tree. What is the probability of each of these four outcomes? _____ What is the probability of drawing a blue block and a red block? _____

b. Using the probability tree diagram, you can see that there are two outcome paths that have a blue block and a red block. Fill in the probabilities of these two outcomes. Do you see a way to obtain the probability of drawing a red block and a blue block, using the probabilities of these two paths? Explain.

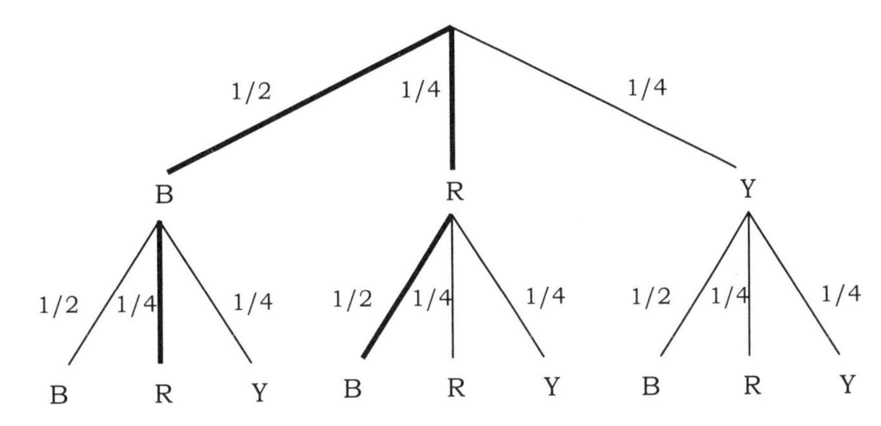

c. As you saw in part (b), in a probability tree, you can find the probability of an event by adding all the probabilities of the outcomes in that event. This is the **additive**

property of probability. Using the multiplicative and additive properties and the probability tree in part (b), find the following probabilities.

P(drawing at least one blue block) = _____

P(drawing 2 blocks of the same color) = _____

3. Draw the probability tree diagram for the following experiment, and then show how to use it to compute the indicated probabilities.

Experiment: Randomly draw two blocks *without* replacement from a set containing 3 green, 1 white, and 2 orange blocks.

3 G
1 W
2 O

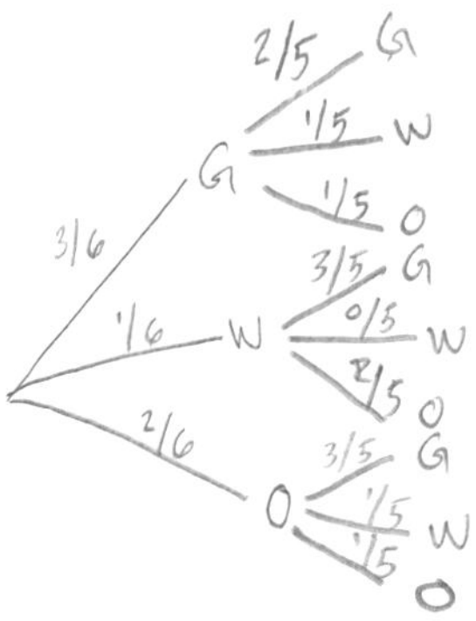

$5 \times 6 = 30$

P(exactly two green blocks) = 6/30

P(GG) 3/6 × 2/5 = 6/30

P(no green blocks) = 6/30

P(WW) = 1/6 × 0/5 = 0/30
P(WO) = 1/6 × 2/5 = 2/30
P(OW) = 2/6 × 1/5 = 2/30
P(OO) = 2/6 × 1/5 = 2/30
6/30

P(two blocks of different colors) = 22/30

P(GW) = 3/6 × 1/5 = 3/30
P(GO) = 3/6 × 1/5 = 3/30
P(WG) = 1/6 × 3/5 = 3/30
P(WO) = 1/6 × 2/5 = 2/30
P(OG) = 2/6 × 3/5 = 6/30
P(OW) = 2/6 × 1/5 = 2/30
22/30

OBJECTIVE: **Use Pascal's Triangle to compute probabilities**

Pascal's triangle, which you first encountered in your study of patterns in Chapter 1, can be useful for determining probabilities in certain types of multi-stage experiments. To discover the connections between Pascal's triangle and probability, first consider experiments that involve coin-flipping.

1. If you flip a coin 2 times, there are four possible outcomes, HH, HT, TH, and TT. These outcomes may be organized according to the number of heads for each one, as shown in the table at the right.

Flip a coin 2 times			
Number of heads	2	1	0
Outcomes	HH	HT, TH	TT
Number of ways	1	2	1

a. Complete the following tables for flipping a coin 1 time, 3 times, and 4 times.

Flip a coin 1 time		
Number of heads	1	0
Outcomes		
Number of ways		

Flip a coin 3 times				
Number of heads	3	2	1	0
Outcomes				
Number of ways				

Flip a coin 4 times					
Number of heads	4	3	2	1	0
Outcomes					
Number of ways					

b. The last row of each table in part (a) can be arranged to show an array called **Pascal's triangle**, as shown at the right. Each row begins and ends with a 1. Describe the pattern that yields each of the other entries in a row from the previous row.

```
      1   1
    1   2   1
  1   3   3   1
1   4   6   4   1
```

c. Extend Pascal's triangle to the row that begins 1 8 ... and ends ...8 1. How can each of the entries in this row be interpreted, in terms of a coin-tossing experiment?

d. Find the sum of each row of Pascal's triangle. What is the sum of the entries in the nth row of Pascal's triangle? _____ What do these numbers represent, in terms of coin-tossing?

2. Use Pascal's triangle to compute the following probabilities. Show your process in each case.

a. P(exactly 4 heads when flipping 6 coins) = _____

b. P(at least 5 heads when flipping 8 coins) = _____

3. The entries of Pascal's Triangle can be used to solve probability problems in which there are two *equally likely* outcomes at each stage (like with a coin that has two such outcomes, a head and a tail). Note that coin outcomes may not be exactly equally likely and that neither are boys/girls. But, for our study, we will assume that they are.

Show how to use Pascal's triangle to find the probability of each of the following events.

a. No girls in a family with 5 children.

b. At least one girl in a family with 5 children.

c. Passing a 10-question true/false quiz, if you randomly choose your answers (passing is at least 7 correct answers).

OBJECTIVE: **Investigate counting techniques**

You Will Need: **Colored blocks from Materials Card 11.1.1**

Many probability experiments have a very large sample space. To find probabilities of an event in such cases, one may find the number of elements in the event and in the sample space rather than listing all the possible outcomes. In this activity, you will develop counting techniques for computing the number of different permutations or combinations of objects. In the problem sets in your textbook, you will apply these techniques to probability problems.

1. Take out 3 of the colored blocks from Materials Card 11.1.1, one red, one yellow, and one blue.

   a. One way to arrange these three blocks in a row is red, yellow, blue (RYB). Systematically list all the other possible ways to arrange these three blocks in order from left-to-right and record your answers below.

   How many arrangements are possible, including RYB? _____

   b. When listing all the possible arrangements for a red, yellow, and blue block, there are ____ choices for the color of the first block. After making a choice for the color of the first block, there are ____ remaining choices for the color of the second block. After the first and second blocks have been chosen, there is ____ remaining choice for the color of the third block. Using the Fundamental Counting Property, you can see that there are ____ × ____ × ____ = 6 possible arrangements of three blocks.

   The product $3 \times 2 \times 1$ is read "3 factorial," and the shorthand notation for this product is written using an exclamation point as 3!. More generally, if n is a counting number, then $n! = n \times (n-1) \times (n-2) \times \ldots \times 1$ is called "n factorial".

   c. Suppose you had a fourth block that is colored green. List all the possible ways to arrange the red, yellow, blue, and green blocks in a row.

   How many arrangements are possible? _____

   d. What factorial is related to your answer in part (c)? Explain.

   e. A **permutation** is an ordered arrangement of objects. When you found all the possible arrangements of blocks in parts (a) and (c), you were listing all the possible permutations of 3 blocks and 4 blocks, respectively. What if you had 5 blocks, how many permutations are there? _____ What if you had n blocks? _____

2. In the examples in problem #1, you considered permutations of n objects, where all n of the objects were used. Now consider permutations of a set of objects taken from a larger set.

   a. As an example, suppose that 5 people are running a race. In order to answer the question, "How many possibilities are there for the first- and second- place finishers of the race?" you could assign a letter to each runner, say A, B, C, D, and E. The following chart shows some of the possible outcomes.

For example, AB represents person A in first place and person B in second place.

Why do we have to list both AB and BA?

Why isn't BB listed?

Possible outcomes for first- and second-place finishers in a race with 5 runners

AB	*BA*
AC	*BC*
AD	*BD*
AE	*BE*

Complete the chart by listing *all* the possible outcomes. How many possibilities are there?_____

b. Notice in the chart there are four possible outcomes with person A in first place, and four with person B in first place. How many groups of four like this do you have altogether? _____ How many possibilities are there for the first- and second-place finishers? _____

c. In part (b), you showed that there are five groups of four, one for each of the runners who could be in first place. So $5 \times 4 = 20$ possible permutations of 5 objects, when we take them 2 at a time. Now consider the problem of finding all the different permutations for the first-, second-, and third-place finishers in a race with 5 people; that is, permutations of 5 objects, taken 3 at a time. The chart at the right shows some of the possible outcomes.

Some possible outcomes for first-, second-, and third-place finishers in a race with 5 runners

A, B, C	*A, C, B*	*A, D, B*	*A, E, B*
A, B, D	*A, C, D*	*A, D, C*	*A, E, C*
A, B, E	*A, C, E*	*A, D, E*	*A, E, D*

All of the possibilities for person A as the first-place finisher are shown in the chart. Notice that there are then 4 possibilities for second place, and for each of these 4 possibilities there are 3 possibilities for third place. Hence there are $4 \times 3 = 12$ possible outcomes with person A in first place. If you were to continue this chart, listing systematically as shown, how many groups of 12 like this would you have? _____ Explain.

So, you see there are $5 \times 4 \times 3 = 60$ permutations of 5 objects, taken 3 at a time.

d. The number of permutations of r objects chosen from n objects is denoted ${}_nP_r$. To develop a general formula for computing ${}_nP_r$, consider the products for ${}_5P_2$ and ${}_5P_3$, from part (c).

You saw that ${}_5P_2 = 5 \times 4$ which is the same as $\frac{5\times4\times3\times2\times1}{3\times2\times1} = \frac{5!}{3!} = \frac{5!}{(5-2)!}$. Similarly, show how to write ${}_5P_3 = 5 \times 4 \times 3$ using factorial notation.

In general, the number of permutations of n objects taken r at a time is: ${}_nP_r = \frac{n!}{(n-r)!}$

3. Suppose a teacher wants to select a team of two students from a group of five volunteers to represent their class at a student council meeting. In order to answer the question, "How many possible teams of two can she choose?" you could assign a letter to each volunteer, say *A*, *B*, *C*, *D*, and *E*. The following chart shows some of the possible teams of two.

Possible teams of two chosen from 5 volunteers for student council meeting

{*A*, *B*}	{*B*, *C*}
{*A*, *C*}	{*B*, *D*}
{*A*, *D*}	{*B*, *E*}
{*A*, *E*}	

a. Why don't you need to list {*B*, *A*}?

How is this problem different from the one in problem #2(a)?

List *all* the possible teams of two chosen from a group of five by continuing the pattern that was begun in the chart above.

A collection of objects *in no particular* order is called a **combination**. The number of combinations of *n* objects, taken *r* at a time, is denoted $_nC_r$. In this case, you found that $_5C_2$ = _____.

b. Since order doesn't matter in the situation involving teams of 2 chosen from 5, listing permutations would yield too many teams. However, comparing the number of possible combinations, $_5C_2$, to the number of permutations, $_5P_2$, can help you discover a rule for computing the number of combinations. In the chart below, note that each combination corresponds to two permutations of the same objects.

Combinations of 2 objects chosen from a set of 5 objects		Permutations of 2 objects chosen from a set of 5 objects	
{*A*, *B*}	⟷	*AB*	*BA*
{*A*, *C*}	⟷	*AC*	*CA*
{*A*, *D*}	⟷	*AD*	*DA*
{*A*, *E*}	⟷	*AE*	*EA*
etc.		etc.	

Use the relationship shown in the chart above to write an expression for $_5C_2$ in terms of $_5P_2$.

c. Now suppose the teacher wanted to make a team of 3 from the group of 5 volunteers. One possible combination is {*A*, *B*, *C*}. List all the possible permutations of *A*, *B*, and *C*. How many are there? _____

d. In part (c), you see that for each combination of 3 objects, there are 6 permutations.

Thus, ${}_5C_3 = \frac{{}_5P_3}{6} = \frac{{}_5P_3}{3\times2\times1} = \frac{{}_5P_3}{3!}$. Also, from part (b) you know ${}_5C_2 = \frac{{}_5P_2}{2} = \frac{{}_5P_2}{2\times1} = \frac{{}_5P_3}{2!}$.
Write a general formula for ${}_nC_r$ in terms of ${}_nP_r$..

4. For each of the following, decide whether the situation involves permutations (ordered arrangements) or combinations (arrangements in no particular order) and explain why. Then compute ${}_nP_r$ or ${}_nC_r$ in each case.

   a. How many different 5-card hands can be dealt from a deck of 52 cards?

   b. How many different 4-letter radio station call letters can be formed from the letters A, B, C, W, X, and Z?

OBJECTIVE: **Do a simulation**

You Will Need: **A coin and a die**

Many times, a simulation of a probability experiment may be more convenient to conduct than the actual experiment. Simulations can also make many interesting and otherwise challenging problems approachable for young students, because they involve models such as dice, coins, numbered slips of paper, spinners, or random number tables.

1. At a party, a friend bets you that at least two people in a group of five strangers will have their birthday in the same *month*. Should you take the bet? That is, do you think it is likely or unlikely that at least two people in a group of five strangers will have their birthday in the same month?

2. One way to determine the experimental probability of the event "at least two people in a group of five strangers will have their birthday in the same month" is to interview a large number of groups of 5 strangers. Conducting a simulation of this experiment is another method and is more convenient.

   In a simulation of an experiment, there is a one-to-one correspondence between outcomes in the original experiment and the outcomes in the simulation. The probability of an outcome in the original experiment is estimated to be the probability of its corresponding outcome in the simulated experiment.

   a. Several different simulation devices would work for this simulation. As an example, you could use 12 numbered slips of paper. For each trial, you would randomly draw a slip of paper 5 times from a container, replacing the slip after noting the number on the slip.

      Why do you need to use 12 slips of paper?

      Why do you need to draw 5 times for each trial?

      Why do you need to replace each slip after it has been drawn?

   b. Explain how you could use the combination of a coin and a die to simulate the birthday experiment.

3. Now take out your coin and die to simulate the birthday problem.

   a. Conduct at least 20 trials and record your results in the following table. Circle the outcomes that represent at least two people in a group of five strangers with their birthday in the same month.

Results of simulation:

b. Show how to use your results in part (a) to estimate the probability that two or more people in a group of five strangers will have their birthday in the same month.

4. The theoretical probability that at least two people in a group of five strangers will have their birthday in the same month can be computed using the Fundamental Counting Property.

   The sample space, *S*, consists of all possible arrangements of birth months for 5 people. The event, *E*, is "two or more people (out of five) have the same birth month".

   a. Counting the number of outcomes in the complement of event E can make this problem easier to do. Describe the outcomes in the complement of E, namely $\overline{E}$.

   b. Recall the Fundamental Counting Property which states, "If an event *A* can occur in *r* ways, and for each of these *r* ways, an event *B* can occur in *s* ways, then events *A* and *B* can occur, in succession, in $r \times s$ ways." This property can be generalized to more than two events; for example, the number of outcomes in *S* above is $12 \times 12 \times 12 \times 12 \times 12$, because there are 12 possible birth months for the first person, and for each of these 12 possibilities, there are 12 possible birth months for the second person, etc.

      Using the Fundamental Counting Property, you can also find the number of outcomes in $\overline{E}$. There are 12 possibilities for the birth month of the first person. Once the birth month of the first person is set, there are only 11 possibilities for the birth month of the second person. Why is that the case?

   c. Find the number of outcomes in $\overline{E}$.

      $$\underset{1^{st}}{\underline{12}} \times \underset{2^{nd}}{\underline{11}} \times \underset{3^{rd}}{\underline{\quad\quad}} \times \underset{4^{th}}{\underline{\quad\quad}} \times \underset{5^{th}}{\underline{\quad\quad}} = \underline{\quad\quad}$$

   d. Lastly, use the fact that $P(E) = 1 - P(\overline{E})$ to find the probability of *E*.

   e. How does the experimental probability you found using a simulation in problem #3 compare to the theoretical probability?

OBJECTIVE: **Compute expected value of games**

You Will Need: **A partner, Materials Card 11.3.2, a paperclip, and two dice (Physical Manipulative option: *Spinners*)**

The concept of expected value has many real-life applications; for example, insurance companies use expected value to determine their premium rates. One common way to introduce elementary students to the ideas of expected value is through the use of games, such as the ones in this activity involving spinners.

1. The spinner shown at the right has four sectors, two labeled R(red) and two labeled Y(yellow). In a two-player game, the players each take a turn spinning the spinner. If the pointer lands on the same color on both spins, the first player gets 1 point. If the spinner lands on a different color each time, the second player gets 2 points.

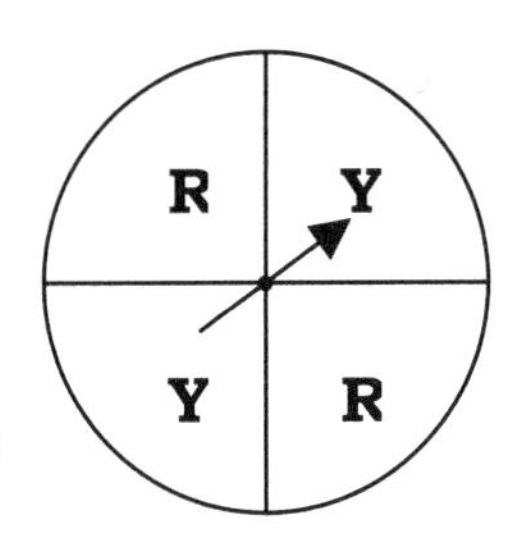

   a. Using the spinner on Materials Card 11.3.2, play 10 rounds of the game with your partner and record the points in the chart below.

Round #	1	2	3	4	5	6	7	8	9	10	Total points
Player #1 (same color)											
Player #2 (different color)											

   Who won the game? ____________________

   b. Compute the theoretical probability of each of the following events.

   Two spins on this spinner land on the same color: _____

   Two spins on this spinner land on different colors: _____

   c. Do you think this is a fair game? __________Why or why not?

   d. How could you change this game to make it a fair game? Explain your answer.

2. A carnival game costs $3 to play. A player spins the spinner, shown at the right, one time. If the spinner lands on "You Lose", the player loses his or her $3.

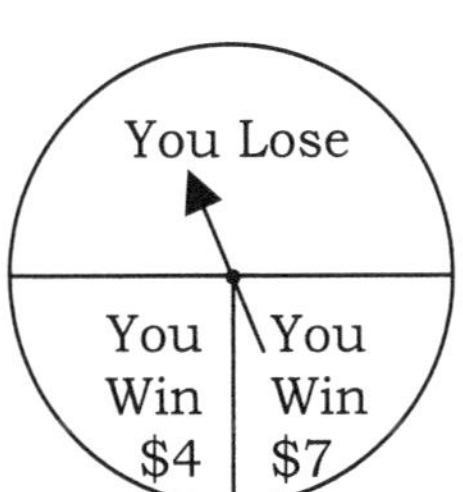

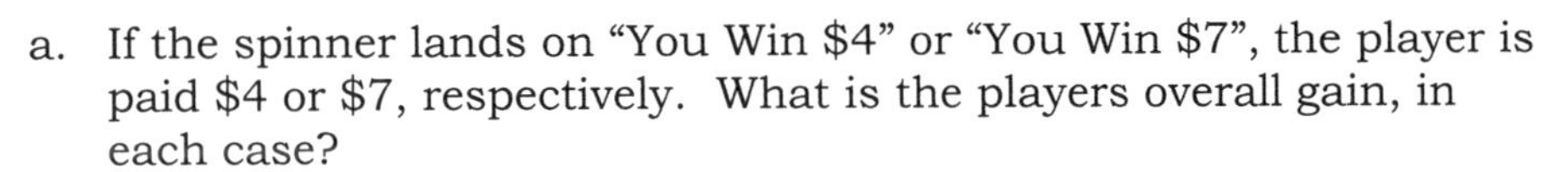

   a. If the spinner lands on "You Win $4" or "You Win $7", the player is paid $4 or $7, respectively. What is the players overall gain, in each case?

   b. Suppose you played this game 100 times. How many times would you expect to lose your $3? _____ How many times, out of 100, would you expect to gain $1?_____How many times, out of 100, would you expect to gain $4? _____ Explain your answers.

c. Theoretically, if you played 100 games, you would lose \$3 in 50 of the games, gain \$1 in 25 of the games, and gain \$4 in 25 of the games. What would be your total gains or losses for these 100 games?

d. On average, how much would you lose *per game*, out of 100 games? _____

e. Do you think this is a fair game? _____ Why or why not?

f. Your theoretical winnings (or losses) per turn, in a game such as this one, is called the **expected value** of the game. The expected value of this game is computed as follows:

$$P(\text{losing } \$3) \times (-3) + P(\text{winning } \$1) \times (1) + P(\text{winning } \$4) \times (4)$$
$$= (1/2)(-3) + (1/4)(1) + (1/4)(4) = -\$0.25$$

Why do we multiply the probability of losing \$3 by "-3"?

What does the answer "-\$0. 25" mean, in terms of the game?

g. In general, the expected value of a probability experiment in which the outcomes are real numbers $v_1, v_2, v_3, \ldots, v_n$ that have probabilities $p_1, p_2, p_3, \ldots, p_n$, respectively, is defined as:

$$E = p_1(v_1) + p_2(v_2) + p_3(v_3) + \ldots + p_n(v_n)$$

A game is a **fair game**, if the expected value of the game is 0. How does this definition of a fair game compare with your answer for problem #1(c)?

3. Consider the following casino game: A player pays \$1 to roll two dice. The player wins \$5 if he or she rolls a sum of 7 or 11 and loses the \$1 if he or she rolls any other sum.

a. Roll your two dice at least 25 times. Use the table below to record your overall gain or loss each time you roll; for example, if you roll a sum of 5, record "-1" in the loss row. Then, compute your total gain or loss for the 25 rolls.

gain		Total gain or loss
loss		

b. Based on your results in part (a), do you think this is a fair game? Why or why not?

c. Compute the expected value for this game. The table you created in Activity 11.1.3 will be helpful for determining the probabilities of the events "sum of 7 or 11" and "not a sum of 7 or 11". Is this a fair game?

EXERCISE

What do you get when you cross an owl with an oyster?

For each problem below, find the answer and then write the letter of that answer on the line above the problem number it matches at the bottom of the page.

B 1/2 **V** 1/4 **A** 1/3 **C** 7/12 **L** 2/3 **J** 3/4 **U** 5/12 **F** 1/6

For problems #1-4, imagine a spinner divided into 12 equal-sized sectors, numbered 1 through 12. If the spinner is spun one time, what is the probability that it will stop on a number that is:

1. a multiple of 2?
2. a multiple of 3?
3. a multiple of 2 *and* a multiple of 3?
4. a multiple of 2 *or* a multiple of 3?

Use the answers below to complete problems #5-18:

H 7/15 **O** 1/3 **I** 2/5 **W** 1/15 **T** 2/15 **P** 3/5 **S** 4/15 **K** 11/15

R 2/3 **M** 8/15 **E** 1/5 **D** 4/15 **G** 1 **Y** 1/2 **N** 0 **Q** 1/4

Consider the shapes shown at the right. If a shape is picked at random from this set, what is the probability that the shape:

5. is hexagonal (six-sided)?
6. is dotted?
7. has more than 4 sides *and* is not white?
8. is striped?
9. is triangular?
10. is a multiple of 3 *or* a square?
11. is not dotted?
12. is even *or* dotted?
13. is odd?
14. is dotted *and* pentagonal (five-sided)?
15. is not circular?
16. is not triangular?
17. is dotted *or* striped *or* white?
18. is odd *and* a multiple of 2?

___   ___ ___ ___ ___   ___ ___ ___ ___   ___ ___ ___ ___ ___
2   1 8 11 15   5 10 2 5   16 7 7 12 9

___ ___ ___ ___ ___ ___ ___ ___   ___ ___ ___ ___ ___ ___
15 11 6 12 12 8 18 17   12 7 2 11 4 9

___ ___   ___ ___ ___ ___ ___ ___!
6 3   14 8 9 15 6 13

CONNECTIONS TO THE CLASSROOM

1. When solving a probability problem related to rolling two dice, a student says, "I don't understand why I need to list both (4, 3) and (3,4) in the sample space. Aren't these the same outcome?" How would you respond?

2. When computing the probability of an event, a student obtains the answer of "2." What could you say to convince this student that his answer does not make sense?

3. An experiment consists of rolling a single die one time. A teacher asked her students to find the probability that the die will land on an even number *or* a number that is a divisor of 6. One student says, "It's 3/6 + 4/6 = 7/6, since three out of the six numbers on the die are even and there are four numbers on the die that go into six." Is the student correct? Why or why not?

4. A student wants to use two dice to simulate the birthday problem in Activity 11.4.1. She says, "I can use two dice because there are 12 outcomes when I roll the two dice, just like there are 12 months in the year."

   a. Is the student correct? Why or why not?

   b. Although there are 36 possible outcomes when rolling two dice, and not 12 as the student claimed, it is still possible to simulate the birthday problem using two dice. Explain how she could do this.

5. A student has flipped a coin 9 times and each time it has landed as a head. He claims that the probability of a tail on the next flip is greater than the probability of another head. Is he correct? Why or why not?

MENTAL MATH

Ella, the school cook, surveyed 20 students to determine their favorite lunch. The results of her survey were:

HAMBURGERS	10	PIZZA	3
HOT DOGS	5	FISHWICH	2

Ella wants to please everyone at least once in a while. What is the probability that student will be smiling when he comes through the lunch line on Hotdog Day?_____ on Hamburger Day?_____on Fishwich Day?_____

What is the probability he will *not* be smiling on Pizza Day?_____

DIRECTIONS IN EDUCATION
Forming Mathematical Connections
Getting It All Together

Being mathematically literate, according to the *Curriculum and Evaluation Standards*, means that students not only possess the ability to use their math skills and understanding in the context of the mathematics classroom, but also that they think to use their math skills and knowledge in contexts which occur outside the mathematics classroom. In addition, being mathematically literate means that students call upon prior math learning to assist and enrich new learning in mathematics. Connecting these previously attained skills and knowledge to new contexts is the focus of curriculum integration.

Curriculum integration may be viewed from the perspective of a single content area such as mathematics. When integration is examined from this point of view, the teacher is looking for ways to form connections among concepts and over time. In mathematics, integration is the means by which students are helped to see that math is not a series of discrete skills, but rather a whole with many interconnected parts.

In the classroom, curriculum integration may also be viewed from the perspective of the total instructional program. The teacher who practices this type of curriculum integration looks for goals from different content areas which overlap or which complement each other. A science activity in which data is collected may incorporate some graphing and statistical manipulation of the data. An art lesson about mosaic art might incorporate the study of tessellations. A health lesson about target heart rates might include calculation of personal heart rates and comparison of classroom average rates. In fact, a single lesson may meet instructional objectives from many individual content areas. This holistic approach to instruction provides the means by which children can form connections between the classroom and the real world.

Benefits of forming connections within the mathematics curriculum include:

- The opportunity to link conceptual and procedural knowledge.
- The ability to relate various representations of concepts or procedures to one another.
- The ability to recognize relationships among different topics in mathematics
- The opportunity to see mathematics as an integrated whole.
- The opportunity to explore problems and describe results using graphic, numerical, physical, algebraic, and verbal mathematical models or representations.
- The ability to use a mathematical idea to further understanding of other mathematical ideas.
- The opportunity to reinforce and expand prior learning in a meaningful context.

Benefits of forming connections between mathematics and other curriculum content areas include:

- The opportunity to see mathematical skills and knowledge as useful in other content areas.
- The ability to apply mathematical thinking and modeling to solve problems that arise in other disciplines, such as art, music, psychology, science, and business.
- Acceptance of the value of mathematics in our culture and society.

Benefits of forming connections between mathematics and real-life settings include:

- The motivation to study mathematics as its usefulness becomes known.
- The opportunity to make sense of the mathematics studied.
- The ability to use mathematics in daily life.
- The opportunity to see the usefulness of mathematics in describing and modeling real-world phenomena and in solving problems in everyday society.

As you think about forming mathematical connections, ask yourself:

1. Must students have mastered every aspect of a skill before they are able to deal with it in an integrated context? Can the integrated lesson provide further motivation for skill attainment or refinement?
2. How does the teacher determine when direct skill or concept instruction is required and when students are ready to encounter the mathematical idea in an integrated setting?
3. How does the teacher provide meaningful instruction in various content areas without giving only superficial treatment to other areas in an integrated lesson?
4. Do I understand and appreciate connections within mathematics in such a way that I can communicate those connections to my students?

12 Geometric Shapes

THEME: Analyzing Geometric Shapes

HANDS-ON ACTIVITIES

The last five chapters of this book are devoted to geometry, the study of shapes and relationships between shapes. Studying geometry is important for many reasons: geometry provides rich and diverse problem-solving opportunities; understanding geometric shapes and their relationships to one another helps students to build spatial reasoning skills; and there are countless applications of geometry, including in science and art.

In this chapter, you will begin your study of geometry by analyzing two- and three-dimensional shapes – their definitions, properties, attributes, and relationships to one another. The following activities involve many of the hands-on materials and tools commonly used to teach children geometry. In your study of symmetry, you will use a geoboard, as well as paper-folding techniques and tracings. To investigate how shapes fit together to form other shapes, you will use Tangram puzzles. You will also create tessellations to explore patterns and discover relationships between angles of polygons.

Children enjoy using shapes to create tessellations, and coloring or painting their designs.

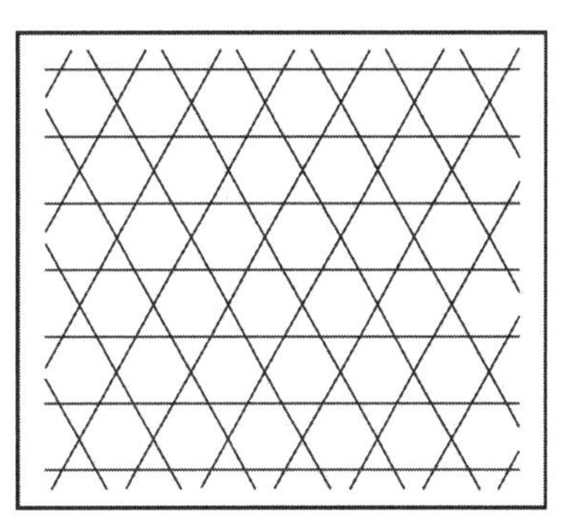

What shapes are used to make this tessellation? Why do these shapes fit together like this?

If you took high school geometry, you likely recall doing lots of two-column proofs. While you will prove some results in your work with geometry in this course, the majority will be of an informal type, accessible to elementary students. These informal proofs will involve hands-on approaches, including the use of paper-folding and tracings.

OBJECTIVE: **Use visualization to reason about shapes**

You Will Need: **Blocks (optional)**
(Physical manipulative option: *Geoboard*)

The abilities to recognize geometric shapes within other shapes, or to visualize how a shape will appear if it is reflected in a mirror, for example, are critical to the study of geometry. Activities such as the ones that follow will help you sharpen your visualization skills. The last problem will also help you practice communicating about shapes.

1. Each shape below was built from two puzzle pieces.

For example:

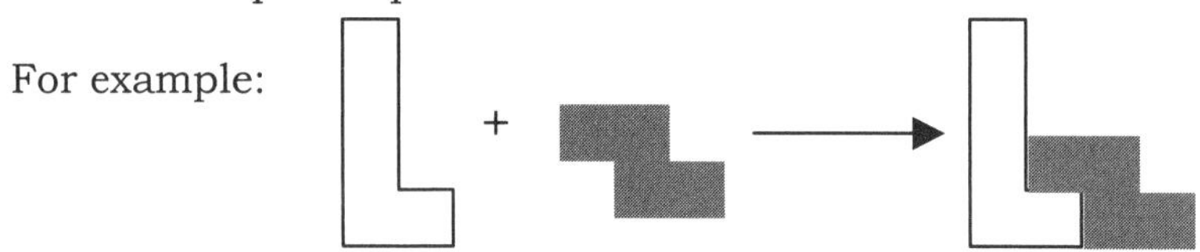

Use shadings to show how the pieces to the left of the arrows fit together to form the shapes on the right of the arrow.

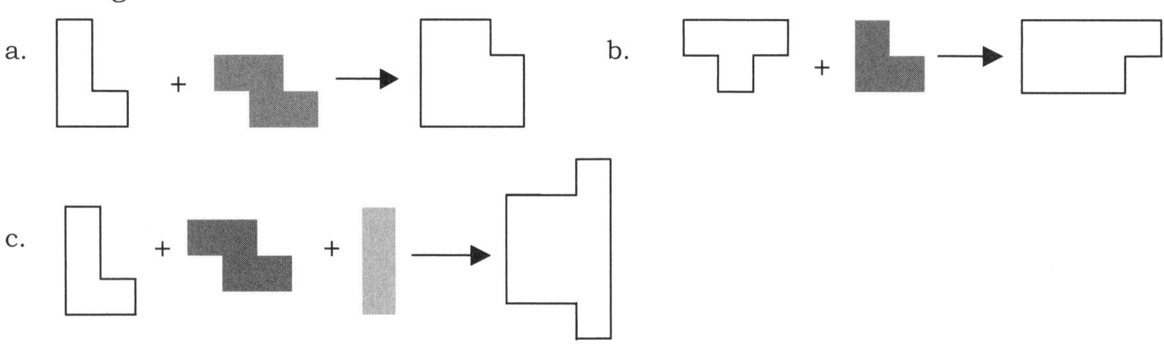

2. In each of the following pictures, the vertical line in the middle represents a mirror. Draw in everything exactly as it would appear in the mirror. Note that things appear reversed in a mirror.

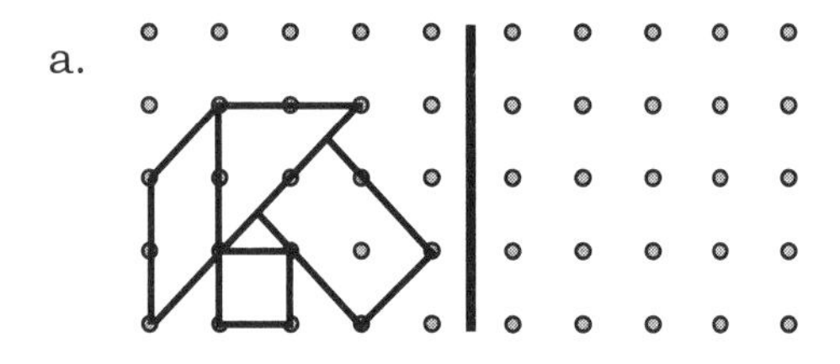

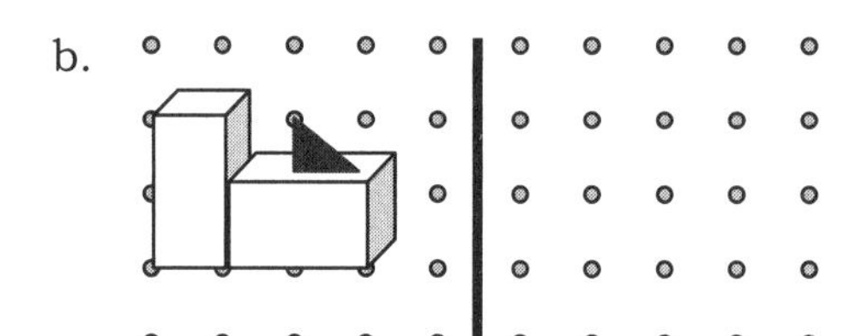

3. The "floor plan" of a building of blocks is shown at the right. The number in each square indicates the number of blocks that will be placed on that square. Draw the silhouette view of the building from the front, right, and left (the back view is shown below). Try to do this using only your visualization skills. To check your answers, draw a sketch or build the building using blocks.

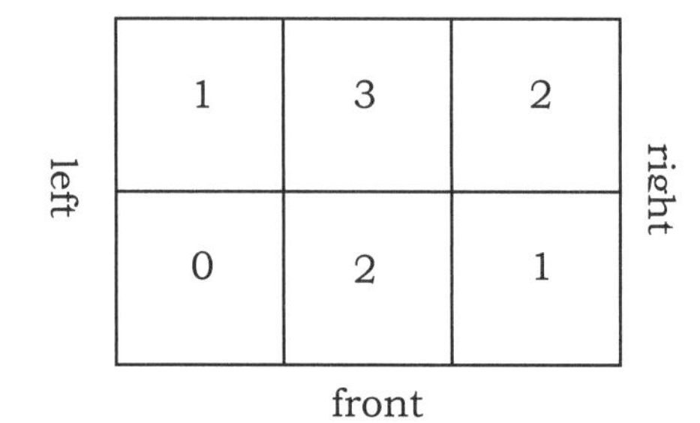

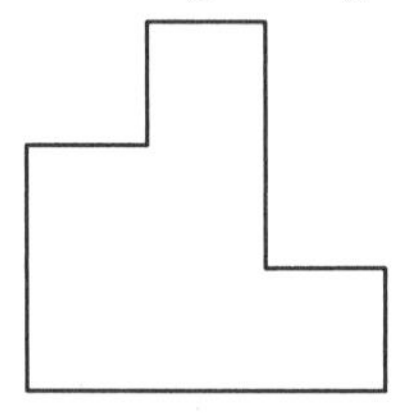

rear view

4. Work with a partner for this activity. Choose one person as the "describer." The describer makes a rubber band shape on the geoboard so the partner cannot see it (use the geopaper on Materials Card 12.1.1 if you do not have a geoboard). Using only words (no hand gestures), the describer will describe the shape and its important features to the partner. The partner should write down the description. Based on this description, the partner will construct a shape on his/her own geoboard. Note: the describer should not see the partner's geoboard shape until he/she is finished re-creating the shape. Also, the describer should not continue to describe the shape once the partner begins making the shape. Repeat this activity with the partners switching roles.

   What are the key elements of a successful description?

OBJECTIVE: **Construct and classify shapes on a geoboard**

You Will Need: **(Physical Manipulatives option: *Geoboard* and *Geoboard Triangular Lattice*)**

From your work with the Pythagorean Theorem in Chapter 9, recall that a geoboard is a square array of pegs around which rubber bands can be stretched to form shapes. In this activity, you will use a geoboard to explore the characteristics of various types of triangles and quadrilaterals, as well as parallel and perpendicular lines.

1. For this activity, the geoboard pegs will be represented by a 5 × 5 square lattice and the rubber bands by line segments. In this problem you will investigate the different types of triangles that can be formed on a geoboard.

   a. A triangle that has two or more sides the same length is called an **isosceles triangle**; for example, the triangle at the right is an isosceles triangle.

   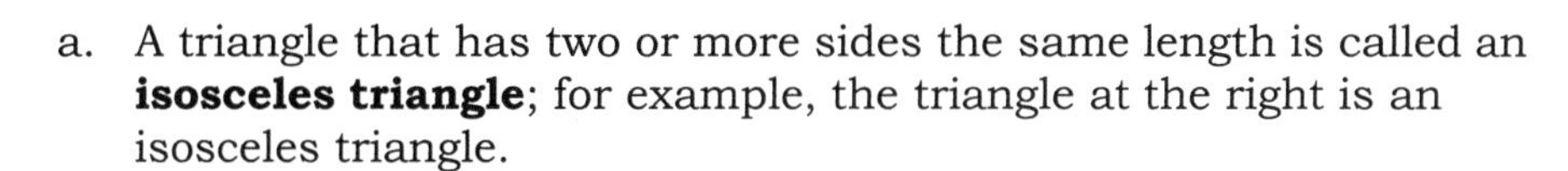

   We will consider two triangles on a geoboard to be the same (congruent) if the sides of the triangles can correspond in such a way that the corresponding sides have the same length. There are 22 **different** isosceles triangles that can be formed on a 5 × 5 geoboard, including the one in the example. Try to list these systematically, so you will be sure to find them all and that no two are the same. You may record more than one triangle on each grid below.

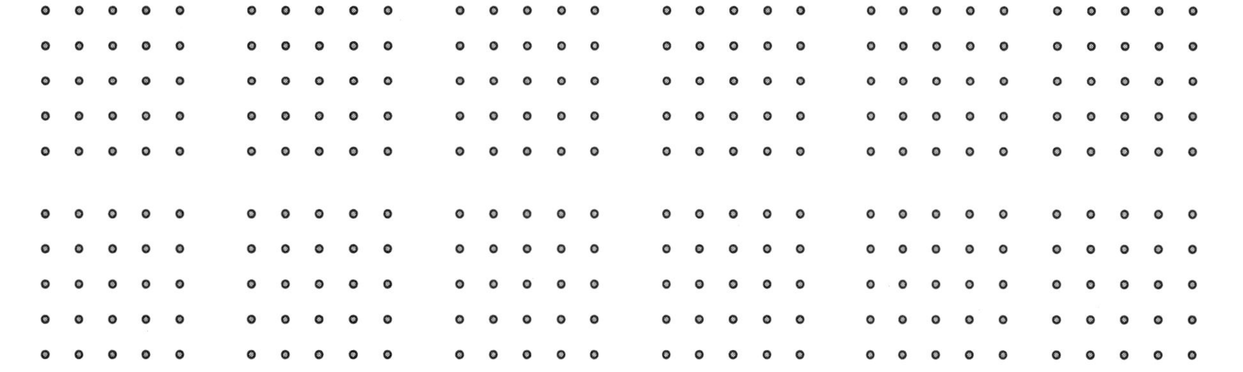

   b. A triangle is called a **right triangle** if it has a right angle (a 90° angle). Circle all the right triangles you drew in part (a). These are called isosceles right triangles.

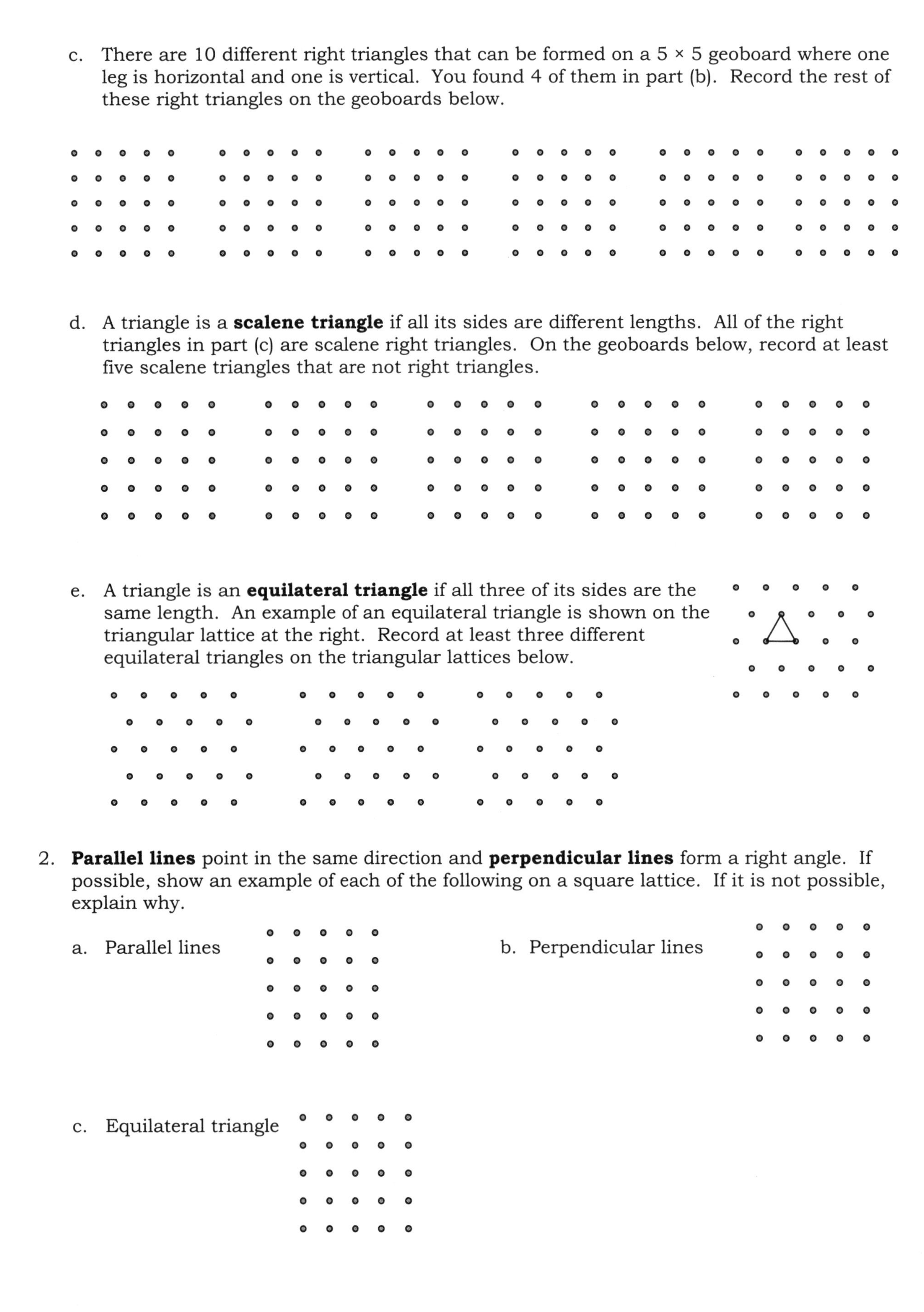

c. There are 10 different right triangles that can be formed on a 5 × 5 geoboard where one leg is horizontal and one is vertical. You found 4 of them in part (b). Record the rest of these right triangles on the geoboards below.

d. A triangle is a **scalene triangle** if all its sides are different lengths. All of the right triangles in part (c) are scalene right triangles. On the geoboards below, record at least five scalene triangles that are not right triangles.

e. A triangle is an **equilateral triangle** if all three of its sides are the same length. An example of an equilateral triangle is shown on the triangular lattice at the right. Record at least three different equilateral triangles on the triangular lattices below.

2. **Parallel lines** point in the same direction and **perpendicular lines** form a right angle. If possible, show an example of each of the following on a square lattice. If it is not possible, explain why.

a. Parallel lines

b. Perpendicular lines

c. Equilateral triangle

3. If possible, show an example of each of the following on a triangular lattice. If it is not possible, explain why.

   a. Parallel lines

   b. Perpendicular lines

   c. Scalene triangle

4. A **quadrilateral** is a shape with four (straight) sides.

   a. Record at least ten different quadrilaterals on the geoboards below.

   b. Quadrilaterals are classified according to their angles and sides. Read the following definitions, and then label each quadrilateral in part (a) with the appropriate name. Each shape may fit more than one classification or it may not fit any of the classifications. Also, if you cannot find an example of one of the shapes, then draw one.

      Square: a quadrilateral with four sides the same length and four right angles.
      Rectangle: a quadrilateral with four right angles.
      Parallelogram: a quadrilateral with two pairs of parallel sides.
      Kite: a quadrilateral with two non-overlapping pairs of adjacent sides that are the same length.
      Rhombus: a quadrilateral with four sides the same length.
      Trapezoid: a quadrilateral with *exactly one pair* of parallel sides.
      Isosceles trapezoid: a trapezoid in which the non-parallel sides are the same length.

   c. In part (b), you classified all the squares also as rectangles. Explain why.

   d. Are all rectangles also squares? Why or why not?

   e. In part (b), you classified all rectangles also as parallelograms. Explain why.

   f. Is a trapezoid also a parallelogram? Why or why not?

OBJECTIVE: **Explore symmetry**

You Will Need: **Materials Card 12.2.1, tracing paper, and squares from Materials Card 3.1.2**

The concepts related to symmetry are usually introduced informally in the elementary classroom, through the use of paper-folding and tracings. This hands-on approach can help students build the foundation they will need for a later, more formal study of these concepts. Symmetry is central to the study of geometry and you will revisit it often in later chapters.

1. A figure that can be folded on a line so that one half of its outline exactly fits on the other half is said to have **reflection symmetry.** As an example, consider the shape shown at the right. The dotted line is called a **line (or axis) of symmetry.** The line of symmetry is where the fold line should be for the left and right halves to match up.

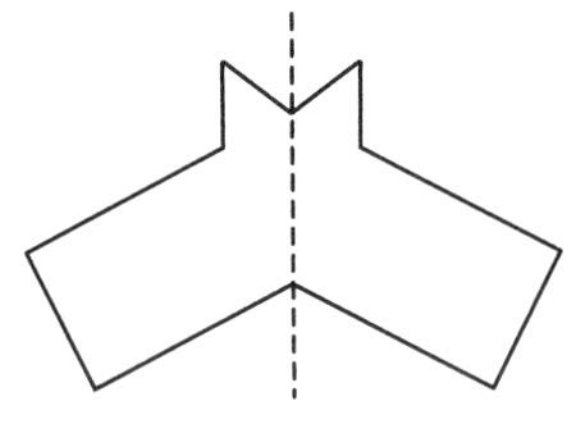

   a. Trace each of the shapes on Materials Card 12.2.1 onto your tracing paper. First, visually determine which of them has reflection symmetry. Then, use paper-folding to demonstrate which shapes have reflection symmetry. Be sure to crease the paper to show all lines of symmetry; this will be a record of your work for the problem. For each shape that has reflection symmetry, sketch it below, along with all of its lines of symmetry. Also, identify the shape by name, if possible.

   b. Sketch and write the names of the shapes on Materials Card 12.2.1 that do not have reflection symmetry.

2. A five pointed star is shown below, along with one of its lines of symmetry. Without using paper-folding, visualize whether this star has any other lines of symmetry. If it does, sketch them as dotted lines. How many lines of symmetry does the 5-pointed star have all together? _____

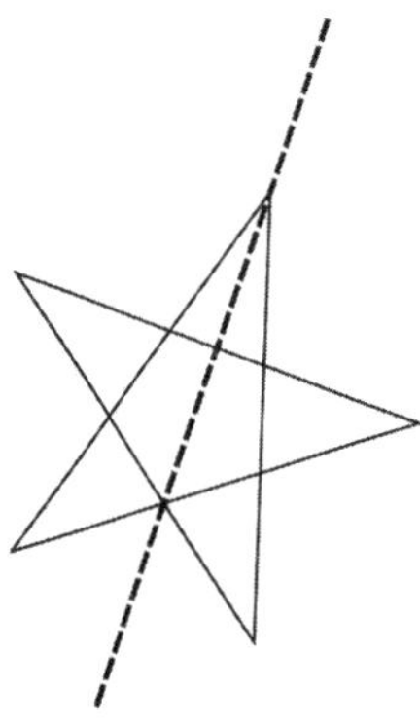

3. A second type of symmetry is rotation symmetry. A shape has **rotation symmetry** if there is a point around which the shape may be rotated (or turned), *less than a full turn*, so that the image exactly matches the original shape.

   The sequence of drawings shown below shows that a square has rotation symmetry since it can be rotated around point *O* at the center of the square so that the tracing (shown as dotted) exactly matches the original.

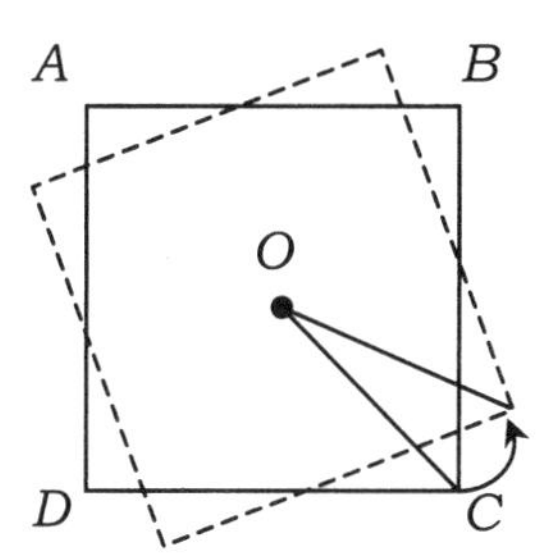

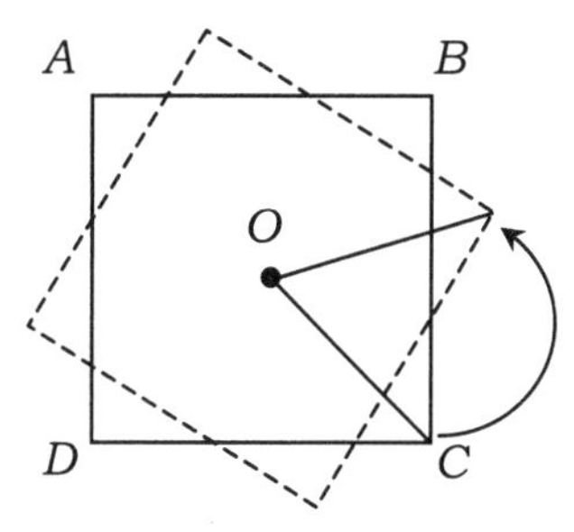

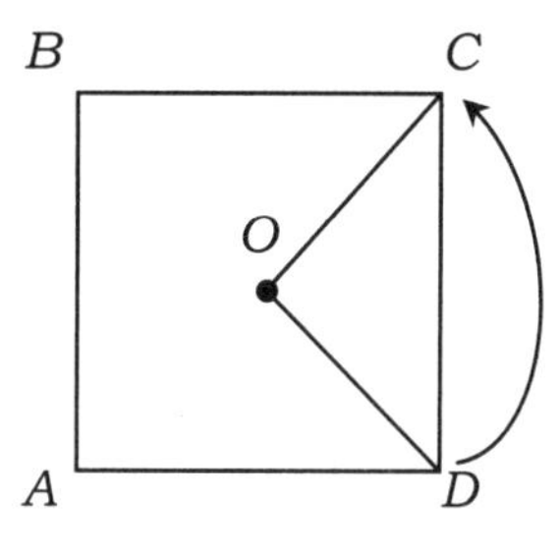

   a. One way to describe the rotation symmetries of a shape is as a fraction of a full turn. What fraction of a full turn is shown for the square above? _____ Another way is to give the number of degrees in the angle of rotation, shown as $\measuredangle COD$ in the final drawing above. How many degrees are in a full turn? _____ Since the square has been rotated 1/4 of a full turn, what is the measure of $\measuredangle COD$? _____

   b. The square has two other angles of rotation symmetry: a 180°-turn and a 270°-turn, Write the corresponding fractions of a full turn that corresponds to each of these three angles of rotation symmetry. _____, _____. Note: One could say that there are infinitly many more possible turns where a square can rotate onto itself: 360°, 450°, 540°, etc. However, here we will only consider turns with measures greater than 0° and less than 360°.

   c. The shape from problem #1is shown at the right. Trace the shape and rotate it about any point. What number of degrees produces a matching image? _____

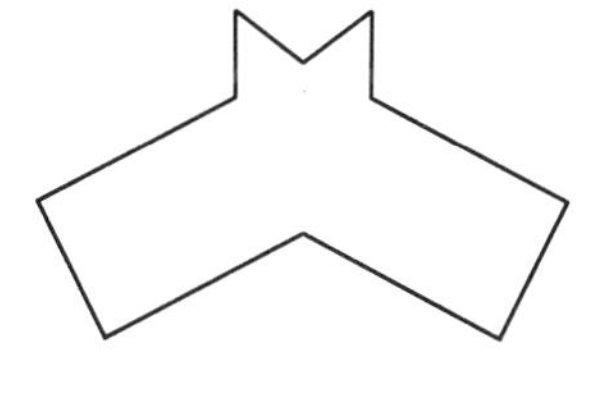

   A shape for which only a full turn produces a matching image, such as the one to the right, *does not have rotation symmetry.*

4. Use tracing and rotating to determine which of the shapes on Materials Card 12.2.1 have rotation symmetry.

   a. Sketch the shapes that have rotation symmetry here, along with the rotation point you used. Indicate how much to turn them by both fraction and degrees.

   b. Sketch the shapes that do not have rotation symmetry.

5. Elementary teachers often introduce their students to the concepts of reflection and rotation symmetry through the use of pentominos. A pentomino is made by five connected squares that overlap only on one complete side. The two shapes shown below on the left are pentominos and the two shapes on the right are not pentominos.

Two pentominos are the considered to be the same if they can be matched by turning or flipping. The two pentominos shown below are the same.

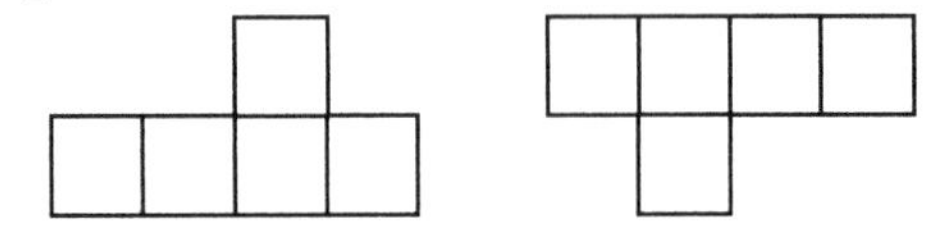

a. Two possible pentominos are shown above. Use your squares from Materials Card 3.1.2 to find all of the other different pentomino shapes and record them below. Do this systematically to be sure you have found them all. Compare your answers with a classmate.

b. Which of the pentominos have reflection symmetry? Indicate the line(s) of reflection.

c. Which of the pentominos have rotation symmetry? Indicate the center and the angles of rotation symmetry.

ACTIVITY 12.2.2

OBJECTIVE: **Discover properties of diagonals of quadrilaterals**

You Will Need: **Quadrilaterals from Materials Card 12.2.1, tracing paper, and a straightedge**

In the next activity, you will use paper-folding and tracing to discover some of the properties of the diagonals of special quadrilaterals. You will also investigate how paper-folding can be used to determine whether two lines are parallel or perpendicular.

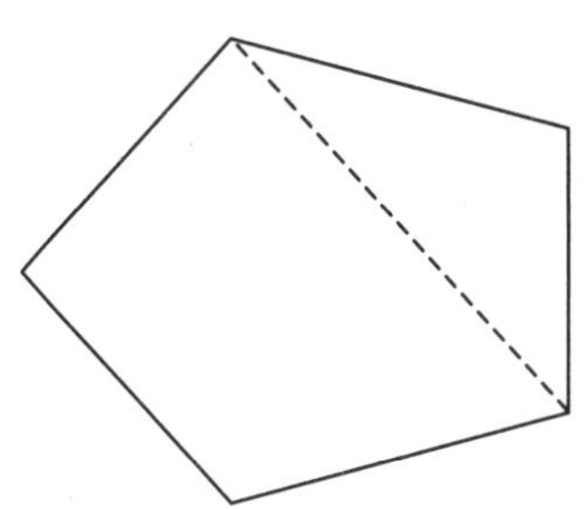

1. A **diagonal** of a polygon is any line segment that connects non-adjacent vertices. **Adjacent vertices** of a polygon have a common side, so diagonals connect vertices that are not adjacent. As an example, consider the pentagon shown to the right. One of its diagonals is shown by the dashed line segment.

   a. Sketch all of the other diagonals of the pentagon. How many diagonals does a pentagon have? _____

   b. How many diagonals does a quadrilateral have? _____ How many diagonals does a triangle have? _____

2. Take out the following shapes from Materials Card 12.2.1: parallelogram, rectangle, square, rhombus, trapezoid, isosceles trapezoid, and kite. Use this set of seven shapes for parts (a) – (d). For each part, find a quadrilateral that has the given property and a quadrilateral that does *not* have the given property, and sketch the shapes in the appropriate column. Then, in the space for justification, explain *how to use paper-folding or a tracing* to justify both of your answers, in each part. The first one is partially done for you as an example.

Property	Quadrilateral with that property	Quadrilateral that does not have that property
a. **Diagonals are congruent**	A square	
Justification: **Draw in the diagonals of the square and trace one of them. Rotate the tracing 180° about the point of intersection of the diagonals - the traced diagonal exactly matches (is congruent to) the other diagonal.**		
b. **Diagonals are perpendicular**		
Justification:		

Property	Quadrilateral with that property	Quadrilateral that does not have that property
c. **Diagonals bisect each other**		
Justification:		
d. **Each diagonal divides the shape into 2 congruent triangles**		
Justification:		

3. Lines can be shown to be parallel or perpendicular, using paper-folding. Consider the following two tests.

   a. **Parallel Line Segments Test:** Given two lines, *l* and *m*, fold the paper so that line *l* folds onto itself. If line *m* also folds on itself, the lines are parallel.

   Carefully trace lines *l* and *m* shown at the right, using a ruler, and determine if they are parallel.

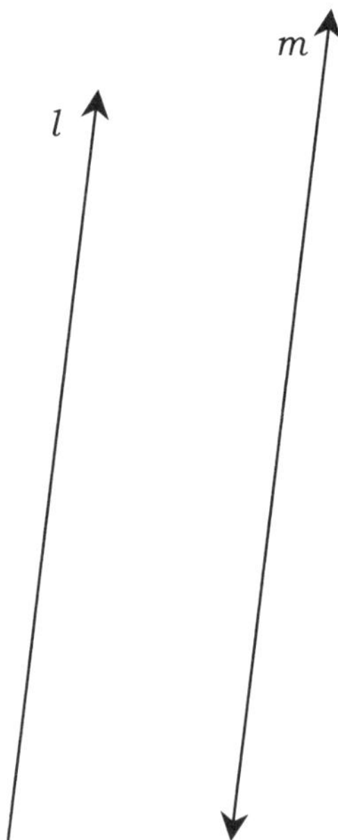

   b. **Perpendicular Line Segments Test**: Given two lines *s* and *t*, and point *P*, the point of intersection of the two lines, fold line *s* onto itself at point *P*. If line *t* is the fold line, the lines are perpendicular.

   Carefully trace lines *s* and *t* shown below, using a ruler, and determine if they are perpendicular.

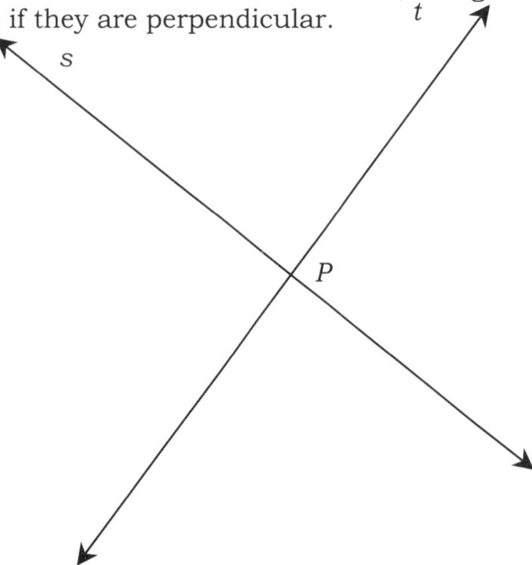

c. Carefully trace the six lines shown below onto your tracing paper, using a ruler. Then apply the Parallel Line Segments Test and record the pairs of parallel lines.

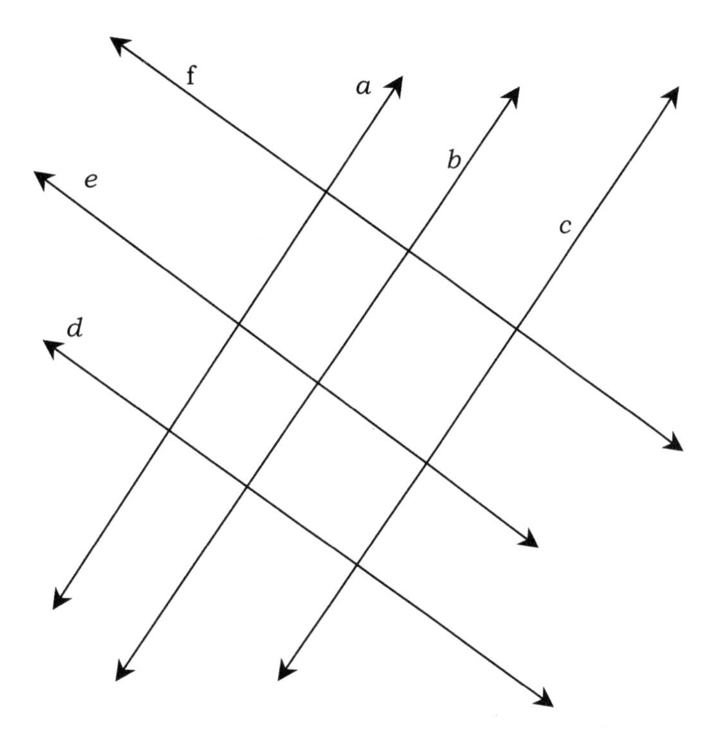

d. Apply the Perpendicular Line Segments Test for these six lines and record the pairs of perpendicular lines.

4. Trace a square and parallelogram from Materials Card 12.2.1 and then use the Perpendicular Line Segments Test to show that the diagonals of a square are perpendicular and that the diagonals of a parallelogram are not perpendicular.

OBJECTIVE: **Use tangram pieces to form polygons**

You Will Need: **Materials Card 12.2.3**

Tangrams are a popular Chinese puzzle commonly used to teach geometry concepts in the elementary classroom. As you will see in the next activity, this versatile manipulative can help students learn about the relationships between polygons as well as develop visualization skills.

1. The tangram pieces on Materials Card 12.2.3 consist of 7 shapes: a square, a parallelogram, and 5 isosceles right triangles (2 small, 1 medium-sized, 2 large). As you can see, the shapes are made so that they fit together to form a square. These shapes can be arranged to form many other shapes, as well. Cut out and use the pieces of the tangram puzzle to complete the following chart; for example, you can make a square from 3 tangram pieces as illustrated in the chart. Many shapes have more than one answer, while others are impossible. Give all the possibilities you can find or explain why you think the shape is impossible.

   Note: the polygons you make might not be regular polygons (with congruent sides and congruent angles).

Number of tangram pieces used	triangle	square	rectangle	parallelogram	trapezoid	pentagon	hexagon
2							
3							
4							
5							
6							
7							

OBJECTIVE: **Find the sum of the measures of the vertex angles of polygons**

You Will Need: **Paper, scissors and a straightedge (such as a ruler)**

You likely recall from your previous study of geometry that the sum of the measures of the vertex angles of any triangle is 180°, but can you explain why? In the following activity, you will see one way to explain why this is the case. You will also develop formulas for the sum of the measures of the vertex angles for other polygons, as well as for the measure of *each* vertex angle of regular polygons.

1. Use a ruler to draw any triangle. If you are working in a group, each member should draw a different type of triangle (isosceles, right, scalene, etc.) so that you will see this works for any type of triangle. Cut out your triangle and tear off the 3 corners (**vertex angles**) of the triangle as illustrated at the right.

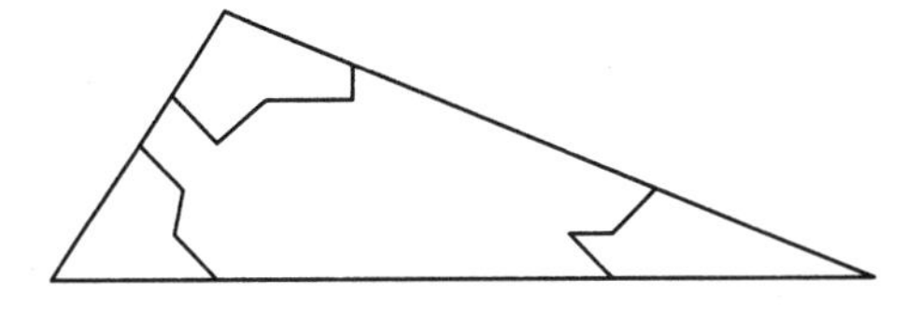

   Rearrange the three angles to show why the sum of the measures of the vertex angles of a triangle is 180°. Make a sketch of your result below and explain.

2. Now draw any quadrilateral and tear off the vertex angles as in problem #1 and rearrange them. Try using several different types of quadrilaterals, such as a square, rectangle, parallelogram, and one such as that shown to the right.

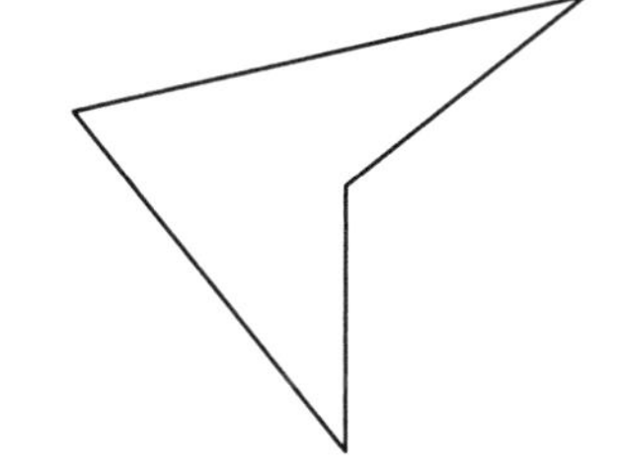

   Make a sketch of your results below.

   What does this show about the sum of the measures of the vertex angles of any quadrilateral?

3. To find the sum of the measures of the vertex angles in a polygon with more than 4 sides, you can divide the shape into triangles in a special way. First, choose one vertex and draw all the diagonals from that vertex. In the pentagon shown at the right, all the diagonals from vertex A are shown.

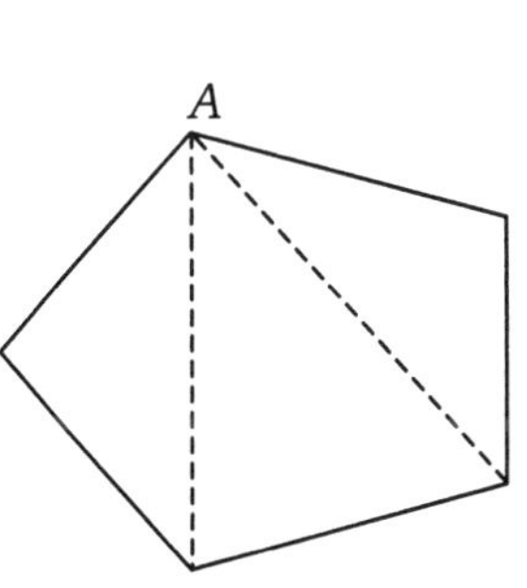

   a. How many triangles are formed within the pentagon? ____

   b. Since the sum of the measures of the vertex angles in any triangle is 180°, the sum of the measures of the vertex angles of the pentagon is ______.

4. Using a method similar to that in problem # 3, find the sum of the measures of the vertex angles of each of the following polygons. In each case, record the number of triangles formed and the sum of the vertex angle.

   a. Number of triangles = _____

   Sum of vertex angles = ________

   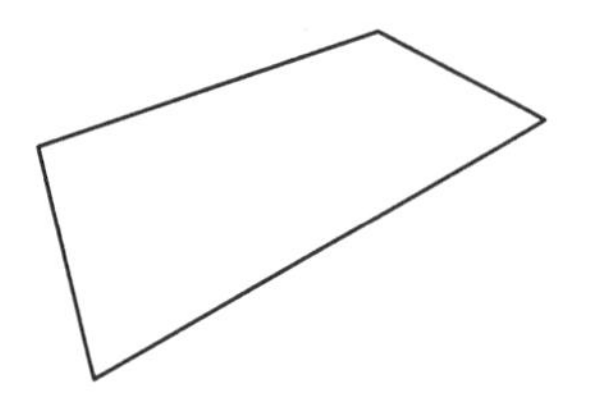

   b. Number of triangles = _____

   Sum of vertex angles = ________

   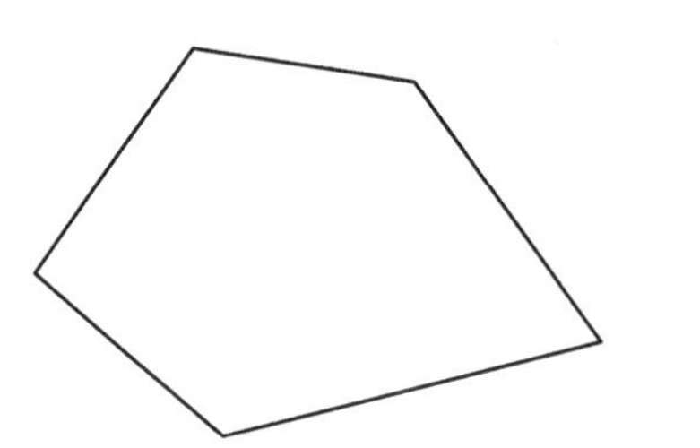

   c. Number of triangles = _____

   Sum of vertex angles = ________

   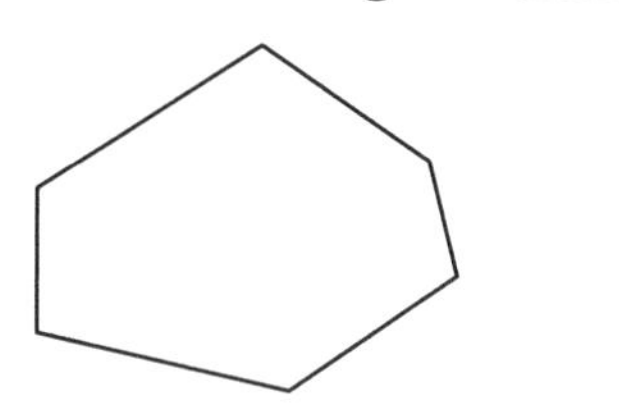

   d. Look for a pattern in parts (a) – (c) and write a formula for the sum of the measures of the vertex angles of a polygon with n sides.

5. A **regular polygon** is a polygon in which all the sides are congruent and all the vertex angles are congruent. The first four regular polygons - equilateral triangle, square, regular pentagon, and regular hexagon - are shown below. Use your results in the previous problems to find the measure of each vertex angle in each of these polygons. Show how you got your answer, in each case.

   a. vertex angle measure = ________

   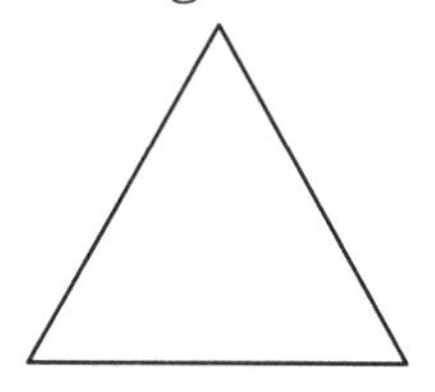

   b. vertex angle measure = ________

   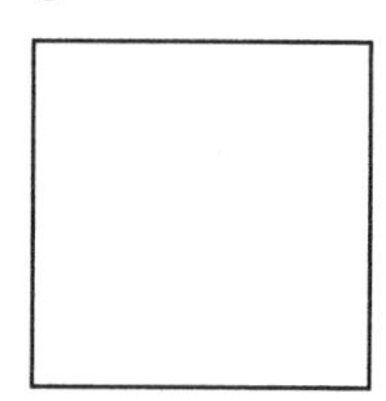

   c. vertex angle measure = ________

   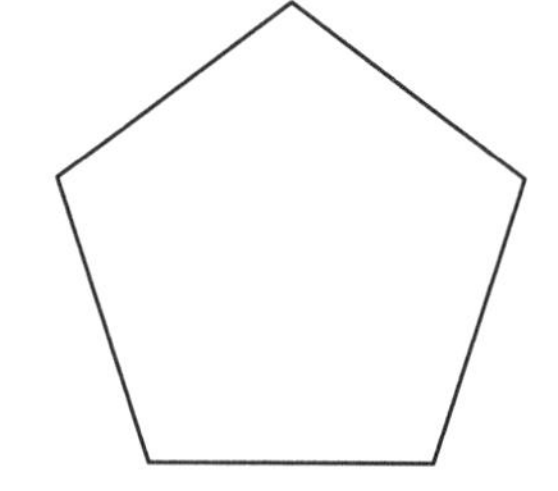

   d. vertex angle measure = ________

   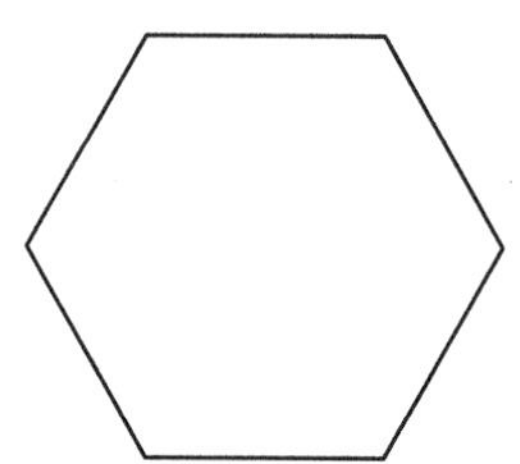

   e. Look for a pattern in parts (a) – (d) and write a formula for the measure of each vertex angle of a regular polygon with n sides.

OBJECTIVE: **Investigate tessellations**

You Will Need: **Paper, scissors, ruler, note card, and Materials Card 12.4.2**

A **tessellation** is a covering of the plane with one or more shapes *with their interiors* so that there are no gaps and no overlaps. (Henceforth, when we talk about shapes in tessellations, we always mean the shape with its interior.) Tessellations appear in artwork throughout the world, from ancient cultures to present day. These geometric designs are eye-pleasing and they provide many interesting and creative activities to help elementary students learn properties of shapes and relationships between them.

1. A tessellation of the plane is shown at the right. (This pattern is assumed to extend indefinitely in all directions.)

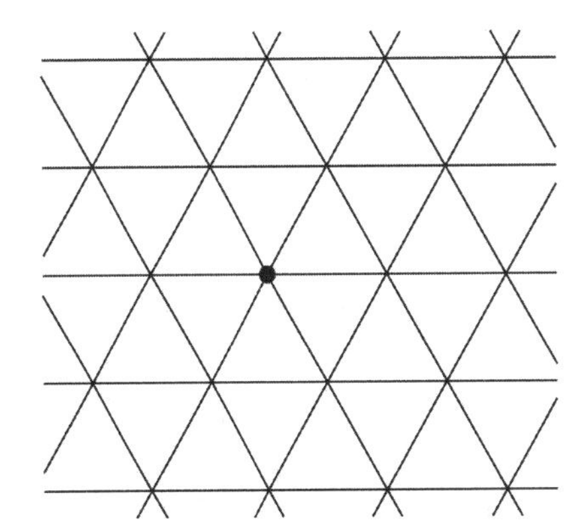

   a. What shape was used to cover the plane?

   b. Consider the enlarged vertex point shown in the tessellation. How many vertex angles of the equilateral triangles meet at this point? _______

   One way to describe a tessellation is to give its **vertex arrangement**, which denotes the number of sides of the polygons that meet at each vertex. In this case, the vertex arrangement is written as (3, 3, 3, 3, 3, 3), because there are six 3-sided polygons that meet at *each* vertex point. This tessellation is called a **regular tessellation** because it is formed with a single regular polygon (an equilateral triangle) and every vertex arrangement is identical (3, 3, 3, 3, 3, 3).

   c. Cut out the regular polygons from Materials Card 12.4.2. Trace around each shape on a separate sheet of paper and try to form a regular tessellation with each one. You do not need to cover the entire page, but be sure to show enough of the tessellation so the reader can see how to extend it in all directions. If a shape does not tessellate the plane, show *why* in your tracing; that is, show that there is a gap or overlap when you try to form a tessellation with that shape.

   You saw in part (a) that it is possible to tessellate the plane with an equilateral triangle. Which of the other regular polygons from Materials Card 12.4.2 tessellate the plane?

   What are the vertex arrangements for these regular tessellations?

   Which of the regular polygons from Materials Card 12.4.2 do not tessellate?

   d. In terms of the measures of the vertex angles, explain why it is possible to tessellate the plane with some regular polygons, but not with others.

e. In parts (a) – (c), you saw the equilateral triangle, square, and regular hexagon tessellate the plane and that a regular pentagon does not tessellate the plane. Explain, in terms of vertex angle measures, how you know there is no regular n-gon, with $n > 6$, that will tessellate the plane by itself.

2. In problem #1, you showed that regular 3-, 4-, and 6-gons are the only regular polygons that will tessellate the plane by themselves. In this problem, you will consider tessellations that involve a combination of two or more regular polygons. A tessellation using two or more regular polygons is called a **semi-regular tessellation**, if every vertex arrangement of the tessellation is identical.

   a. The vertex arrangements of two semi-regular tessellations that use an equilateral triangle and a square are: (3, 3, 3, 4, 4) and (3, 3, 4, 3, 4). On a separate sheet of paper, trace shapes from Materials Card 12.4.2 to form these two tessellations. Show enough of each one so the reader can easily see how to extend the pattern in any direction.

   How are these two tessellations similar? How are they different?

   b. Using the regular polygons from Materials Card 12.4.2, create another semi-regular tessellation. Record your tessellation on a separate sheet and give its vertex arrangement.

   c. Using at least two of the regular polygons from Materials Card 12.4.2, create a tessellation that does not have the same vertex arrangement at each vertex; that is, one that is *not* semi-regular Record your tessellation on a separate sheet and give each of its different vertex arrangements.

3. A tessellation may also be formed with polygons that are not regular.

   a. Using a ruler, draw any triangle (except equilateral) on a note card. Cut it out to make a template for tessellating. If you are working in a group, each person should make a different type of triangle (isosceles, scalene, right, etc.) Trace your shape on a separate sheet of paper to form a tessellation of the plane.

   b. On your triangle template, label the interior of each of the three angles x, y, and z (here x, y, and z represent the respective angle measures). Then, choose one vertex of your tessellation and write the appropriate label, *either* x, y, *or* z, for each of the angles meeting at that vertex. What do you notice?

   c. Use your observation in part (b) to explain why *any* triangle will form a tessellation of the plane.

4. Every quadrilateral will tessellate the plane also. Use your ruler to draw an arbitrary quadrilateral, such as those shown at the right. Cut it out and use it for a template to for tessellating.

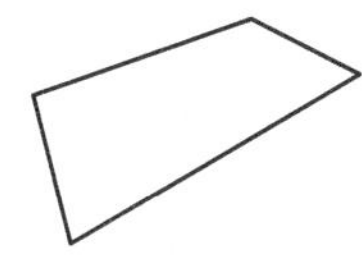

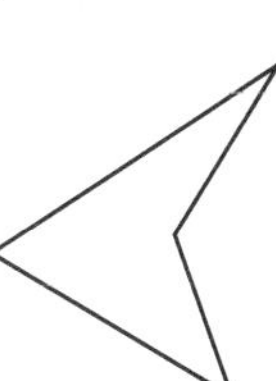

   a. What is the sum of the measures of the vertex angles of any quadrilateral? _______How does this fact, together with the ideas from problem #3b, help you to decide how to arrange the quadrilateral for a tessellation?

   b. Trace your quadrilateral on a separate sheet of paper to form a tessellation of the plane.

   c. Why will every quadrilateral tessellate the plane?

OBJECTIVE: **Build models of polyhedra**

You Will Need: **Paper, scissors, paper clips or tape or glue, and Materials Card 12.4.2**

Three-dimensional objects are everywhere we look – rooms, cars, our bodies, etc. are all three-dimensional. In geometry, our study of three-dimensional objects includes shapes such as pyramids, cylinders, cones, cubes, and spheres. In this activity, you will build polyhedra, special types of three-dimensional objects. Building these objects will help you see the relationship between the three-dimensional objects and their two-dimensional counterparts.

1. A **polygonal region** is the union of a polygon and its interior. A **polyhedron** (plural: polyhedra) is a closed surface made up of polygonal regions. The cube shown at the right is an example of a polyhedron.

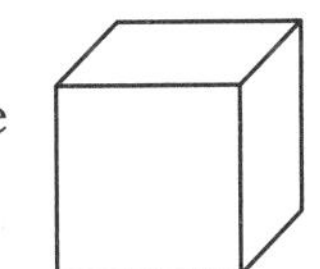

   a. Remove the circle patterns from Materials Card 12.5.1 and find the pattern that has a square on it. To build a cube, using copies of this pattern, you will fold on the dotted lines to create flaps. You can use paper clips, glue, or tape to put the square regions together. How many square regions do you need to make a cube? _____ Make the copies you need and build the cube.

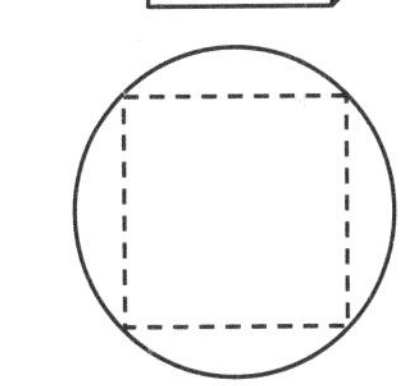

   b. It is possible to build the polyhedron shown at the right using the patterns on Materials Card 12.5.1. Which shapes and how many of each do you need to build this figure?

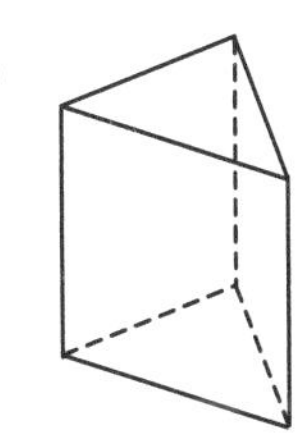

   Make the copies of the patterns you need and build this polyhedron.

   c. The two polyhedra you built in parts (a) and (b) are examples of prisms. A **prism** is a polyhedron with two identical parallel faces, called the **bases**. The vertices of the bases are joined to form the **lateral faces** of the prism. The prism you built in part (b), has triangular bases and the lateral faces are squares.

The following are three examples of prisms. For each one, shade the bases. Then write the name of the shape of the base and the shape of the lateral faces.

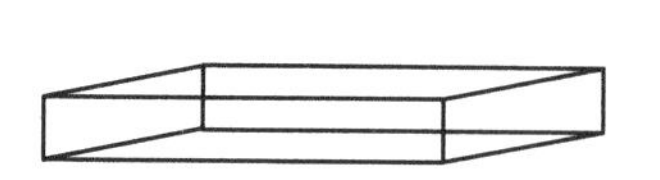 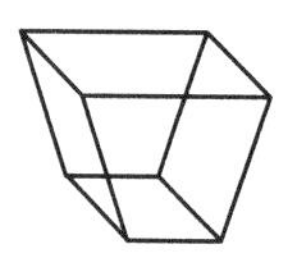 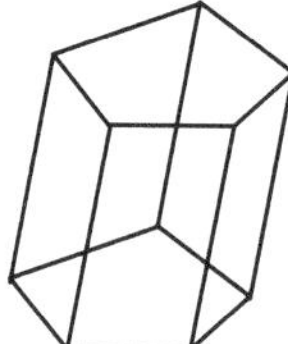

bases: ________________ bases: ________________ bases: ________________

lateral faces: ________________ lateral faces: ________________ lateral faces: ________________

d. Notice the shape of each lateral face of a prism is always a parallelogram. If the lateral faces are rectangular, the prism is called a **right prism**. If the lateral faces are not all rectangular, the prism is called an **oblique prism**. Prisms are named according to the shape of their base and whether they are right or oblique; for example the prism in part (b) is called a "right triangular prism." Write the names of the prisms in parts (a) and (c).

2. A pyramid is another type of polyhedron. A **pyramid** is a polyhedron with a polygonal base and triangular lateral faces that all share a vertex (the **apex**) that is not on the base.

   a. Make the copies of the patterns from Materials Card 12.5.1 that you need to build the pyramid shown at the right. Then build it. This is called a "right square pyramid" because the shape of the base is a square and its lateral faces are isosceles triangles.

   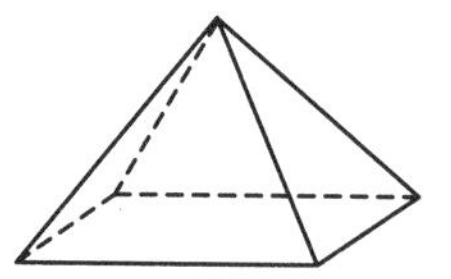

   The pyramid shown below is an example of an "oblique square pyramid" – the triangular faces are not isosceles.

   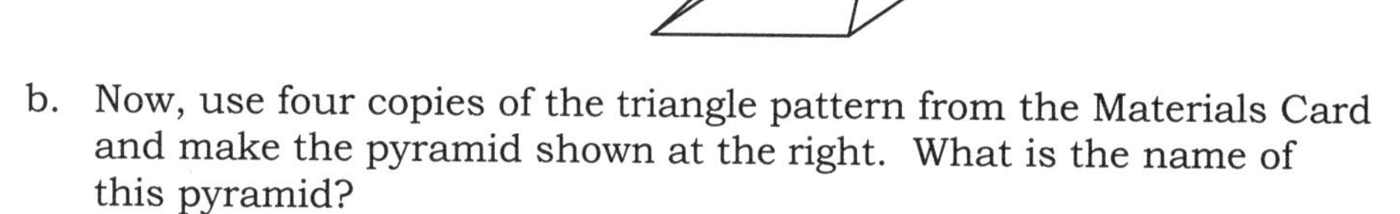

   b. Now, use four copies of the triangle pattern from the Materials Card and make the pyramid shown at the right. What is the name of this pyramid?

   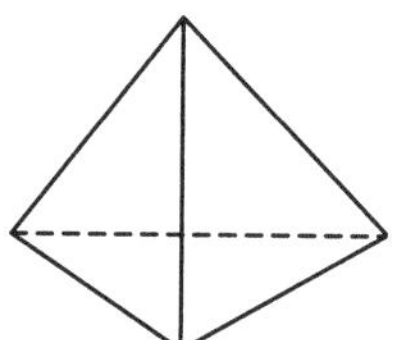

3. A **regular polyhedron** (or **Platonic solid**) is one in which all the faces are identical regular polygonal regions and each vertex is surrounded by the same arrangement of polygons. It has been proven that only five such regular polyhedra exist. You have already built two of these, the cube in problem #1(a) and the right triangular pyramid (also called a tetrahedron) in problem #2(b).

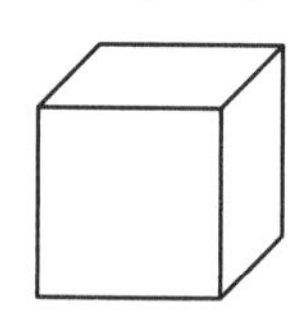

Cube

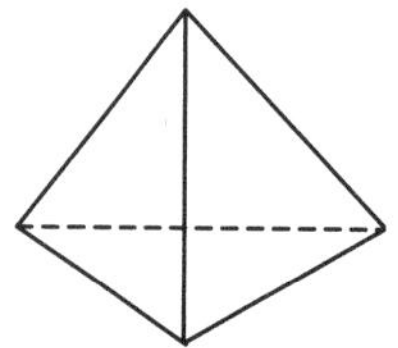

Tetrahedron

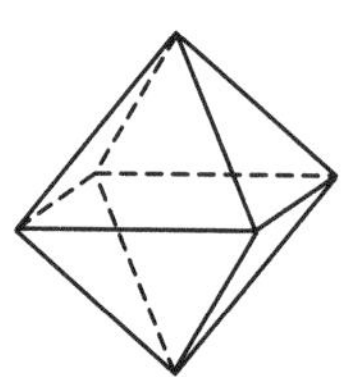

Octahedron

a. Using the patterns from the Materials Card, build the octahedron (to make this easier, you may use one of the polyhedron you already built).

b. Another regular polyhedron is called a dodecahedron – it is formed from 12 pentagonal regions. Using the patterns from the Materials Card, build a dodecahedron.

You will not build the fifth of the Platonic solids, the icosahedron (made up of 20 equilateral triangles) in this activity, but a picture of one appears in your textbook.

4. A **semi-regular polyhedron** is one made up of regular polygonal regions of more than one type such that each vertex is surrounded by the same arrangement of polygons. Make the following with 8 equilateral triangles and 6 squares: each vertex is surrounded by a triangle-square-triangle-square arrangement. This semi-regular polyhedron is called a "cube octahedron."

5. An interesting relationship exists between the number of faces, vertices, and edges of a polyhedron. Using the models you have made, complete the following table.

	Faces (F)	**Vertices (V)**	**Edges (E)**
Cube			
Triangular prism			
Square pyramid			
Tetrahedron			
Octahedron			
Dodecahedron			
Cube Octahedron			

Write an equation involving F, V, and E. ______________________This is called Euler's formula and it holds for any polyhedron.

Plane Slices

Each figure below is sawed in two by the plane indicated. Match the figure with the resulting cross section. Put the letter from the answer in the blank below that is numbered the same as the problem.

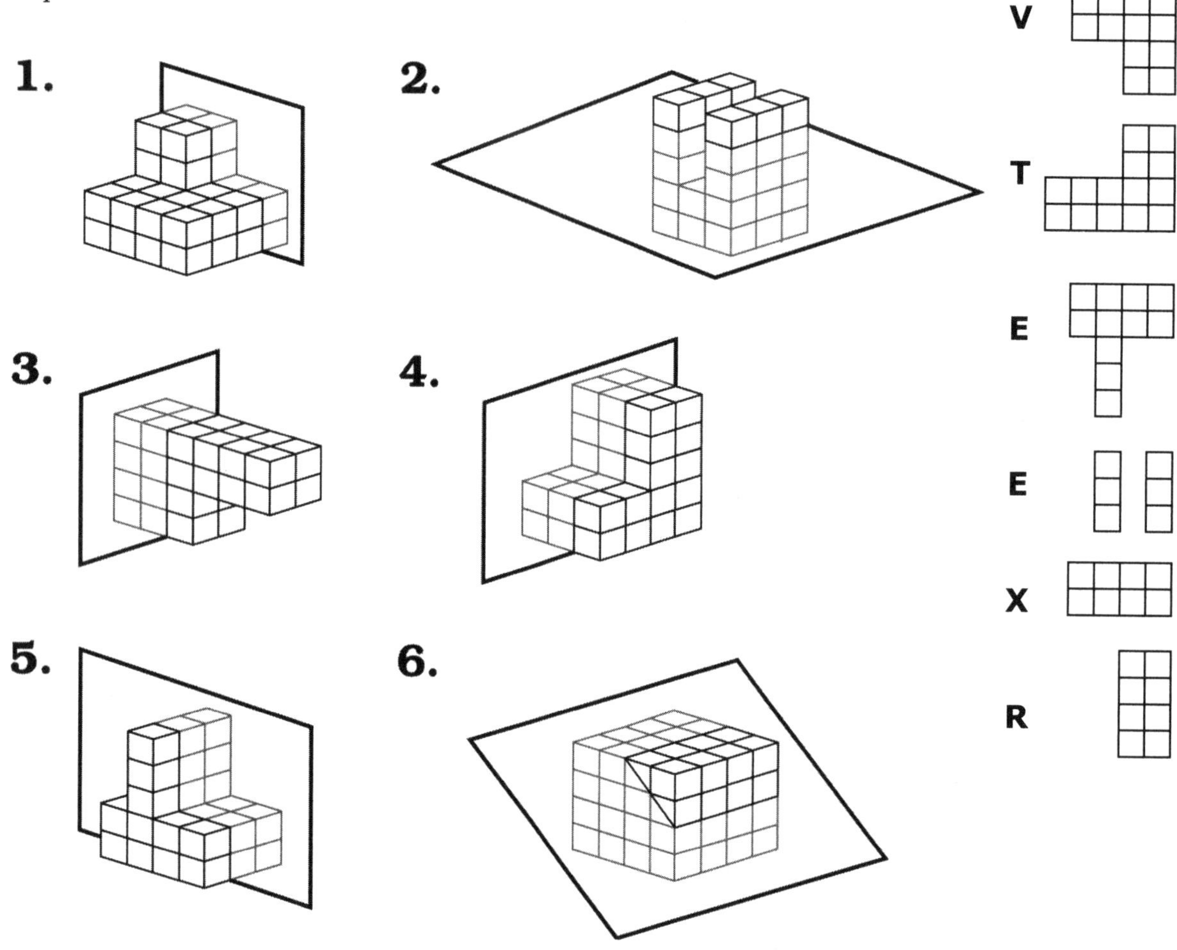

What the heroine said when the cowboy rescued her just as she was about to be sawed in two?

If it weren't ___ ___ ___ ___ ___ ___ I would have been bisected.
1 2 3 4 5 6

CONNECTIONS TO THE CLASSROOM

1. When asked whether the following triangles are isosceles triangles, a student says, "The one on the left is isosceles, but the one on the right isn't. I marked the base angles and I see that they are not congruent."

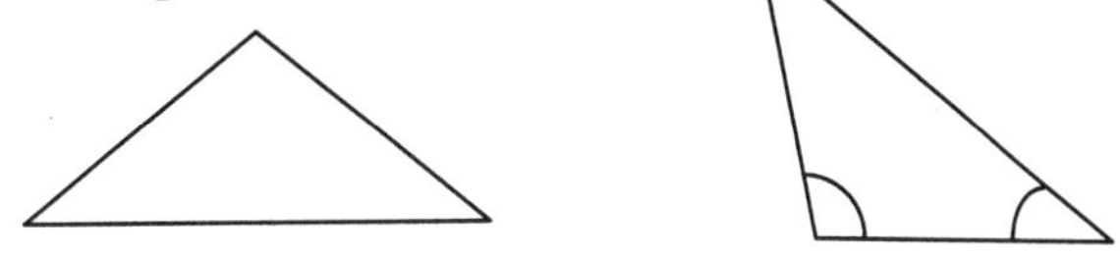

   a. What part of the definition of isosceles triangle does this student not understand?

   b. What questions could you ask this student to help her understand the term "base of an isosceles triangle"?

   c. If the student traced one of the two triangles above and rotated her tracing, what surprising result would she find?

2. A student constructed the geoboard triangle shown at the right and claimed that it is a right triangle. Is the student correct? Prove your answer (it is not enough to say, "it looks like a right triangle" or to measure with a protractor).

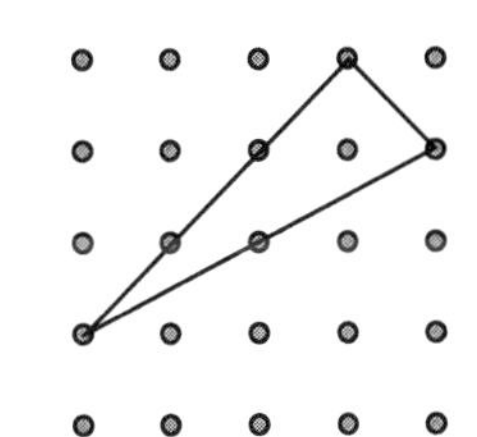

3. When asked whether a parallelogram has reflection symmetry, as student answers, "Yes, because you can cut it on this dashed line and get two pieces that match." (The student's picture is shown at the right.)

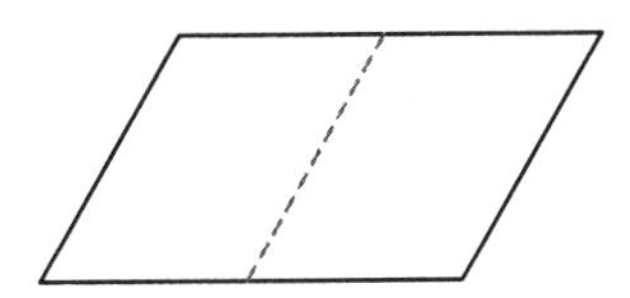

   a. Is the student correct? Why or why not?

   b. What activity would you suggest to the student to help her check her answer and to help her understand the concept of reflection symmetry?

4. A student claims that the polygon shown at the right is a regular dodecagon and justifies his answer by saying, "The twelve sides are all the same length and the 12 angles are all congruent." What is wrong with what this student said?

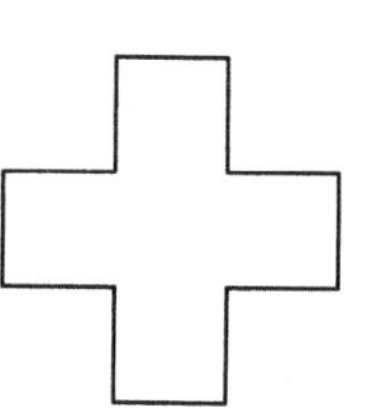

5. **Vertical angles** are pairs of angles formed by intersecting lines. In the diagram at the right, angles 1 and 4 are a pair of vertical angles. Angles 2 and 3 are another pair of vertical angles. Explain how one could use paper-folding or a tracing to help a student see that pairs of vertical angles are always congruent?

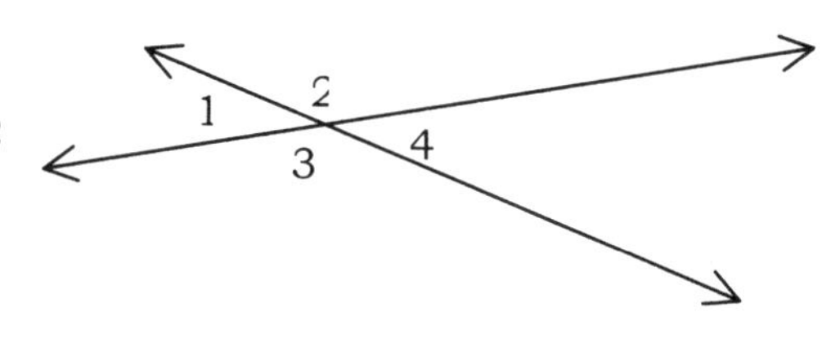

MENTAL MATH

It is possible to make one straight cut in each of these shapes so that the two pieces will form a square.

For example:

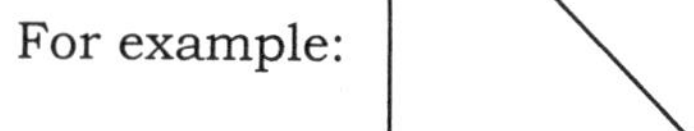

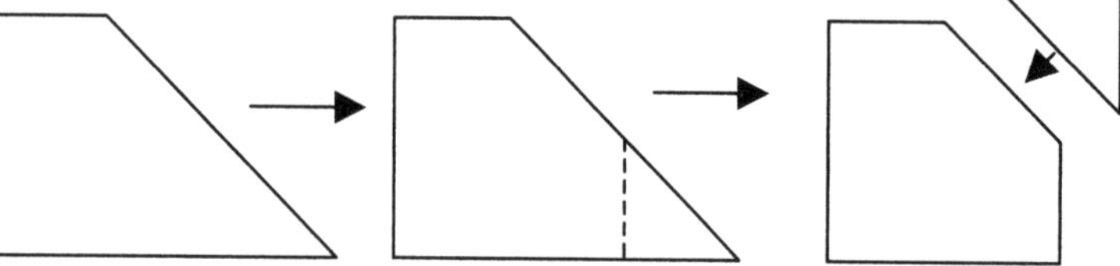

Draw a line on each figure to show where the cut should be made:

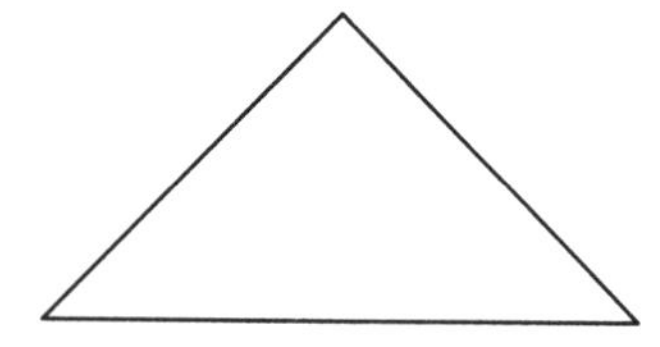

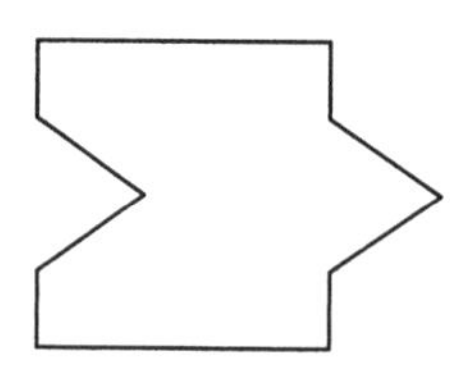

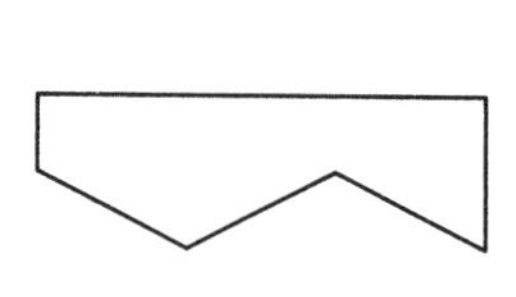

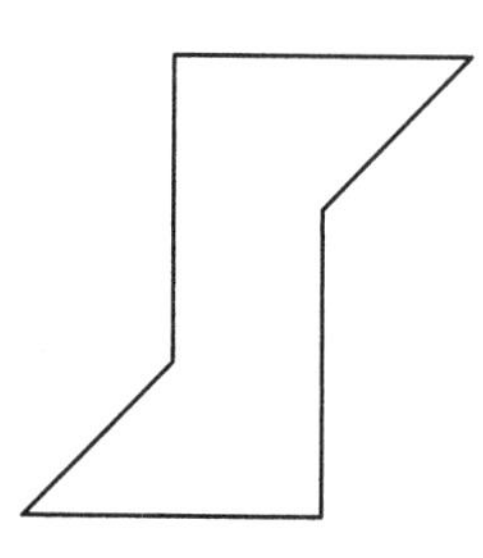

DIRECTIONS IN EDUCATION
Assessment Techniques
Inspect What You Expect

Testing can have a major impact on the content of classroom instruction. Results of testing may be used to evaluate the performance of individual schools or of individual classroom teachers. For that reason, it is imperative that assessment techniques used in the classroom examine student progress in the full range of skill acquisition, conceptual understanding, problem solving, and applications of knowledge expected in that classroom. Tests which do not match classroom instruction give feedback which is of extremely limited use for evaluation of individual student progress or for prescribing further instructional needs.

Standardized tests, which may be useful in comparing schools or school districts, have limited value in the individual classroom because the results are not received for several weeks after the testing session. In addition, the goals are drawn from popular national textbooks and are designed to be general enough to meet the needs of a full array of classrooms across the nation. Individual test items may match only a limited portion of the curriculum in an individual classroom and may test that portion of the curriculum in a format different from the one used for instruction.

In the *Curriculum and Evaluation Standards* published in 1989 and the *Assessment Standards for School Mathematics* published in 1995, the NCTM called for changes in mathematics assessment, including assessing what students know and how they go about thinking in mathematics; having assessment be an integral part of teaching; taking a holistic view of mathematics and focusing on a broad range of mathematical tasks; developing problem situations that require the application of a number of mathematical ideas; using multiple assessment techniques, including written, oral and demonstration formats; and using calculators, computers, and manipulatives in assessment. The NCTM further clarified and supported their original calls for assessment changes in the 2000 *Principles and Standards for School Mathematics*. In general, the recommended changes in assessment matched the NCTM's suggested changes in mathematics instruction.

What criteria do the Assessment Standards establish for judging the quality of mathematics assessment?

- Assessment should reflect the mathematics that all students need to know and be able to do (Mathematics Standard).
- Assessment should enhance mathematics learning (Learning Standard).
- Assessment should promote equity (Equity Standard).
- Assessment should be an open process (Openness Standard).
- Assessment should promote valid inferences about mathematics learning (Inferences Standard).
- Assessment should be a coherent process (Coherent Standard).

What assessment strategies can be used in the mathematics classroom?

- **Pencil and paper tests** still play an important role in assessing student understanding of concepts and skills. Even traditional tests can include a wide range of items including open-ended and essay-like questions.

- **Teacher observations** can provide information about mathematical or affective characteristics. Teachers may observe students either individually or in groups.

- **Interviews and conferences** with students provide information about their thoughts, understandings, and feelings about mathematics. They also provide opportunities to offer encouragement and evaluate students' progress.

- **Performance assessment** involves presenting students with a mathematical task, project, or investigation to complete. As a result of observing, interviewing, and looking at the resulting product, the teacher can assess what the students know and can do.

- **Journal writing** provides students with the opportunity to reflect on and communicate about what they are learning. Such writing provides insights for teachers on changing student attitudes and understandings.

- **Mathematics portfolios** provide a way of showcasing student work and documenting progress made over time. Many types of assignments and examples of work may be collected in a portfolio, including projects, reports, and writing assignments.

As you think about classroom assessment, ask yourself:

1. What can I do to learn more about these new assessment techniques?

2. How might students benefit from a wide range of classroom assessment techniques?

13 Measurement

THEME: Using Real Numbers to Analyze Geometric Shapes

HANDS-ON ACTIVITIES

The concept of measurement is fundamental to the study of geometry. The process of measuring an object involves comparing a chosen unit, or fixed reference amount, to the attribute being measured. A shape may have many measurable attributes, including width, area, or volume. You also have many choices for the units of measure you can use - standard units (inches, square centimeters, cubic yards, etc), and non-standard units (such as such as finding the length of a room in "paces"). When elementary students use such natural units to measure, they are able to focus on the attribute they are measuring. Additionally, they will develop a sense for the need for standardized units, especially when they report their measurements and compare theirs with others.

The first several activities in this chapter will help you deepen your understanding of the *meanings* of perimeter, area, surface area, and volume. Your focus will be on understanding answers to questions like, "What are we measuring when we find the area of a shape?" and "What is a square unit?"

In the later activities you will explore measurement formulas. As a teacher, it will be important for you to be able to apply these formulas, as well as explain *why* they work. To that end, you will *develop* the perimeter, area, surface area, and volume formulas for the common 2- and 3-dimensional shapes. These are developed through hands-on techniques commonly used in elementary classrooms; for example, students often develop surface area formulas through the use of nets. Consider the prism shown on the left below. The shape on the right is a two-dimensional version of the prism. This 2-dimensional pattern, called a net, can be folded to form the prism.

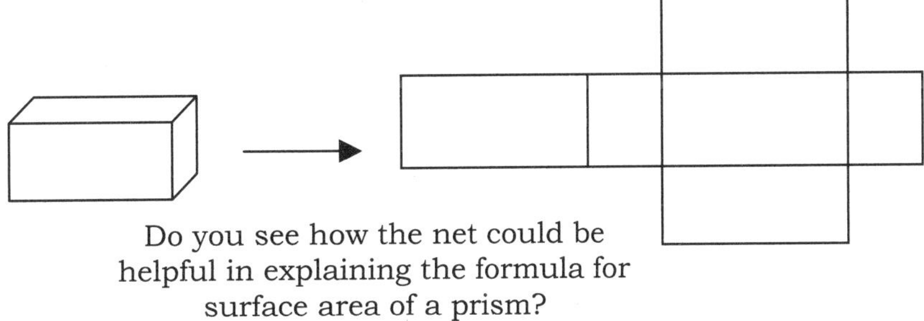

Do you see how the net could be helpful in explaining the formula for surface area of a prism?

OBJECTIVE: **Use body parts to investigate measurement**

You Will Need: **Materials Card 13.1.1 or a tape measure with centimeters**

In the first activity, you will investigate the measurement process, using your body parts as non-standard units for measuring length. You will also develop mental images of the standard metric units for linear measurement, by comparing these to your body parts.

1. Early measurements were often reported in terms of body parts such as thumbs, hands, and feet. For this problem, you will use the width of your thumb, your hand span (the distance from the tip of your thumb to the tip of your little finger when your hand is outstretched), and the length of your foot as units for measuring lengths of objects in your room.

   a. The measurement process involves three steps:

      First, choose an object, such as this page, and an attribute to measure, such as its width.

      Second, select an appropriate unit for measuring the object, such as the width of your thumb.

      Third, lay the unit side-by-side to determine the number of units required to measure the object.

      Measure the width of this page in thumbs, and record your measurement: ____________

   b. Measure the width of this page in hand spans and record your measurement: ________

   c. Which of thumbs or hand spans do you think is the more appropriate unit for measuring the width of this page? __________________ Explain.

   d. Compare your answers for parts (a) and (b) to those of several of your classmates. Did you obtain the same measurements? What factors might cause your measurements to be different?

2. For each of the following objects in your room, choose an appropriate unit (thumb, hand span, or foot) for measuring. Estimate each distance and then measure.

	UNIT	ESTIMATE	MEASUREMENT
a. The length of a table	______________	_________	_________
b. The height of a door	______________	_________	_________
c. The length of a pencil	______________	_________	_________
d. The width of a hallway	______________	_________	_________

3. Because you are familiar with standard English units such as inches and yards, you likely have a mental image of the approximate length of each of these. For most Americans, this is not the case when it comes to metric units, because the metric system is not widely used in the United States. In the next problem, you will compare the common metric units to body parts in order to develop mental images of the units of linear measure.

   Certain body measurements correspond very closely to metric units for measuring distance. This information is very useful when no metric measuring device is available or when an estimate of the metric length is all that is needed.

   a. For example, the width of the little finger is approximately one centimeter.

      Check your little finger: 1 cm. |——|

   b. The width of your hand (with fingers together), including your thumb, may be close to one **decimeter**.

      Check your hand. |————————————————|

      1 dm.

   c. About how many centimeters does it appear are equal to one decimeter?________

   d. A **meter** is approximately the distance from your navel to the floor or the distance from your nose to the end of the hand when your head is turned to the opposite direction from your hand (for people with average builds).

4. Using your personal metric references, estimate the following measurements. When you have completed your estimates, tape together the meter tape from Materials Card 13.1.1 and use it to measure each distance.

	ESTIMATE	MEASUREMENT
a. Height of the door in decimeters	________	________
b. Length of your pencil in centimeters	________	________
c. Width of the room in meters	________	________
d. Length of your foot in centimeters	________	________
e. Length of your table in decimeters	________	________

OBJECTIVE: Use dimensional analysis

Dimensional analysis is the term used to describe the process of changing from one unit of measure to another. There are many correct methods for solving a dimensional analysis problem and this activity will introduce one of them.

1. If you buy 4 yards of fabric for $7.00 per yard, what is your total cost? _________ Explain how you found your answer.

   Notice when you multiply 4 yards by $7.00 per yard by including the units (as shown at the right), the yards "cancel," and you are left with dollars as the unit. So the answer of 28 means $28.00.

   $$4 \text{ ~~yards~~} \times \frac{7 \text{ dollars}}{\text{~~yard~~}}$$

   Canceling units in this way is one part of the process of dimensional analysis.

2. Two other fraction ideas also play a role in dimensional analysis. Recall from your work with fractions that any amount over itself is equal to 1. You also know from the identity property of fraction multiplication that any amount multiplied by 1 leaves the amount unchanged.

   a. As an example, consider the problem of converting 13 feet to yards.

      Since 1 yard equals 3 feet, you can think of $\frac{1 \text{ yard}}{3 \text{ feet}}$ as a fraction equivalent to 1. Thus, in the multiplication problem shown at the right, you leave the *value* of 13 feet unchanged.

      $$13 \text{ feet} \times \frac{1 \text{ yard}}{3 \text{ feet}}$$

      Why is it useful in this problem to use yards/feet rather than feet/yards?

   b. In the multiplication problem in part (a), what does the answer $4\frac{1}{3}$ mean? Explain.

3. Some dimensional analysis problems are multi-step problems because there are several intermediate conversions that must be made.

   For example, suppose Alonso has saved a large collection of nickels in a jar. He put the nickels into rolls of 25 coins each. He then put all the rolls of nickels into three shoe boxes, each containing 30 rolls. He wants to know how much money he has, in dollars.

   One way to solve this problem is to first convert the 3 shoeboxes to rolls, then rolls to nickels, and lastly nickels to dollars. The first step is shown below.

   $$3 \text{ boxes} \times \frac{30 \text{ rolls}}{1 \text{ box}} = 3 \text{ ~~boxes~~} \times \frac{30 \text{ rolls}}{1 \text{ ~~box~~}} = 90 \text{ rolls}$$

   a. Why can you multiply by $\frac{30 \text{ rolls}}{1 \text{ box}}$?

b. Show the remaining two steps for this problem and answer Alonso's question.

Once you thoroughly understand dimensional analysis, you can do problems such as this one without the need to write out each step individually, as shown below.

$$3 \text{ \sout{boxes}} \times \frac{30 \text{ \sout{rolls}}}{1 \text{ \sout{box}}} \times \frac{25 \text{ \sout{nickels}}}{1 \text{ \sout{roll}}} \times \frac{1 \text{ dollar}}{20 \text{ \sout{nickels}}} = 3 \times 30 \times 25 \div 20 = 112.50 \text{ dollars}$$

c. Use dimensional analysis to solve the following problem. Show your process and how you cancel the units.

A farmer planted 75 acres of wheat. He expects to harvest 35 bushels per acre and, for each three bushels, he hopes to be paid \$20. What are his total anticipated earnings?

4. Some dimensional analysis problems involve a given quantity that is a rate (a ratio of two units) where one or both of the units must be converted to different units.

For example, consider this problem: Jordia spends 95¢ every day for a cup of coffee (the rate here is 95 cents over 1 day). How many dollars per year does she spend on coffee?

To solve this problem, you must convert cents to dollars and days to years. The process could be shown as follows.

$$\frac{95 \text{ \sout{cents}}}{1 \text{ \sout{day}}} \times \frac{365 \text{ \sout{days}}}{1 \text{ year}} \times \frac{1 \text{ dollar}}{100 \text{ \sout{cents}}} = 95 \times 365 \times 1 \div 100 = 346.75 \text{ dollars/year}$$

Show how to solve each of the following dimensional analysis problems involving rates.

a. A runner averages 1 mile every 8 minutes. How fast is the runner traveling in feet per second? Note: 1 mile equals 5280 feet.

b. If you spend an average of 8 hours per day sleeping, how many weeks sleeping is this per year?

OBJECTIVE: **Use a geoboard to find perimeter and area of geometric figures**

You Will Need: **(Physical Manipulative option: *Geoboard*; eManipulative option: *Geoboard*)**

In the next activity, you will use a geoboard to investigate the *meanings* of perimeter and area, as well as learn about the units used for these measurements. In a later activity, you will develop formulas for computing the perimeters and areas of common polygons and circles. So, even if you know and are able to apply area formulas, put aside your knowledge of these for this activity. Try to approach the problems from the perspective of a student learning these concepts for the first time.

1. The unit used to measure length on a geoboard is the horizontal (or vertical) distance between two adjacent pegs as show to the right.

   What is the diagonal distance between two adjacent pegs on a geoboard as marked by the question mark to the right? Show how you got answer.

   _____ units

   1 unit ?

   1 unit

2. The **perimeter** of a polygon on a geoboard is found by adding the lengths of all its sides. Find the perimeter of each of the following geoboard figures and record your answer in each case. (You will need to apply the Pythagorean Theorem to find some lengths.)

   a. 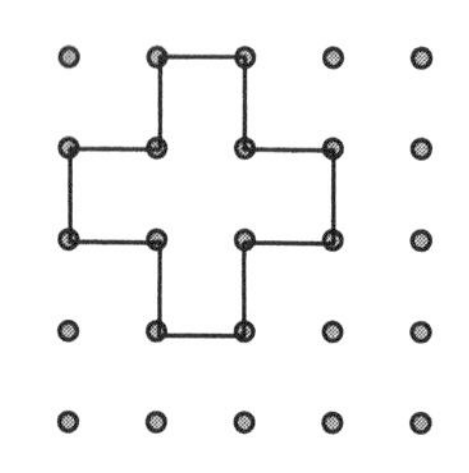

   _______ units

   b. 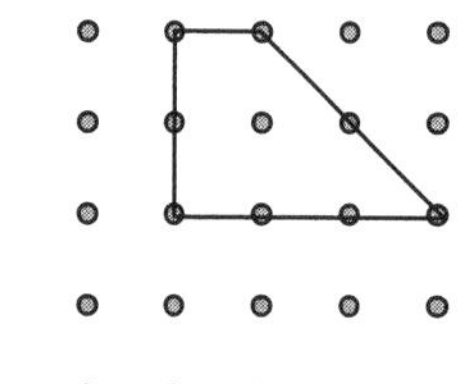

   _______ units

   c. 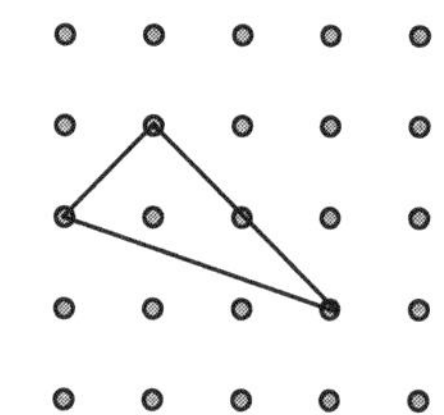

   _______ units

3. Linear units are useful for measuring one-dimensional attributes such as width, height, or perimeter, but they are not sufficient for reporting the two-dimensional measure of area.

   For this activity, use the smallest square that can be formed on the geoboard as the unit of area. One is shown at the right. Why do you think this is called this a "square unit?"

   1 square unit (denoted unit^2)

   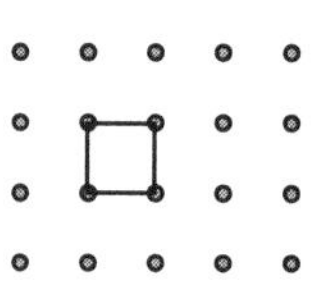

4. The **area** of a figure is the number of square units required to cover the figure. The example at the right shows a rectangle with an area of 6 square units.

   Area = 6 square units

   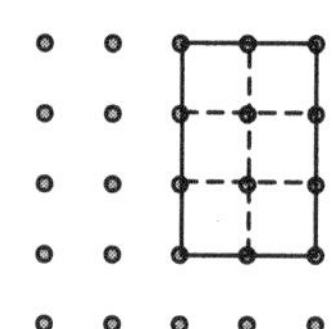

Find the area of each of the following figures.

a. 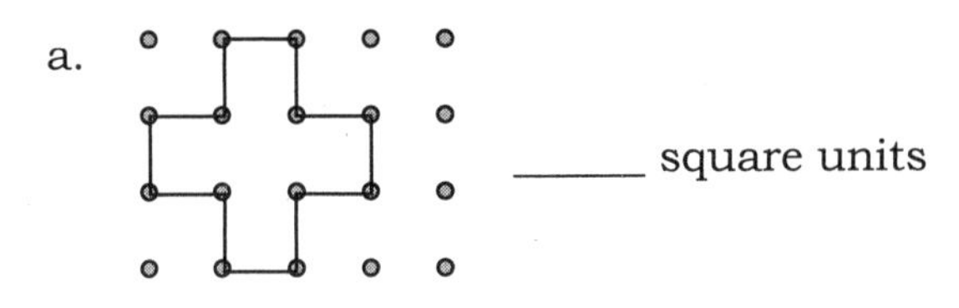 ______ square units

b. 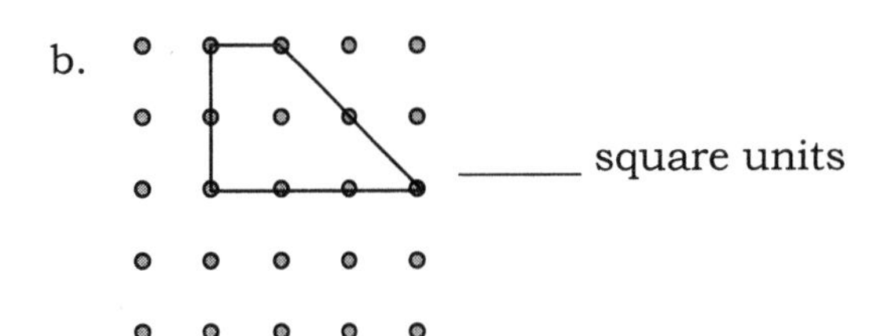 ______ square units

5. Consider the following method for finding the area of a *right* triangle.

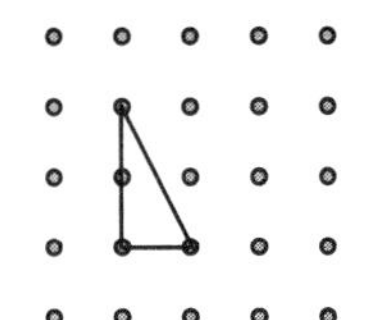

To find the area of the triangle to the left, you can form the rectangle to the right that is comprised of two such triangles.

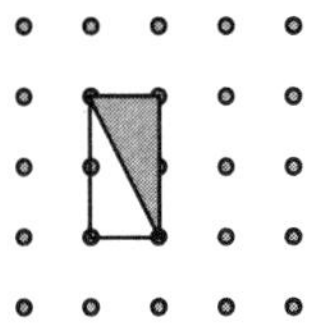

a. What is the area of the resulting rectangle? ______ square units

Explain how to use the area of the rectangle to find the area of the original triangle.

Area of the right triangle = ______ square units

b. The idea in part (a) is useful for finding the areas of many geoboard figures; for example, polygons may be subdivided into pieces that are rectangles and right triangles.

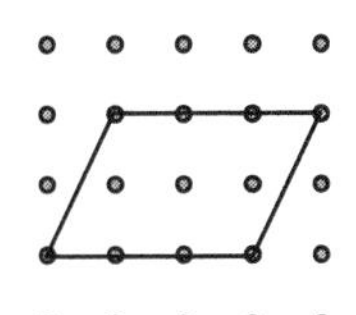

To find the area of the parallelogram to the left, you can form the square and two right triangles shown to the right.

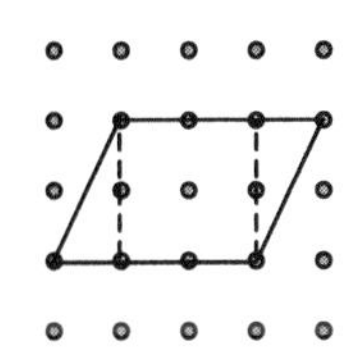

Show how to use the areas of the three polygons in the subdivided shape to find the area of the original parallelogram.

Area of the parallelogram = ______ square units

c. The next example shows another way to use right triangles to find the area of a polygon.

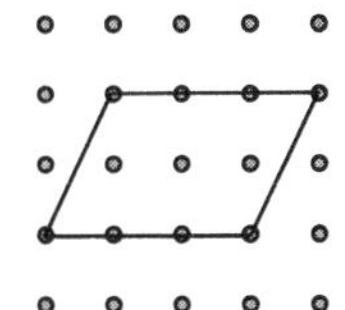

To find the area of the parallelogram to the left, you can surround it by right triangles to form the rectangle shown to the right.

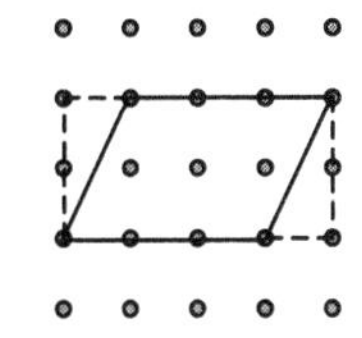

What is the area of the resulting rectangle? ______ square units

What is the total area of the triangles surrounding the parallelogram? ____ square units

What is the area of the parallelogram? ______ square units

d. Show how to find the area of each of the following polygons using two methods: (1) subdivide the shape into rectangles and/or right triangles; (2) surround the shape by right triangles to form a rectangle.

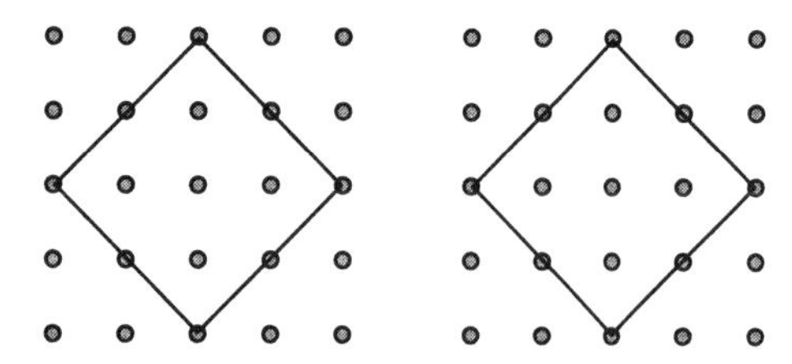

Area = ______ square units

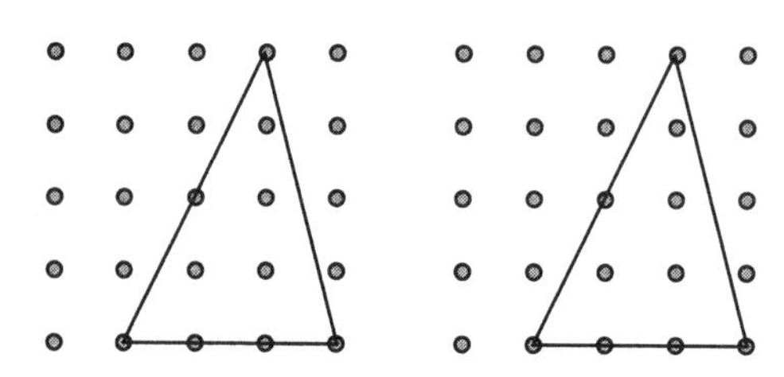

Area = ______ square units

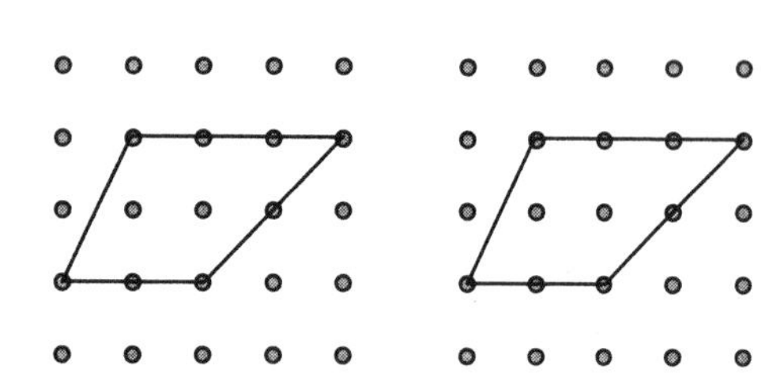

Area = ______ square units

6. For each of the following polygons, find the area and perimeter.

a.

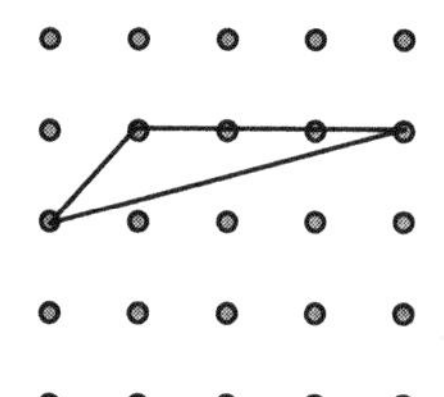

Area = ______ square units

Perimeter = ______ units

b.

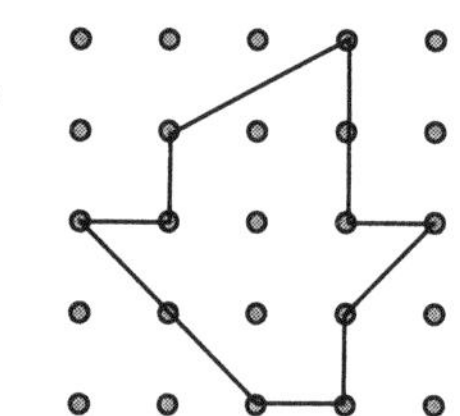

Area = ______ square units

Perimeter = ______ units

OBJECTIVE: **Develop area formulas for polygons**

You Will Need: **Geoboard (optional)**
(Physical Manipulative option: *Geoboard*;
eManipulative option: *Geoboard*)

Many students have memorized and are able to apply area formulas, but do not understand where these formulas come from. In this activity, you will *develop* formulas for areas of certain polygons; that is, you will gain a deeper understanding of the answers to the question: "Why do these formulas work?"

1. Consider the parallelogram formed on the geoboard shown below.

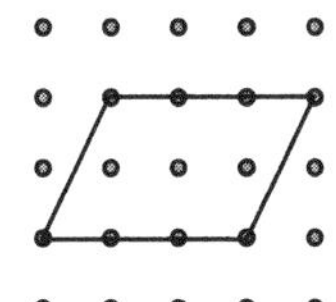

The parallelogram on the left can be "rearranged" to form the rectangle shown at the right.

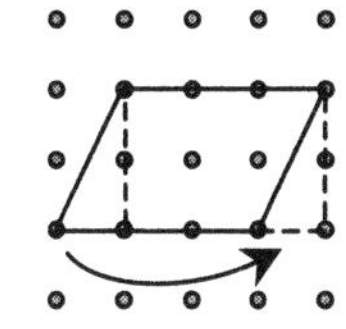

   a. The length of the base of the resulting rectangle is 3 units and the height of the rectangle is 2 units. What is the area of the rectangle? _____ square units

   b. Explain how to use the area of the rectangle to find the area of the original parallelogram.

      Area of original parallelogram = _____ square units

   c. What is the length of the base of the original parallelogram? _____ units

      What is the height of the original parallelogram? _____ units

   d. Complete the following statement: If a rectangle and a parallelogram have the same length base and same height, the two shapes have the same _____________.

   e. Write a formula for the area of a parallelogram with base b and height h.

      Area of a parallelogram = _____________

2. To develop the formula for area of a triangle, use what you discovered about parallelograms in problem #1.

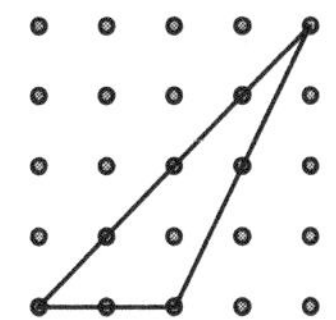

Two copies of the triangle at the left can be put together to form the parallelogram at the right

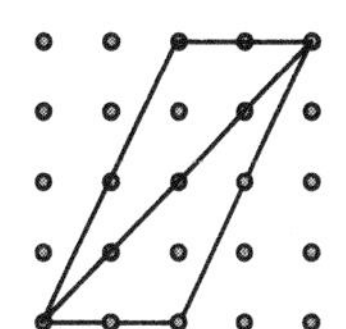

   a. Explain how to use the resulting parallelogram to find the area of the original triangle.

   b. Write a formula for the area of a triangle with base b and height h.

      Area of a triangle = _____________

c. Use your formula from part (b) to find the area of each of the following triangles.

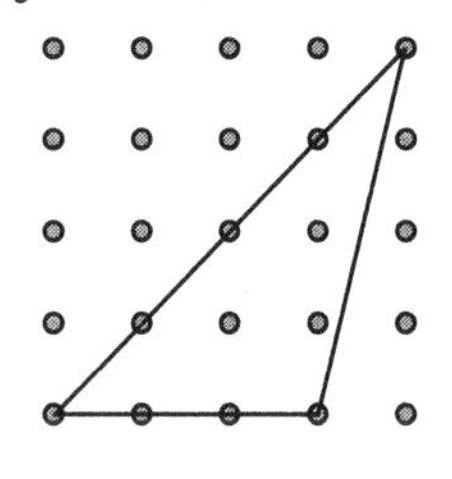

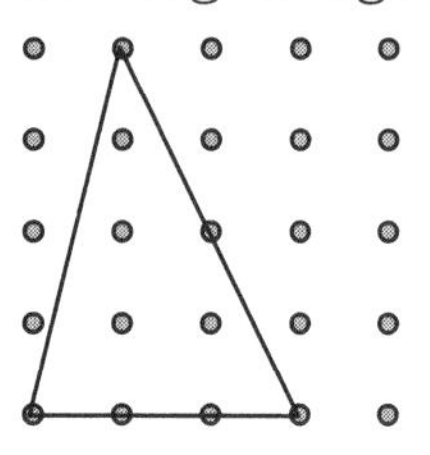

Area = ________units^2 Area = ________units^2 Area = ________units^2

d. Examine the three triangles in part (c). What traits do these triangles share?

Based on this example, what can you infer about the areas of triangles whose bases have the same length and whose heights are equal?

Test your conclusion by drawing three more triangles and finding their areas.

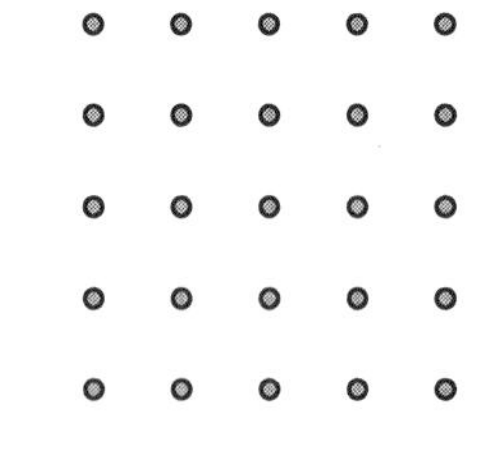

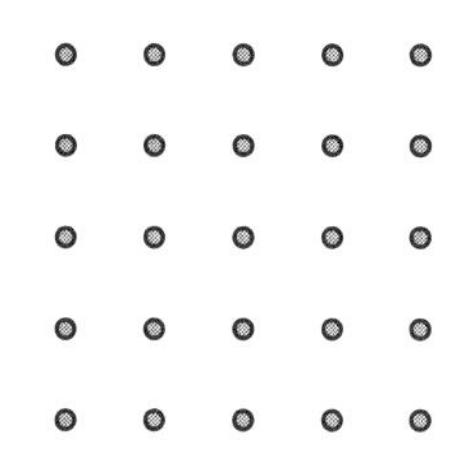

Area = ________units^2 Area = ________units^2 Area = ________units^2

3. One way to develop the formula for area of a trapezoid is to use what you know about the area of a parallelogram.

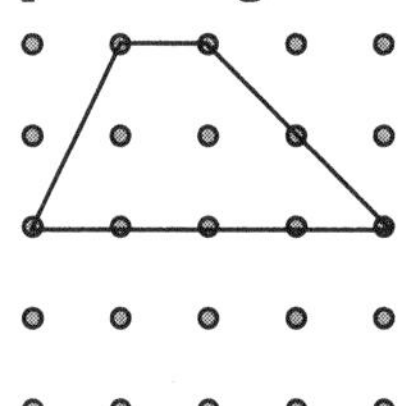

The trapezoid on the left can be rearranged to form the parallelogram on the right.

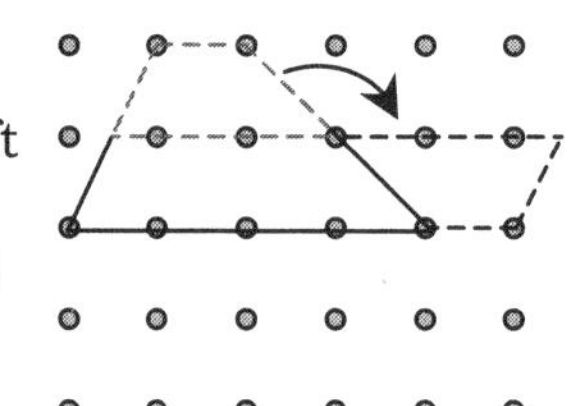

a. Do the resulting parallelogram and the original trapezoid have the same area? _____ Explain your answer.

b. What is the relationship between the length of the base of the parallelogram and the length of each of the two bases of the trapezoid? (Recall: the bases of a trapezoid are the two parallel sides.)

c. How does the height of the resulting parallelogram compare to the height of the original trapezoid?

d. What is the area of the parallelogram? ________units^2

What is the area of the trapezoid? ________ units^2

e. Based on what you found in parts (a) – (d), write a formula for the area of a trapezoid with bases whose lengths are a and b and whose height is h.

Area of a trapezoid = ____________

4. The method for developing the formula for area of a trapezoid in problem #3 relied on your knowledge of the area of a parallelogram. In addition, this method required you to go outside the 5 by 5 geoboard.

A different approach to developing this formula relies on the area of a triangle formula. This method also has the added advantage of staying inside the 5 by 5 geoboard.

Write a short paragraph explaining how to use the geoboard trapezoid shown at the right to develop the area formula for a trapezoid with bases whose lengths are a and b and whose height is h.

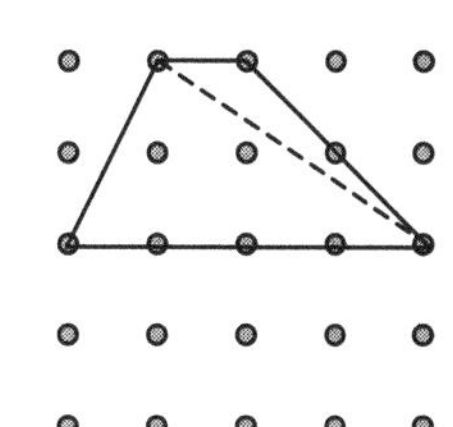

OBJECTIVE: **Develop circumference and area formulas for circles**

You Will Need: **Circles of various sizes (such as can, lid, cd), ruler, and Materials Cards for 13.2.3**

The formulas for circumference and area of a circle are a mystery to many students. As a teacher, you will likely hear the question: "What does the number π have to do with circles?" As you work through the problems in the next activity, think about how you will respond when your future students ask you a question such as this.

1. For this problem, you will need to gather circles of various sizes – these could be items such as the base of a soup can, the lid of a yogurt cup, a coin, a music cd, etc. Also, cut out the strips of paper from Materials Card 13.2.3.

Follow the directions in parts (a) and (b) to measure the circumference and diameter of each of your circles. *Take careful measurements – precision and accuracy are important here.*

a. First, use a strip of paper from the Materials Card to measure the circumference C of each of your circles; for example, if you are using a soup can, carefully wrap a strip of paper around the base of the can and mark the strip. Tape strips end-to-end, if needed for larger circles.

Lay the strips on the table, and use your ruler to measure the circumference of each circle. Record your measurements at the right, along with the name of each item (such as "soup can").

Circle #1: ______________________

C = ________ d = ________

Circle #2: ______________________

C = ________ d = ________

Circle #3: ______________________

C = ________ d = ________

b. To measure the diameter d of each of your circles, use your ruler. Then record your measurements in the space provided in part (a).

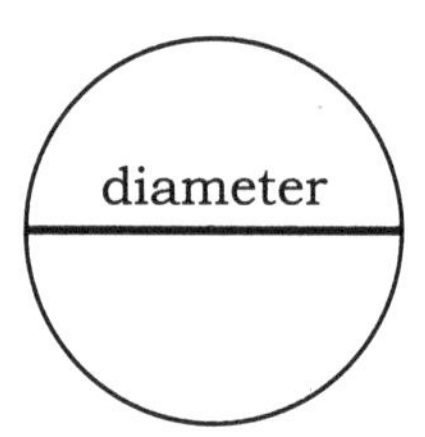

If the center of a circle is not marked, how can you be sure to measure the distance across the circle through the center point?

c. For each of your circles, compute C/d. What do you notice?

Circle #1: $C/d.$ = ______ Circle #2: $C/d.$ = ______ Circle #3: $C/d.$ = ______

d. One way to state the relationship between the circumference and diameter of a circle is to say, "The ratio of the circumference to the diameter is always equal to π." Use this idea to write a formula for the circumference of a circle with diameter d.

Circumference of a circle = ________

2. Consider the following method for developing the formula for area of a circle.

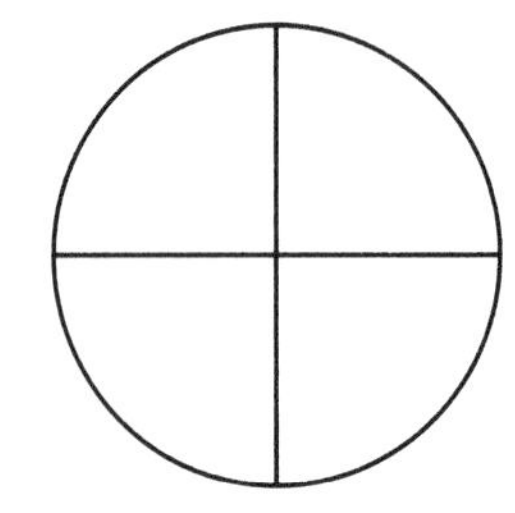

The sectors of the circle on the left can be rearranged to form the shape shown on the right.

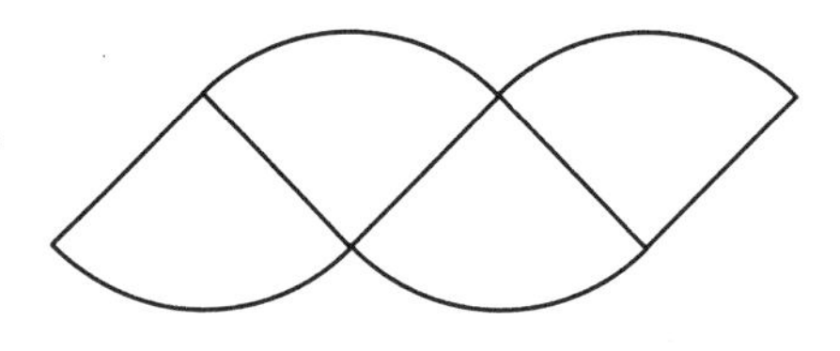

a. Cut the circles on Materials Card 13.2.3 into sectors. For each circle, rearrange the sectors as shown in the example above.

Trace the sixteen 1/16 sectors below to show your new arrangement of the circle.

What shape does your arrangement resemble? ________________________

b. Write an expression for the length of the base of the parallelogram-like shape you drew in part (a), in terms of the radius of the circle, r. Explain your answer.

Use your expression to label the base of the figure in part (a).

c. What is the height of the parallelogram-like shape? _______ Explain your answer.

Sketch an altitude of the parallelogram-like figure you drew in part (a) and label it.

d. What is the area of the parallelogram-like figure? _________ square units

What is the area of the original circle of radius *r*? _________ square units

---

ACTIVITY 13.3.1

OBJECTIVE: **Use nets to investigate surface area of polyhedra and cylinders**

You Will Need: **Materials Cards for 13.3.1, tape, and ruler**

A **net** is a 2-dimensional pattern that can be folded to create a 3-dimensional shape. By using nets of certain polyhedra in this activity, you will see the relationship between these 3-dimensional shapes and the 2-dimensional measure of surface area. You will also practice measuring, choosing appropriate units, and applying area formulas.

1. Cut out the three nets lettered *A*, *B*, and *C* from Materials Cards for 13.3.1.

a. Make a sketch of each of the shapes formed by these nets.

net *A* net *B* net *C*

b. The **surface area** of a 3-dimensional figure is the area of its surface.

To find the surface area of each of the prisms formed by nets *A*, *B*, and *C*, use your ruler to find any measurements necessary for computing the areas of the polygons that make up its surface (this may be easier if you flatten out the net again). Then apply your area formulas.

To record your work, write your measurements directly on the nets. Record your computations and answer, *including units*, below.

Surface area of net *A* = __________________

Surface area of net *B* = __________________

Surface area of net *C* = __________________

c. In general, how do you find the surface area of a prism?

2. A student, Scott, found pictures in his textbook of a prism and a regular hexagon shown below. He wants to make a net for a prism with a base that is the same size as the hexagon shown and a height of 5 inches. He plans to trace the hexagon when drawing the net.

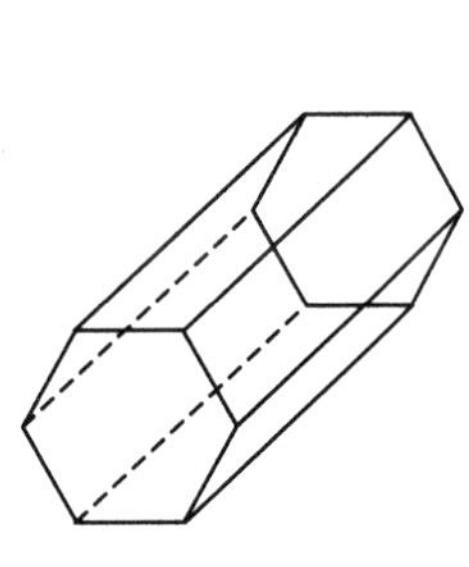

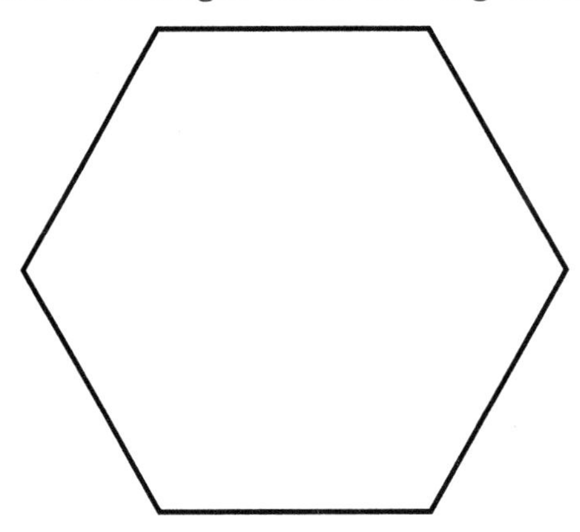

a. Make a sketch of a net for this prism.

b. In your sketch in part (a), label all the dimensions Scott will need to measure when drawing the net, so that the prism will have the desired dimensions.

c. Show how to find the surface area of Scott's prism in square inches. To find the area of the hexagonal bases, it may help if you think of dividing the hexagons into six equilateral triangles, as shown at the right.

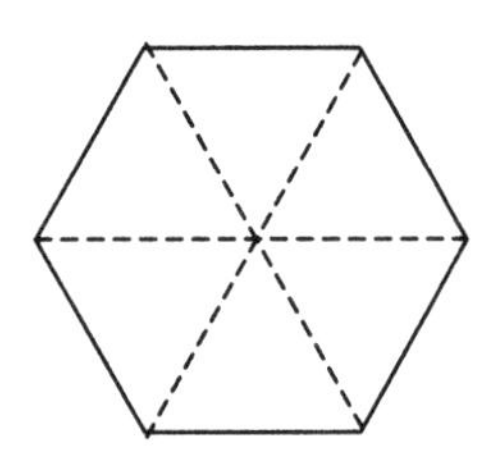

Surface area of prism =

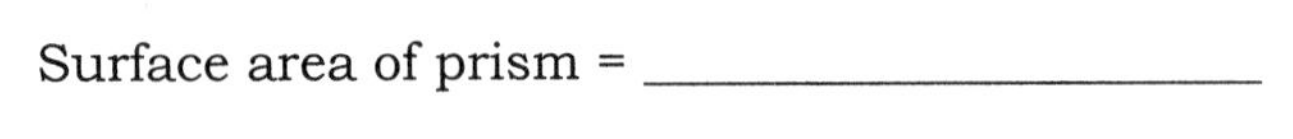

3. Now cut out, fold and loosely tape together the two nets lettered D and E.

   a. Make a sketch of each of the pyramids formed by nets D and E.

   net D net E

   b. Find the surface area of each of the square pyramids formed by nets D and E. Again, record your measurements directly on the nets and show all your computations below.

   Surface area of net D = ____________________

   Surface area of net E = ____________________

   c. In general, how do you find the surface area of a pyramid?

4. A student, Dianne, has an oatmeal container (in the shape of a cylinder) and wants to cover it with construction paper.

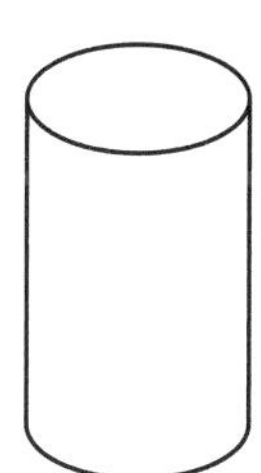

   a. What 2-dimensional shapes will Dianne need to cut from the construction paper to cover the cylinder? Sketch them below.

   b. Dianne wants the construction-paper cover she is making to exactly cover the oatmeal container with no overlap. What dimensions of the oatmeal container will Dianne need to measure? Label these in your drawing in part (a).

   c. The formula for surface area S of a cylinder is often stated as $S = 2(\pi r^2) + (2\pi r)h$. Use your drawing in part (a) to explain why this formula makes sense.

OBJECTIVE: **Explore volume concepts and develop volume formulas for prisms and pyramids**

You Will Need: **Materials Card 13.4.1, cm. cubes (optional), and salt**

In the previous activities in this chapter, you investigated linear measures, such as length and perimeter, as well as the two-dimensional measurements of area and surface area. The last two activities in this chapter introduce you to the three-dimensional measure of volume. These explorations will help you understand the concept of volume, as well as develop formulas for the volumes of polyhedra, cylinders, and cones. In this activity, begin by considering prisms and pyramids, and the relationship between their volumes.

1. Recall that linear units are used for measuring one-dimensional attributes such as width, height, or perimeter. The shape of a linear unit is a straight line segment, as shown in the example at the right.

   1 cm.

   For the two-dimensional measure of area, square units are commonly used. The example to the right shows a **square centimeter** which is a square that measures 1 cm. on each side.

   $1\ cm^2$

   a. **Volume** is a measure of the amount of space that a three-dimensional object occupies. Volume is commonly measured in cubic units, written $units^3$. What do you think is meant by the term "cubic unit?"

   b. Make a sketch of a cubic centimeter and label the lengths of its edges:

   c. Why do you think cubic units are used to measure volume?

2. Throughout this activity, use cubic centimeters as the unit of volume.

   a. Materials Card 13.4.1 contains a net for a right rectangular prism like the one shown below. Cut it out, fold and tape the prism together, leaving one end open.

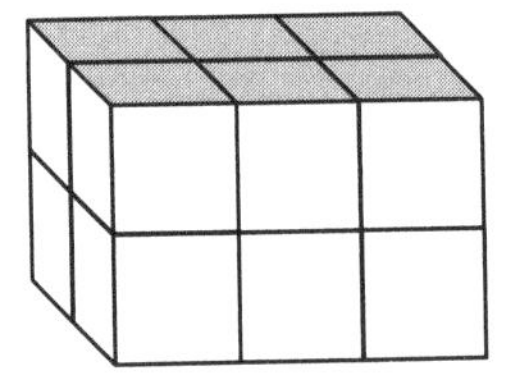

   If you have centimeter cubes, fill the prism with cubes or build a prism that is the same size and shape. If you do not have cubes, sketch a picture of the bottom layer of cubes in the prism here:

   How many cubes are in the bottom layer? ______

   How many layers of cubes will this prism contain? ______

   What is the volume V of the prism? V______ cubic centimeters

b. If you have centimeter cubes, build the following prisms. If not, visualize how they are built.

In each case, find the number of cubes in the bottom layer of the prism, the number of layers, and the volume of the prism.

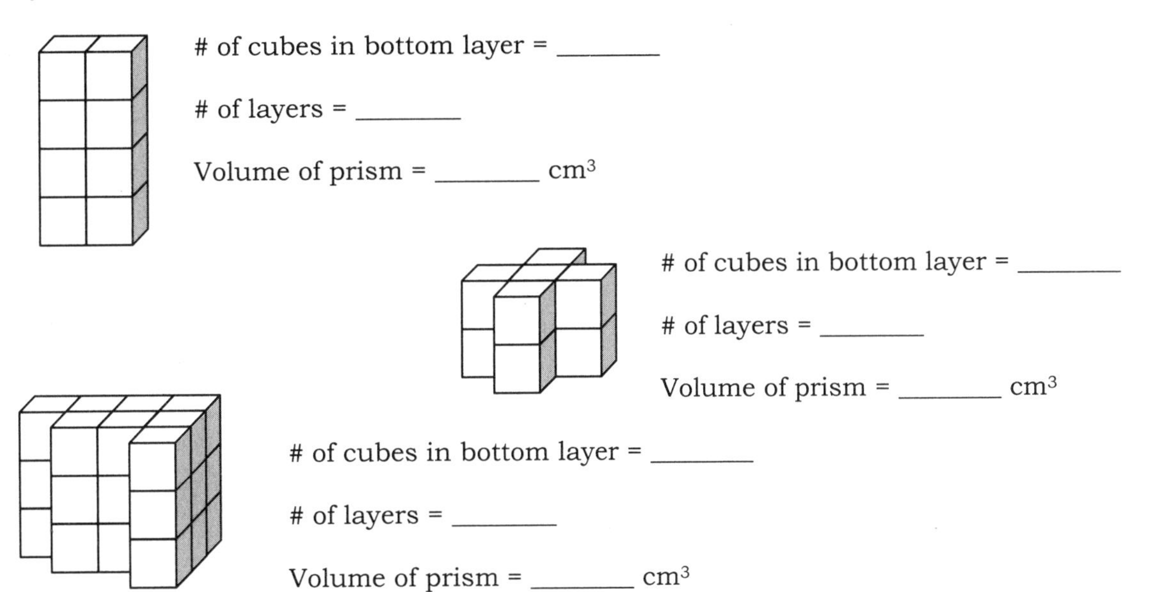

# of cubes in bottom layer = _______

# of layers = _______

Volume of prism = _______ cm^3

# of cubes in bottom layer = _______

# of layers = _______

Volume of prism = _______ cm^3

# of cubes in bottom layer = _______

# of layers = _______

Volume of prism = _______ cm^3

c. The formula for volume V of a prism is often stated in the following way:

$$V = (\text{Area of the base}) \times (\text{height})$$

Referring to your work in parts (a) and (b), explain why this formula makes sense.

3. Materials Card 13.4.1 also contains a net for a right square pyramid, without the base. Cut it out, fold it, and securely tape the flap on the outside.

a. Compare the prism from problem #2 to the pyramid. What is the same about these two shapes?

b. For this part, you are going to fill the pyramid with salt and use it to fill the prism you constructed in problem #2. Before you start, predict the number of pyramids full of salt you will need: _______

Check your prediction (be sure your prism is securely taped together). How does the volume of the pyramid compare to the volume of the prism?

c. Using the formula for the volume of a prism given in problem #2(c), write a formula for the volume of a pyramid.

Volume of a pyramid = ______________________________

OBJECTIVE: **Investigate volume formulas for cylinders and cones**

You Will Need: **Materials Cards for 13.4.2, ruler, rice or oats, and two 8.5 by 11 in. overhead transparencies or cardstock**

As with prisms and pyramids, the volumes of a cylinder and a cone are related. The last activity will help you see where the formulas used for computing volume of these shapes come from.

1. Consider the circle pictured on the centimeter grid at the right.

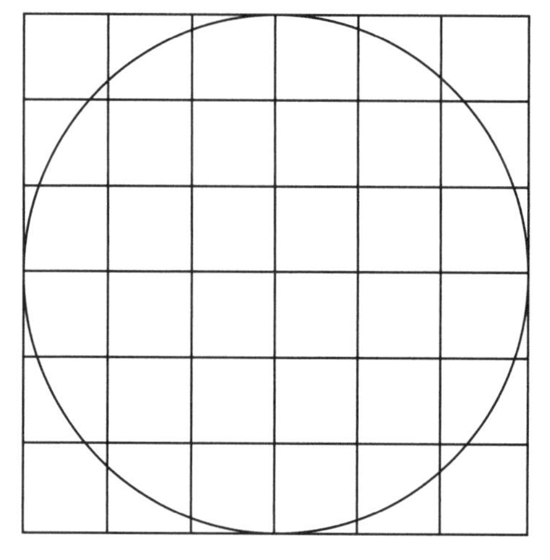

   a. What is the area of the circle? Be sure to include units with your answer.

      Area of circle = ___________________

   b. Imagine forming a layer of centimeter cubes and parts of centimeter cubes to exactly cover the circle. Estimate the volume of this layer of cubes *restricted to the circle.* Explain your answer.

      Volume of one layer of cubes = ___________________

   c. Cut out the rectangle on Materials Card 13.4.2 and tape it together to form the lateral surface of a cylinder with radius equal to the radius of the circle in part (a).

      How many layers of cubes like the one in part (b) are needed to equal the height of the cylinder ? _____

      What is the volume V of the cylinder? $V =$ ___________________

   d. The formula for volume V of a cylinder is:

$$V = (\text{Area of the base}) \times (\text{height}) = (\pi r^2) \times (\text{height})$$

      Referring to your work in parts (a) and (b), explain why this formula makes sense.

2. Roll up your two overhead transparencies or cardstock *in two different ways.* Tape them securely, *without overlap,* to form the lateral surfaces of two cylinders, as shown at the right.

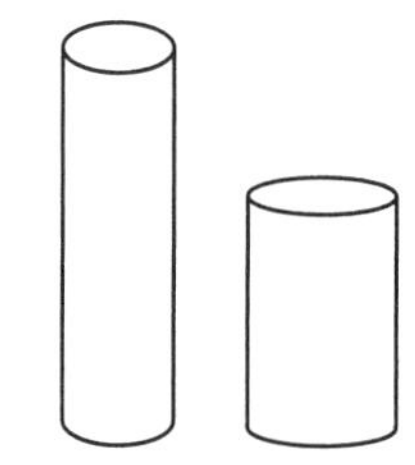

   a. Which cylinder do you think has the greater volume, the one with the long side of the transparency as its height, or the one with the short side of the transparency as its height, or will the volumes be the same? Explain your reasoning.

b. Place the tall cylinder on a paper plate or box lid (to catch the spillage) and fill it with rice or oats. Now place the short cylinder around the tall one, as shown at the right. Predict what will happen when you pull the tall cylinder out and let the rice pour into the short cylinder.

c. Check your prediction in part (a). Slowly pull the tall cylinder up while holding the shorter cylinder in place. Was your prediction correct?

d. Since the two cylinders you worked with in parts (a) – (c) were formed from the same size sheets (that is, they have the same lateral surface area), most students expect them to have the same volume. Explain why the shorter cylinder has a greater volume than the taller cylinder.

3. Materials Card 13.4.2 continued contains a net for a cone, without the base. Cut it out, fold it, and then securely tape the flap on the outside *with no overlap.*

a. Compare the cylinder from problem #1 to the cone. What is the same about these two shapes?

b. Predict how many cones full of rice or oats will fill the cylinder. ________ cones

c. Place the cylinder on a paper plate or inside a box lid to catch the spillage.

Now fill the cone with rice or oats and pour them into the cylinder. Record how many times you must fill the cone and empty it into the cylinder in order to fill the cylinder:________

How does the volume of the pyramid compare to the volume of the prism?

d. Using the formula for the volume of a cylinder given in problem #1(d), write a formula for the volume of a cone.

Volume of a cone = ________________________________

Milli-Golf

Estimate the distance in millimeters from the tee to hole #1 and record it on the appropriate blank in the score card below. Now use a metric ruler to mark off your estimate from the tee (within the boundaries of the fairway). Record that as one stroke. If your line ended in the hole, proceed to tee #2. If you missed the hole, estimate from your current position to hole #1 and tally a second stroke. Continue until the ball is in the hole, and then proceed to tee #2. If you land in a sand trap or water hazard, add one stroke to your score. Continue until you have completed all 6 holes of the Milli-Golf Course.

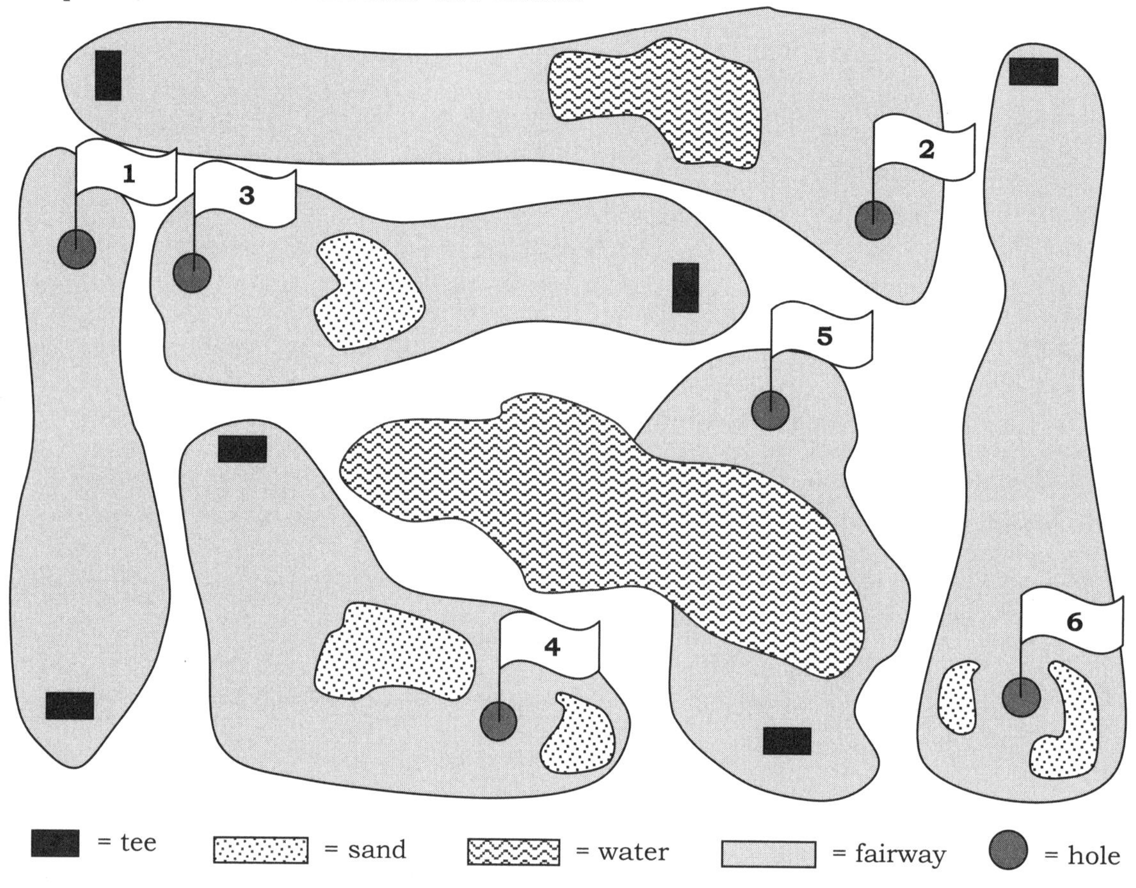

SCORE CARD									
	Estimates			Strokes		Estimates			Strokes
Hole #1 – Par 2					**Hole #4** – Par 3				
Hole #2 – Par 3					**Hole #5** – Par 4				
Hole #3 – Par 2					**Hole #6** – Par 2				

CONNECTIONS TO THE CLASSROOM

1. A teacher asked his class to find the perimeter of a given rectangle, 3 units by 4 units. A student, David, shows the work at the right.

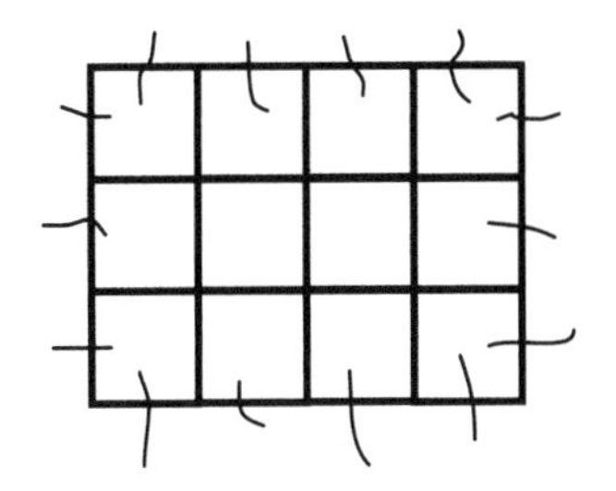

   a. Using his work, what answer will David obtain for the perimeter of the rectangle?

   b. Is David's solution correct? Why or why not?

2. Two students are discussing perimeter and area of rectangles.

   a. Latisha, says, "If the perimeters of two rectangles are the same, the areas must also be the same." Draw two geoboard rectangles that prove or disprove Latisha's statement. Explain what your pictures show.

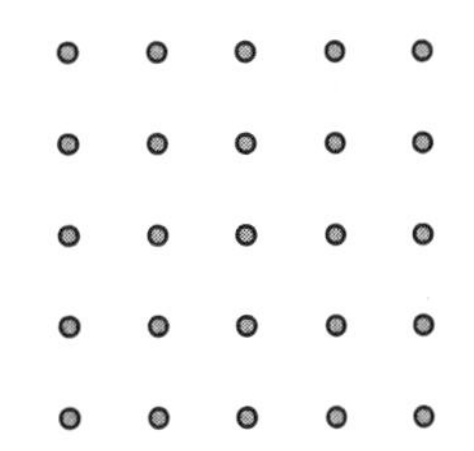

   b. Bobby says, "If you have a rectangle and you double its perimeter, then the area of the new rectangle is doubled, too." Use the geoboards to prove or disprove Bobby's statement. Explain.

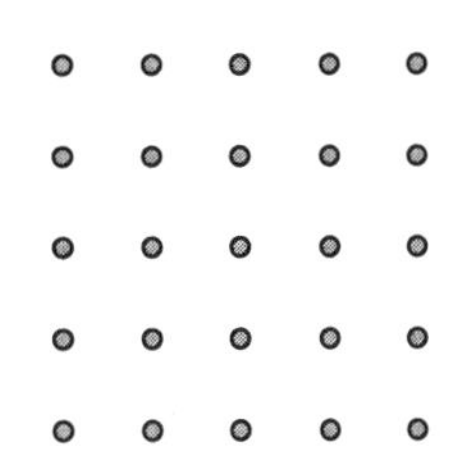

3. In Activity 13.2.2, you found two different ways to develop the area formula for a trapezoid. It may be that, as a teacher, you ask your students to develop the formula using any method they chose. The following two parts show the beginnings of descriptions that students gave to describe their method. Using the trapezoid pictured to the right, complete their process and see if you end up with the same formula as in the activity.

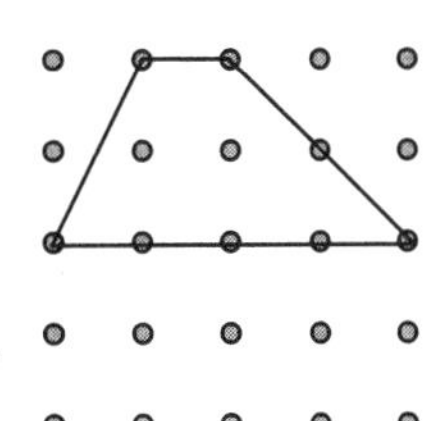

   a. I surrounded the trapezoid with another rubber band to form a rectangle that had the base of the trapezoid as one side and the top row as its opposite side. This formed two new right triangles on the outside of the trapezoid. To find the area formula for the trapezoid, I found the area of the rectangle and subtracted the areas of the two new right triangles.

b. From the ends of the top base of the trapezoid, I put rubber bands straight down to the bottom base. This formed a right triangle on the left, a rectangle in the middle, and a right triangle on the right. I found the areas of these three figures and added them.

4. Two students are discussing cones.

   a. Karen wants to make a cone-shaped birthday hat, out of construction paper. She says, "I need to start by cutting out a triangle then roll it up and it will be a cone." Will Karen's method work? Why or why not?

   b. Sharma says, "I've seen a cone flattened out before and it has a curved edge, like this." How do you think Karen could use Sharma's idea to create her cone? What will she need to do to create a right circular cone?

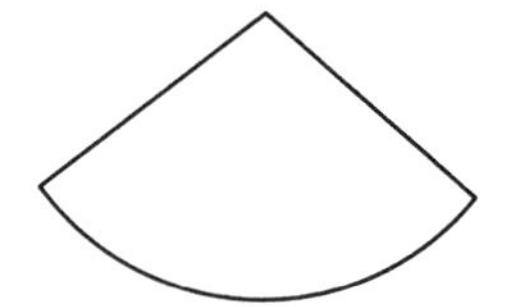

5. a. Juan was asked to find the lateral surface area of the square pyramid to the right (lateral surface area is the sum of the areas of the faces, not including the area of the base). Juan computes the lateral surface area as follows: 1/2(5*6)*4. Is he correct? Explain.

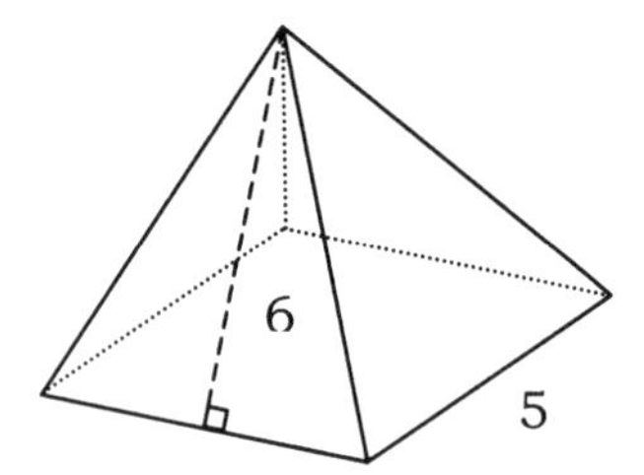

   b. Natalie says "Hey, if you rearrange your numbers as 1/2(5*4)*6, you have 1/2 times the perimeter of the base times the slant height." Does her method always work? Explain.

   c. Can Natalie's method be used to find the surface area of a cone? Explain.

MENTAL MATH

Match each lettered shape with the numbered shape that has the same area.

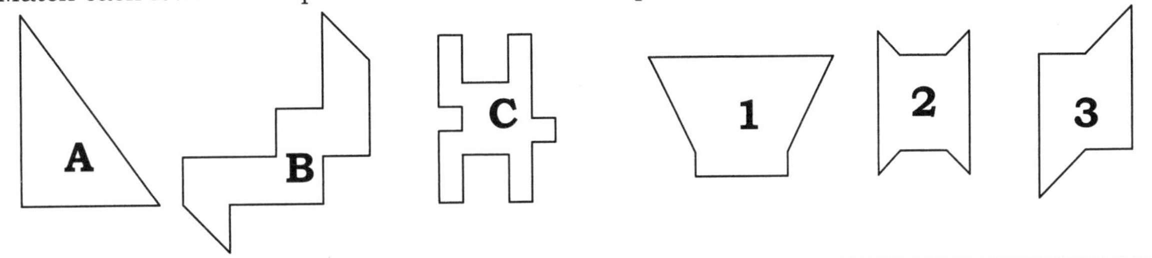

DIRECTIONS IN EDUCATION
Questioning Strategies
The Answer Is...But What Was The Question?

Questions posed by the classroom teacher allow the teacher to probe and extend student understanding, to promote divergent thinking, and to allow students to hear divergent responses. Questions also provide the teacher with immediate feedback on student progress which helps determine the appropriate course for the lesson. Both the quality of the teacher's questions and the teacher's behavior set the stage for thoughtful responses by the students. By explaining, modeling, and encouraging appropriate responses, the teacher can establish a classroom climate in which students are actively involved in the questioning process. Students can learn to consider answers to questions posed by the teacher or by others. When this thoughtful climate for questioning is established, the quality of student responses is enhanced, thus providing the teacher with rich and meaningful glimpses of student thinking.

What kinds of questions may be posed by the teacher?

- Questions may ask for simple recall of factual information – "What is 4 times 5?" or "What are the factors of 12?"
- Conceptual understanding can be checked with questions such as – "How is multiplication like addition?" or "Describe the relationship between fractions and decimals."
- Questions can be posed in such a way that divergent responses are encouraged as well – "If the answer is 7, what is the question?" or "How would our use of mathematic be different if we did not have multiplication or division?"
- Student responses can provide the starting point for further questions – "Craig says that decimals are fractions written in a different way. Do you agree or disagree? Why?" or "What did you do to arrive at that answer?"

- Questions may be used to assess the group's readiness for instruction and to prescribe specific instructional needs – "Tell me all the ways I can make a mistake in working this problem." or "What strategies might be useful in solving this problem?"

- Students can be encouraged to think about their own responses through the use of questions which ask them to examine their thinking – "Why did you decide to multiply?" or "You drew a picture. How did that help you solve the problem?"

Teacher behaviors can enhance the quality of student responses.

- Use wait time to encourage students to think about their responses. Wait time may vary according to the type of response desired. For example, simple recall may require wait time of 3 seconds or less. Questions that require divergent responses should be given wait time of 15 to 20 seconds depending on the age of the students and the complexity of the question. Older students may learn to accept even longer wait time. **It is essential that teachers avoid calling on or recognizing any student until after the wait time has been given. This encourages all students to engage in thinking about an appropriate response to the question.**

- Value student responses b accepting them, clarifying them when necessary, probing for further information when appropriate, and using the response as the basis for further questions when possible. The teacher's use of student responses will be determined by factors such as the level of thinking required by the question, the needs of the individual student for certain types of feedback, and the overall goals of the questioning. **Do not treat the response as the end point of the question/answer process, but rather as an integral part of an on-going discussion.**

- Model careful consideration of student questions by avoiding impulsive responses on your part. Create a classroom climate in which students are encouraged to pose questions for the teacher or for other students. Value student questions by answering them thoughtfully and by extending the questions to other students in the classroom. **Help students to see that questioning is a two-way activity and not something reserved for teachers.**

- Teach and reinforce student behaviors which enrich the questioning process in the classroom. Student can learn to clarify their own responses and to ask for clarification from others. They can learn to probe for more information and they can learn to encourage the questions of others. **Create a climate which encourages comments such as, "That's an interesting question. Let me think about it." Or "When you say...do you mean...?"**

As you think about questioning strategies, ask yourself:

1. What kinds of questions do I use most often in my lessons?

2. Can I develop a series of questions which range from recall to divergent thinking on a single topic?

3. How much time elapses between my question and the acceptance of a response?

4. Can I list factors such as teacher behaviors or student behaviors which enhance/inhibit questioning in the classroom?

14 Geometry Using Triangle Congruence and Similarity

THEME: **Exploring Triangle Congruence and Geometric Constructions**

HANDS-ON ACTIVITIES

When we say that two shapes are the "same", we may mean that they are the same in both shape and size; that is, one of the shapes exactly matches the other when the two are superimposed. On the other hand, two shapes might be described as the same if they have the same shape, even if they are of different sizes. Informally, this is the difference between *congruent* shapes versus shapes that are *similar.*

The two regular pentagons shown below are congruent. A tracing of one of them will exactly match the other.

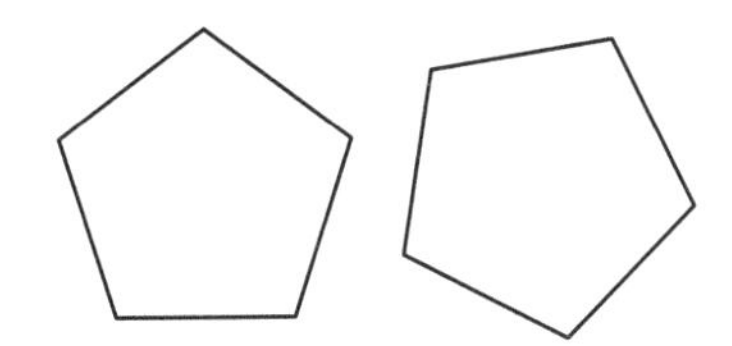

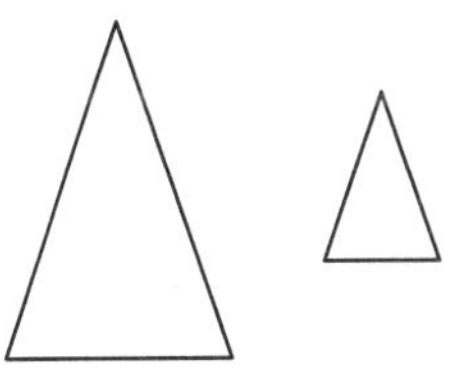

The two isosceles triangles shown above are similar. The lengths of the sides of the smaller triangle are half the lengths of the sides of the larger one.

In the first several activities in this chapter, you will investigate the concepts of congruence and similarity as they relate to triangles. Using the hands-on methods of the elementary classroom, such as tracing and measuring, you will develop the formal mathematical notions of congruent and similar triangles, discover properties of congruence and similarity relationships, as well as apply the concepts of similar triangles in a practical way.

The activities in this chapter also introduce you to compass and straightedge constructions, ancient techniques for creating precise geometric shapes, using simple tools. Using only a compass for making circles and arcs, and a ruler for making straight lines, you will be able to do many challenging constructions, such as parallel lines, regular polygons, and line segments of irrational number lengths.

OBJECTIVE: **Recognize correspondences between congruent triangles**

You Will Need: **Tracing paper**

Informally, we say that two shapes are congruent if they have the same size and shape. In this activity, you will consider the formal mathematical definition of congruent triangles, which involves the concept of corresponding angles and sides.

1. Two triangles are **congruent** if there is a correspondence between vertices such that all corresponding sides are congruent and all corresponding angles are congruent. Two triangles are given to the right. Carefully trace ΔMTO onto your tracing paper and label the vertices as shown.

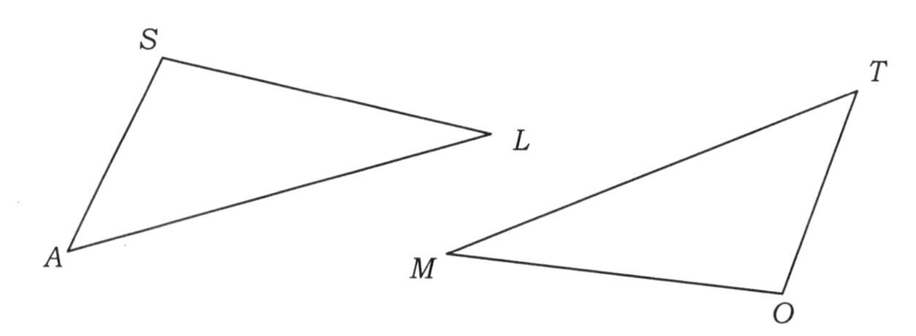

   a. Can you fit the copy of ΔMTO on top of the other triangle so that they match? _____

   b. By laying ΔMTO on top of the other triangle, you can see the correspondence between the pairs of sides and angles. Complete the following correspondence statements.

   Vertex $M \leftrightarrow$ ____________ Vertex $T \leftrightarrow$ ____________ Vertex O $\leftrightarrow$ ____________

   $\overline{MT} \leftrightarrow$ _____ $\overline{TO} \leftrightarrow$ _____ $\overline{OM} \leftrightarrow$ _____

   c. Since you found a correspondence between ΔMTO and ΔLAS such that all the corresponding angles are congruent and all the corresponding sides are congruent as shown by the tracing, by definition, the triangles are congruent. This may be written as: $\Delta MTO \cong \Delta LAS$

   The congruence relationship between two triangles may be written in several ways. Of the following statements, circle the one(s) that are true.

   $\Delta TOM \cong \Delta ASL$ $\Delta MOT \cong \Delta ALS$ $\Delta MOT \cong \Delta LSA$

2. For the three pairs of congruent triangles pictures below, identify the correspondence between the vertices (you may trace one of the triangles, if needed) that indicates that the triangles are congruent. Then write a congruence statement for the triangles.

   a.

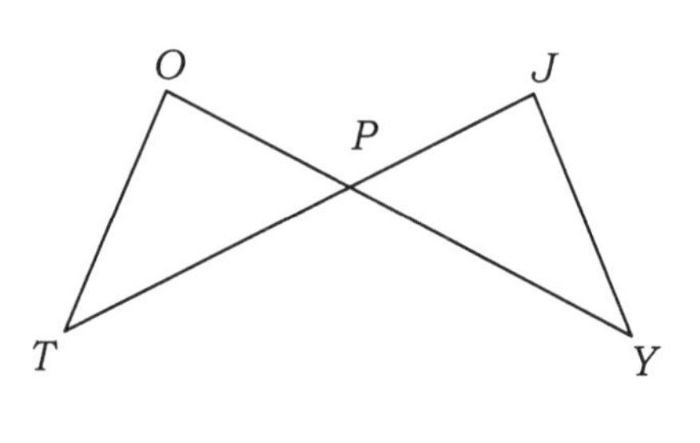

   b.

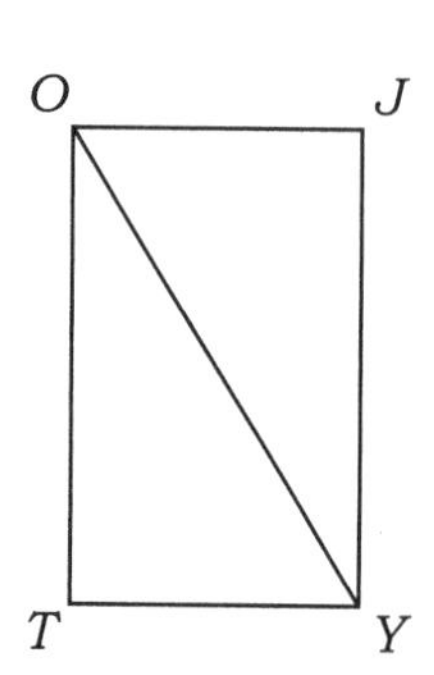

   c.

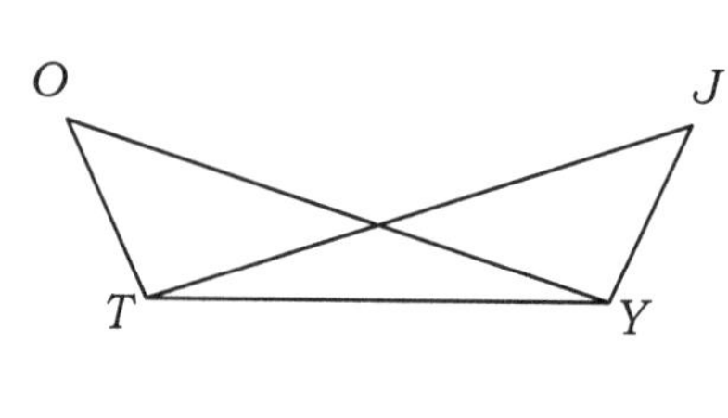

ACTIVITY 14.1.2

OBJECTIVE:	**Investigate triangle congruence properties**
You Will Need:	**Compass, straightedge, tracing paper, and protractor from Materials Card 10.1.1 (eManipulative Option: *Congruent Triangles*)**

In this activity, you will investigate triangle congruence properties which consist of sufficient conditions that may be checked in order to verify that two triangles are congruent. These properties allow us to verify the congruence of two triangles without having to check all six pairs of information, namely, three pairs of corresponding angles, and three pairs of corresponding sides. You will also use a compass as a tool for copying lengths.

1. For the first problem, you will consider the situation where three side lengths of one triangle and the corresponding three side lengths of another triangle are congruent. To discover whether this is enough information to conclude that the two triangles are congruent, follow the steps in parts (a) – (d).

   a. Given at the right are three line segments with lengths l, m, and n.

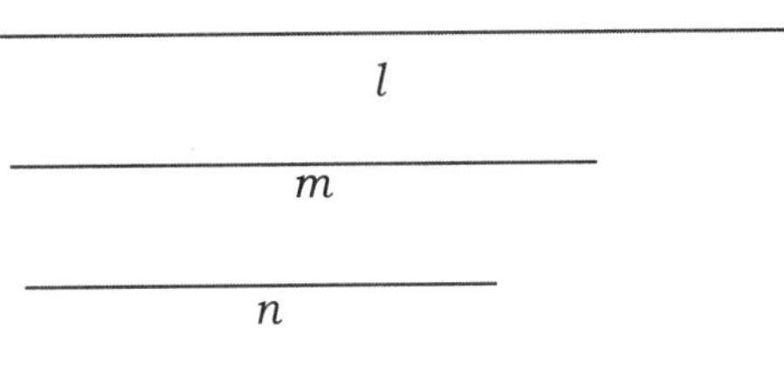

   To construct a triangle having sides congruent to these three line segments, begin by opening your compass to the length of one of the line segments, say l. Then, lay off this length on the line segment below by placing the sharp point of your compass at point P and marking an arc that intersects the segment. Label this intersection point Q. $\overline{PQ}$ is the first side of the triangle.

   P ______________________________

   b. Open your compass to the length of one of the other two given line segments, say m. Place the sharp point of your compass on one endpoint of $\overline{PQ}$, say P. Mark an arc the length of m above the line. The distance from point P to any point on this arc is _____.

   c. Lastly, construct an arc the length of n from the other endpoint, Q, in this case. Label the point where the two arcs meet R, the third vertex of your triangle. Complete your triangle by drawing the sides using a straightedge.

   d. Next, in the space next to ΔPQR, construct another triangle using the lengths l, m, and n.; for example, you may want to use the lengths in a different order.

   e. Trace one of the triangles. Does it fit exactly on top of the other by some combination of flipping, turning or sliding? ________ Are the triangles congruent? ________

   f. Write a sentence summarizing what you found. This is called the **Side-Side-Side (SSS) Congruence Property**.

2. Now consider the situation where two sides and the *included* angle of one triangle are congruent to the corresponding two sides and the included angle of another triangle. Do you think this is enough information to conclude that the two triangles are congruent?

   a. Given at the right are two line segments with lengths x and y, and a 40° angle. Follow the steps below to construct a triangle having two sides congruent to these segments and the angle *between* the two sides congruent to the 40°angle.

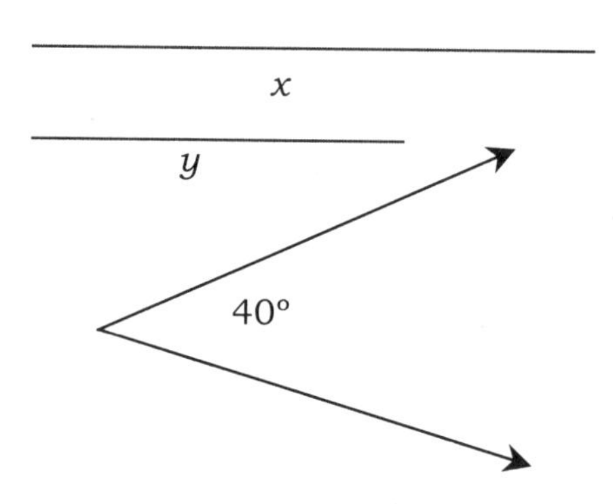

   Using a protractor, draw an angle with a measure of 40°. Use the line segment below as one side of the angle and point R as the vertex.

   R ________________________________

   b. Use your compass to mark lengths x and y along the rays of the angle R. Label the endpoints of these segments S and T. Connect points S and T to form the third side of ΔRST.

   c. Next, in the space next to ΔRST, construct another triangle using these given components. However, switch the segments on the sides of the angle.

   d. Trace one of the triangles. Does it fit exactly on top of the other by some combination of flipping, turning or sliding? ________Are the triangles congruent?________

   e. Write a sentence summarizing what you found. This is called the **Side-Angle-Side (SAS) Congruence Property**.

3. In problems #1 and #2, you showed that SSS and SAS are congruence properties. For each of the following, determine whether the given information is enough to determine that the two triangles are congruent. If it is, show that you can construct only one size triangle with the given information. If not, show two triangles which have the given information, but that are not congruent.

   a. Side-Side-Angle (Two sides and a non-included angle of one triangle are congruent to the corresponding two sides and non-included angle of another triangle)

b. Angle-Side-Angle (Two angles and the included side of one triangle are congruent to the corresponding two angles and the included angle of the another triangle)

c. Angle-Angle-Side (Two angles and a non-included side of one triangle are congruent to the corresponding two angles and non-included side of another triangle)

d. Angle-Angle-Angle (Three angles of one triangle are congruent to the corresponding three angles of another triangle)

e. Summarize what you found in parts (a) – (d).

OBJECTIVE: **Investigate triangle similarity properties**

You Will Need: **Millimeter ruler or Materials Card 14.2.1**

In this activity, you will discover how to determine triangle similarity by checking as few conditions as necessary. As you work through these problems, think about how the mathematical definition of "similar" compares to the everyday usage of this term.

1. Two triangles with the measures of two angles are shown at the right.

   a. What is the measure of the third angle in each triangle?

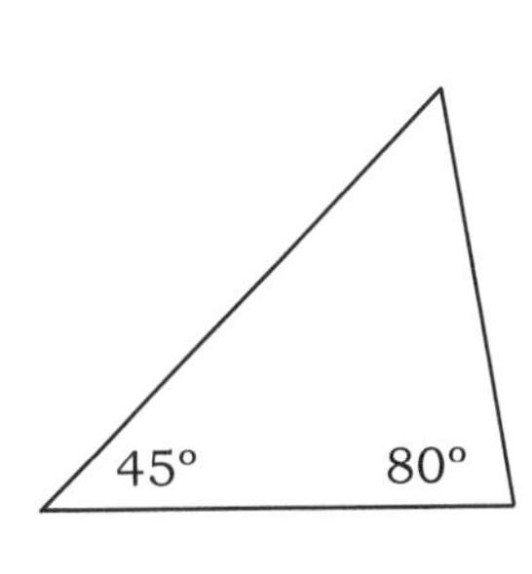

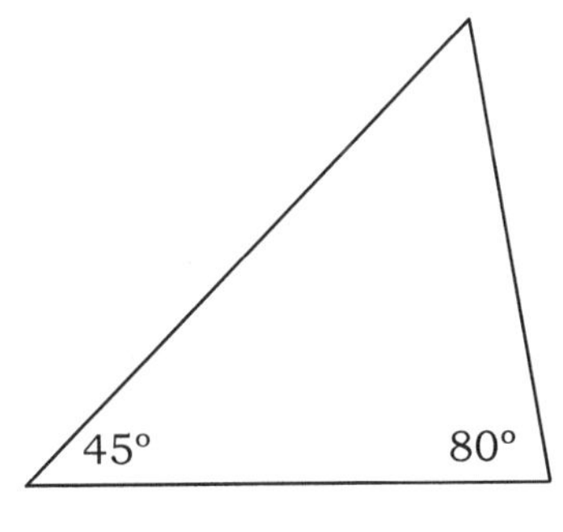

b. Are the two triangles in part (a) congruent? ________ Why or why not?

c. Label the vertices of the triangles in part (a) so that the corresponding angles are identified as A and D, B and E, and C and F. Using a millimeter ruler, measure the lengths of the sides and complete the following ratios.

$\frac{AB}{DE} = \frac{\quad}{\quad} =$ $\qquad$ $\frac{AC}{DF} = \frac{\quad}{\quad} =$ $\qquad$ $\frac{BC}{EF} = \frac{\quad}{\quad} =$

d. Two triangles are **similar** if there is a correspondence between the vertices such that all corresponding sides are *proportional* in length and all corresponding angles are congruent. Visually, this means that the two triangles have the same shape. Does it appear that the two triangles in part (a) are similar? Explain.

Note that you began with only a correspondence between the vertices such that two angles of one triangle are congruent to two corresponding angles of the other triangle. This is enough information to determine that the triangles are similar. This is called the **Angle-Angle Similarity Property (AA similarity)**.

2. When you apply the Side-Angle-Side (SAS) *Congruence* Property, you check whether two sides and the included angle of one triangle are congruent to the corresponding two sides and included angle of another triangle. If so, you can conclude that the triangles are congruent.

a. If you have two triangles and want to apply the **Side-Angle-Side Similarity Property (SAS similarity)**, what conditions must you check?

b. If the conditions you wrote in part (a) are met, what can you conclude?

c. Draw two sides of a triangle; in other words, draw one angle. Next to it, draw two sides of another triangle so that the corresponding side lengths in the two triangles have the same ratio (such as 1/2) and the included angle is congruent to your original angle. Is there a way to draw the third sides so the two triangles are a different shape? Why or why not?

3. In problems #1 and #2, you showed that AA similarity and SAS similarity are similarity properties. For each of the following, determine whether the given information is sufficient to determine that two triangles are similar. If so, use reasoning similar to that in problem #2(c) to explain why. If not, show two triangles which satisfy the given information, but are not similar.

   a. Side-Side-Angle (Two pairs of corresponding sides are proportional and their corresponding non-included angles are congruent)

   b. Side-Side-Side (All three pairs of corresponding sides are proportional.)

OBJECTIVE: **Make indirect measurements using similar triangles**

You Will Need: **A partner, measuring stick, mirror, and a sunny day (optional)**

For this activity, you will use similar triangles to measure lengths *indirectly*; that is, without comparing a unit to the object you wish to measure. There are many applications of indirect measurement; for example, you may need to know the height of a tree that is to be cut down so you are sure it won't hit any buildings.

1. Follow these steps to measure the height of the ceiling in your classroom *indirectly*.

   a. Measure a reasonable distance (1-2 meters) from the base of one wall in your classroom and place the mirror on the floor at that point. Stand beyond the mirror at the point where you can see the image of where that wall meets the ceiling in the mirror. Your partner should mark this point on the mirror, and then measure and record the following distances:

      Your eye level = __________

      Distance between a point on the floor directly below your eye and the point in the mirror where the ceiling is sighted = ________

      Distance between the wall and the point in the mirror where the ceiling is sighted = ____

b. Draw a diagram of this situation that shows the two triangles involved. Label the distances from part (a).

c. *Justify* why the two triangles you've drawn are similar. That is, name the appropriate triangle similarity property that allows you to determine the triangles are similar and give the corresponding parts of the two triangles.

d. Using a proportion, find the height of the wall.

e. Now measure the actual height of the wall. How close was your indirect measurement? Name at least two reasons why your indirect and direct measurements may be different.

2. (Sunny day required) Choose a partner who is not the same height as you. Find a place where your shadows are visible and measure the following distances.

Your height = ________ Length of your shadow = ________

Your partner's height = ________ Length of your partner's shadow = ________

Compute the following ratios:

$\frac{\text{Your height}}{\text{Partner's height}} = \frac{\quad}{\quad} =$ $\frac{\text{Your shadow length}}{\text{Partner's shadow length}} = \frac{\quad}{\quad} =$

$\frac{\text{Your height}}{\text{Your shadow length}} = \frac{\quad}{\quad} =$ $\frac{\text{Partner's height}}{\text{Partner's shadow length}} = \frac{\quad}{\quad} =$

What do you observe? What do you think would happen at a different time of day?

3. Choose an object outside (building, tree, lamp post, etc.) and make the following measurements.

a. Length of the shadow the object casts = ________

Length of your shadow = ________ Your height = ________

b. Draw a picture of two triangles involved in part (a), labeling these distances.

c. Justify why the two triangles you've drawn are similar.

d. Use your similar triangles to find the height of the object.

OBJECTIVE: **Practice and apply basic compass and straightedge constructions**

You Will Need: **Compass and straightedge**

In this activity, you will work through the seven basic compass and straightedge constructions, also called Euclidean constructions, which will serve as the basis for all of your work with constructions. You will also construct line segments of a certain length, the medians of a triangle, and the altitudes of a triangle, using combinations of these seven basic constructions.

1. When you perform compass and straightedge constructions, the only tools you may use are your compass and straightedge. If you are using a ruler for your straightedge, you may not use it to *measure* anything, only for making straight lines.

   The compass and straightedge properties and the step-by-step directions for the seven basic constructions appear in your textbook. After reading through those directions, try to do each of the constructions below without referring to those directions. Work toward understanding what you are doing rather than simply following the steps. You will use these constructions a lot in this chapter and it will be helpful if you are able to recall these without having to refer to the directions.

   a. Construction #1: Copy line segment $\overline{AB}$ onto line l, using point P on the line as one endpoint of the new line segment.

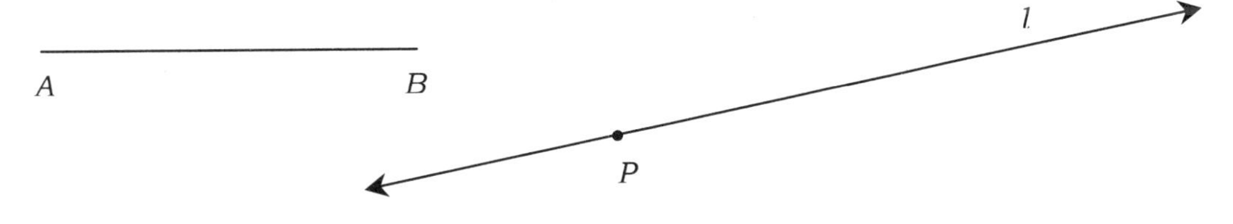

b. Construction #2: Copy $\angle C$ using ray $\overrightarrow{AB}$ as one side of the new angle.

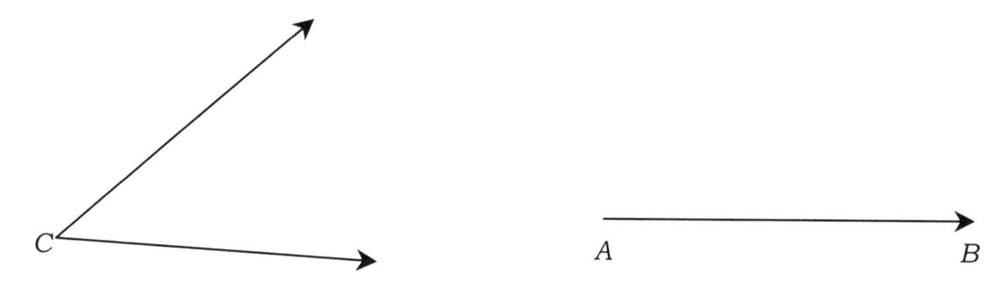

c. Construction #3: Construct the perpendicular bisector of line segment $\overline{AB}$.

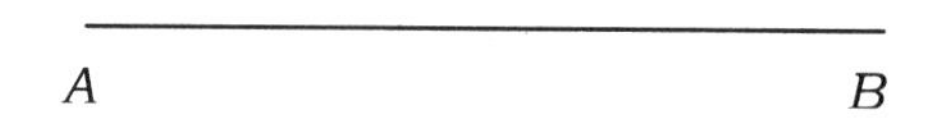

d. Construction #4: Bisect $\angle B$.

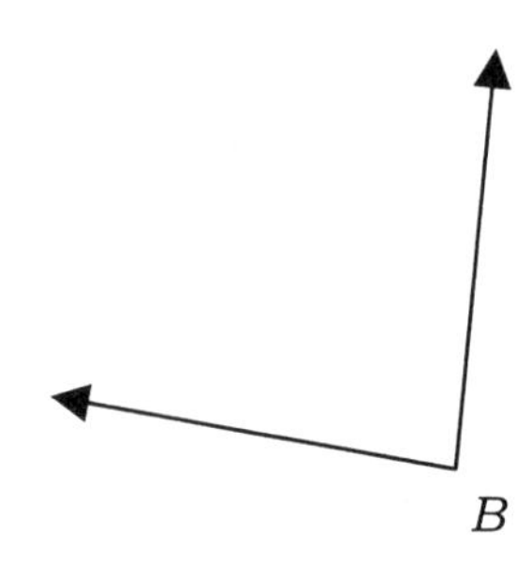

e. Construction #5: Construct a line perpendicular to a line l through a point P on the line.

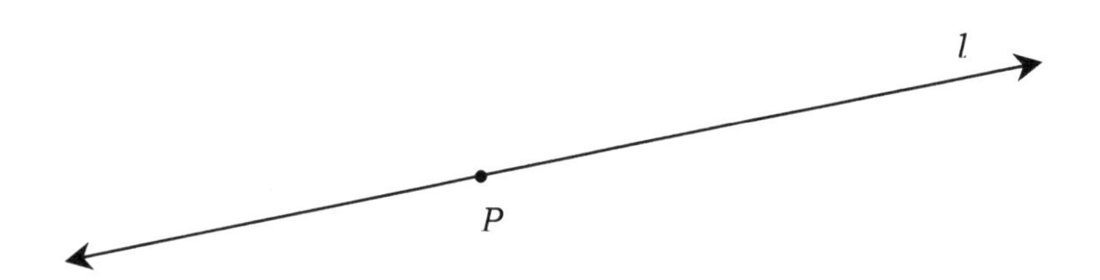

f. Construction #6: Construct a line perpendicular to a line l through a point P not on the line.

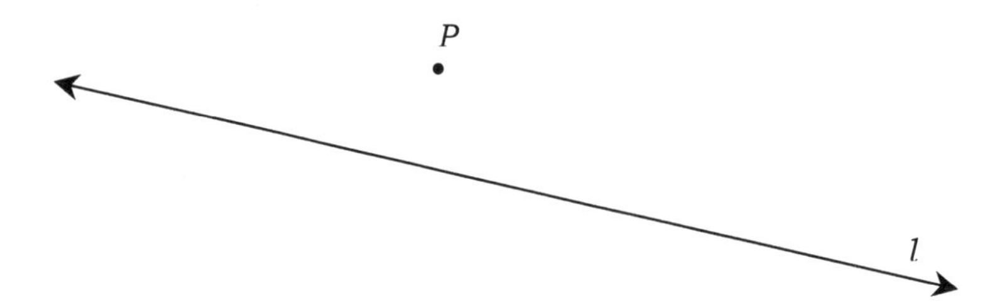

g. Construction #7: Construct a line parallel to line *l* through point *P* not on the line. (The construction directions in your textbook give one way to do this. Can you see a way to use Constructions #6 and #5 instead?)

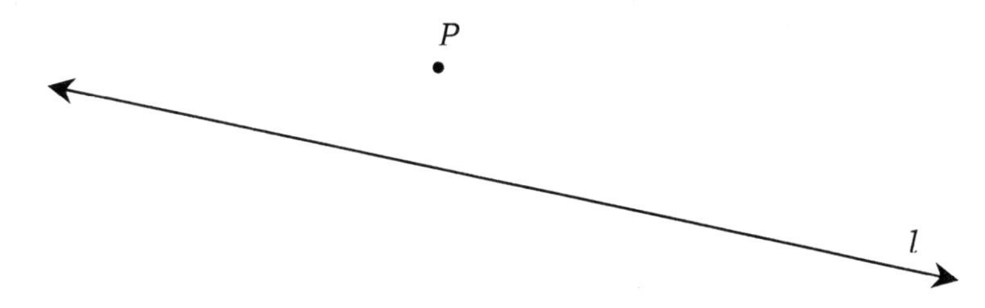

2. By combining some of the basic constructions and applying the Pythagorean Theorem, it is possible to construct line segments with irrational lengths. Show how to construct a line segment of length $\sqrt{5}$ units, using the segment below as the *unit* length. Briefly describe your process so the reader will be able to follow your steps.

1 unit

3. Recall that an **altitude** of a triangle is a line segment perpendicular to a side of the triangle through the opposite vertex.

   a. Which of the seven basic constructions do you need to use to construct an altitude of a triangle? Explain.

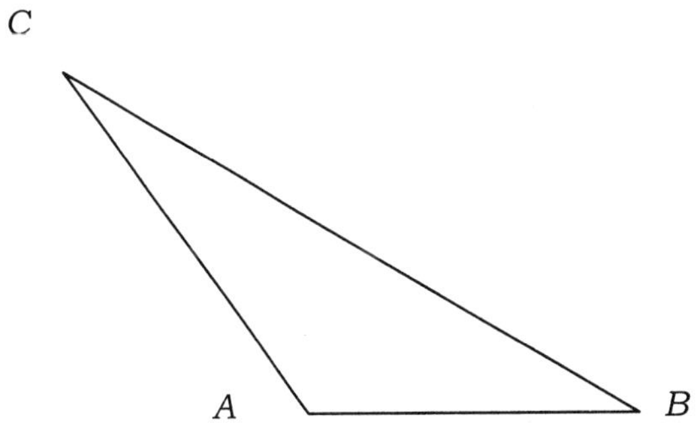

   b. Construct the altitude from vertex A in the triangle shown at the right.

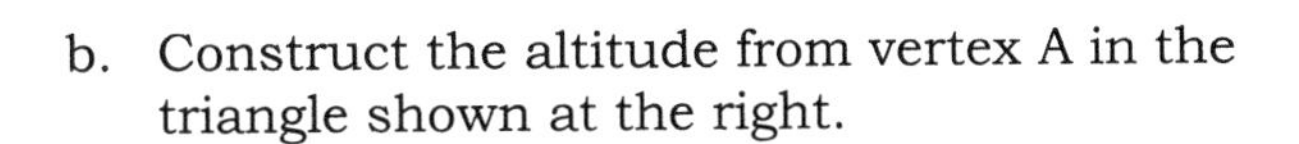

   c. Construct the altitude from vertex *B*.

   d. Construct the altitude from vertex *C*.

4. A **median** of a triangle is a line segment connecting the midpoint of a side of the triangle to the opposite vertex.

   a. Which of the seven basic constructions do you need to use to construct a median of a triangle? Explain.

   b. Draw any scalene triangle ΔABC and then construct the median through each vertex.

   c. Under what circumstances will the altitudes and medians of a triangle coincide?

   d. What happens to the altitudes in a right triangle?

OBJECTIVE: **Perform constructions by paper folding**

You Will Need: **Several sheets of tracing or wax paper, Mira (optional)**
(Physical Manipulative option: *Georeflector*)

Many of the basic constructions may be performed using paper-folding techniques. These techniques can help you and your students understand the constructions from a variety of perspectives.

1. To construct the perpendicular bisector of a line segment, follow these steps:

   a. Draw a segment $\overline{PQ}$ on tracing or wax paper.
   b. Fold the sheet over so that the points P and Q are superimposed on each other.
   c. While holding those two points together with one hand, crease the fold with the other hand.

   d. What does the fold line appear to be? How is this construction related to the Perpendicular Line Segments Test (Chapter 12)?

   e. This construction of the perpendicular bisector of $\overline{PQ}$ can also be performed with a Mira or a Georeflector by superimposing the image of point P on point Q. If you have either of these, try the construction with this method.

2. Draw an angle $\angle RPS$ on your tracing paper.

   a. Explain how you could construct the angle bisector of $\angle RPS$ by paper-folding.

b. How do you know the fold line from your construction in part (a) is the angle bisector?

c. You can also find the angle bisector by fixing the Mira or Georeflector at point *P* and adjusting it until the rays match. If you have one of these tools, try this construction.

3. Follow these paper-folding steps:

   a. Draw a line *l* and choose any point *P* on line *l*.
   b. Fold the paper without creasing it so that the line is superimposed on itself. Slide the paper, keeping the line coinciding with itself, until the fold passes through point *P*. Crease the fold.
   c. What have you constructed?

   d. Repeat the process above, with a point *not* on the line *l*.

   e. These two constructions can be done by fixing the Mira or Georeflector at point *P* and pivoting it until the line is superimposed on itself. If you have one of these tools, try this construction.

4. On your tracing paper, draw an acute triangle.

   a. Use paper-folding or a Mira to construct the perpendicular bisectors of each side. These should meet at a single point, called the **circumcenter** of the triangle.
   b. Repeat the construction in part (a) using an obtuse and a right triangle. This point is the center of the **circumscribed circle** that contains all the vertices of the triangle.
   c. Describe where, in general, the circumcenter is located for each kind of triangle.

5. On your tracing paper, draw an acute triangle.

   a. Use paper-folding or a Mira to construct the angle bisectors of each angle. These meet at a single point, called the **incenter** of the triangle.
   b. Repeat the construction in part (a) using an obtuse and a right triangle. This point is the center of the **incribed circle** that is tangent to each of the sides of the triangle.
   c. Describe, in general, where the incenter is located for each kind of triangle.

6. On your tracing paper, draw an acute triangle.

   a. Recall that an **altitude** of a triangle is the line through a vertex perpendicular to the opposite side (or the line containing the opposite side). Construct the altitudes from each vertex of the triangle you drew. These meet at a single point, called the **orthocenter** of the triangle.
   b. Repeat the construction in part (a) using an obtuse and a right triangle.

   c. Describe, in general, where the orthocenter is located for each kind of triangle.

OBJECTIVE: **Construct the circumscribed and inscribed circles of a triangle, and find the Euler line**

You Will Need: **Compass, straightedge, paper**

In the next activity, you will use some of the basic constructions to perform several more challenging and interesting constructions. You will see how to construct two circles related to a triangle, the circumscribed circle and the inscribed circle. You will also construct the Euler line, which contains three special points for a triangle.

1. The triangle given below will be used for all of the constructions in parts (a) - (e). Using a different colored pencil or pen for each part may be helpful.

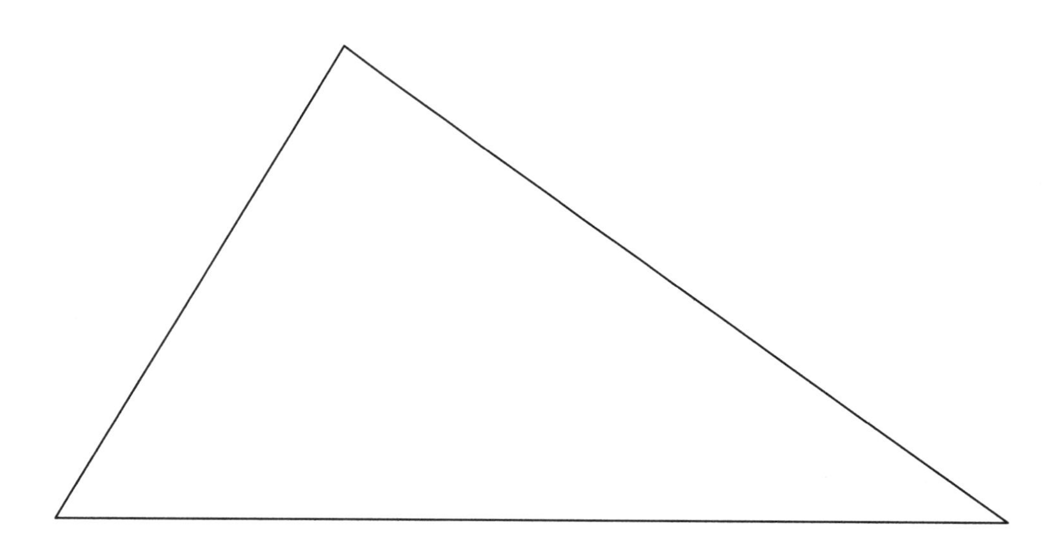

   a. The **circumscribed circle** of a triangle is a circle that contains all the vertices of the triangle. To construct the circumscribed circle, first construct the perpendicular bisectors or each side.

      The point of intersection of the bisectors is called the **circumcenter**, the center of the circumscribed circle. Label this point C.

      The radius of the circle is the distance from any vertex of the triangle to the circumcenter. Find the radius of the circumscribed circle for this triangle and construct the circle.

      Explain why C qualifies as the circumcenter.

b. The **inscribed circle** of a triangle is a circle that intersects each side of the triangle in exactly one point. To construct the center of the inscribed circle, the **incenter**, first construct the angle bisectors of each angle.

The point of intersection of the angle bisectors is the incenter. Label this point I.

To find the radius of the circle, construct a line from I perpendicular to one side of the triangle. Label the point of intersection of this perpendicular line and the side of the triangle as R. Construct a circle with center I and radius IR.

Explain why I qualifies as the incenter.

c. Recall that a **median** is a segment connecting a vertex to the midpoint of the opposite side. Construct the median from each vertex. You already found the midpoints in part (a).

The medians meet at a single point, called the **centroid**. This point is the center of gravity or point of balance of the triangle. Label the centroid D.

d. Construct the altitudes from each vertex.

The **orthocenter** is the point of intersection of the altitudes. Label this point O.

e. Which three of the four centers circumcenter, incenter, centroid, and orthocenter appear to be collinear?

The line containing these three points is called the **Euler line**, named for the Swiss mathematician Leonard Euler.

OBJECTIVE: Construct regular n-gons of the families $n = 3$, 4, and 5

You Will Need: Compass, straightedge, paper

The seven basic constructions, combined with the circumscribed circle construction, give you the tools you need construct many regular n-gons. In this activity, you will construct regular polygons of the families $n = 3$ (equilateral triangle, regular hexagon, etc.), $n = 4$ (square, regular octagon, etc.) and $n = 5$ (regular pentagon, regular decagon, etc.)

Use a separate piece of paper for each of the following constructions.

1. In this problem, you will construct regular n-gons of the family $n = 3$.

   a. Follow these steps to construct an *equilateral triangle.*
      - Start with a line segment the length of the segment shown at the right. Label the endpoints A and B.
      - With your compass open to the length AB, place the compass point on A and mark an arc above (or below) the segment.
      - With the same compass opening, place the compass point on B and make an arc that intersects the previous arc. Label the point of intersection C.
      - With your straightedge, construct segments $\overline{AC}$ and $\overline{BC}$.

      How do you know this is an equilateral triangle?

   b. The following steps outline one way to construct a *regular hexagon.*
      - Construct the circumscribed circle of ΔABC in part (a).
      - Bisect all three angles of the triangle. Construct your angle bisectors so they extend to meet the circle. Label the points of intersection of the bisectors with the circle as D, E, and F.
      - Connect A, B, C, D, E, and F in an appropriate way so that you have a hexagon.

   c. How do you know that your hexagon is regular?

   d. How could you proceed to construct a regular 12-gon? 24-gon?

2. Next, consider regular n-gons of the family $n = 4$.

   a. Using the definition of a *square*, apply the basic Euclidean constructions to construct a square whose sides are congruent to the line segment shown at the right.

   b. How many times did you construct a right angle?

c. The following steps outline one way to construct a *regular octagon.*

- Label the vertices of the square you constructed n part (a) as *A*, *B*, *C*, and *D*. and label the point of intersection of the diagonals as *P*.
- Construct a circle of radius *PA* and center *P*.

  What is this circle called?

- The angles formed by the intersecting diagonals should be right angles. Why?

  Construct the angle bisectors of the 90° angles formed by the diagonals of the square. Extend the bisectors to meet the circle.
- Use your straightedge to connect the consecutive points on the circle to form a regular octagon. How do you know that this is a regular octagon?

d. How can you construct a regular 16-gon?

3. The construction of the following regular polygon is accomplished via a series of intermediate constructions. Use great care in each step.

a. Using *AB* as the unit length, construct a segment $\overline{CD}$ of length $\sqrt{5}$.

A B

b. Using $\overline{AB}$ and $\overline{CD}$, construct a segment $\overline{EF}$ of length $\sqrt{5}-1$.

c. Bisect $\overline{EF}$, forming $\overline{EG}$. What is the length of $\overline{EG}$?

d. Construct a circle whose radius is *AB*.

e. With the compass open to a length equal to *EG*, mark arcs around the circle (starting from any point on the circle). Connect the consecutive points. What regular polygon did you form?

f. How could you utilize the polygon you just formed to construct a regular pentagon? Show how to do this.

g. How could you construct a regular 20-gon?

Constructible Designs

The following designs can be constructed using only compass and straightedge. How many of these designs can you construct? It may be easier to construct versions larger than those shown.

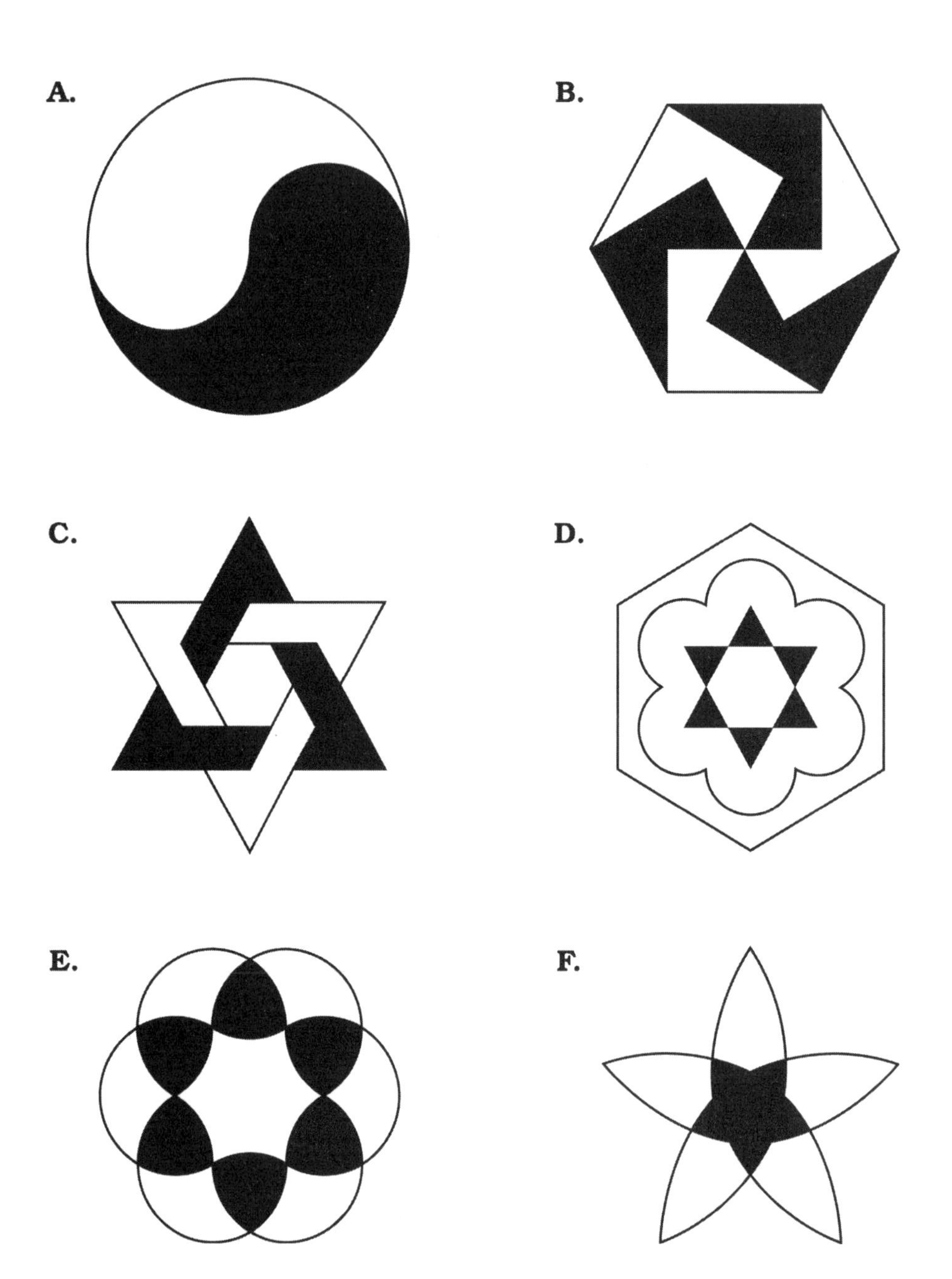

CONNECTIONS TO THE CLASSROOM

1. A student claims that ΔABC is congruent to ΔDEF. Is the student correct? Explain your answer.

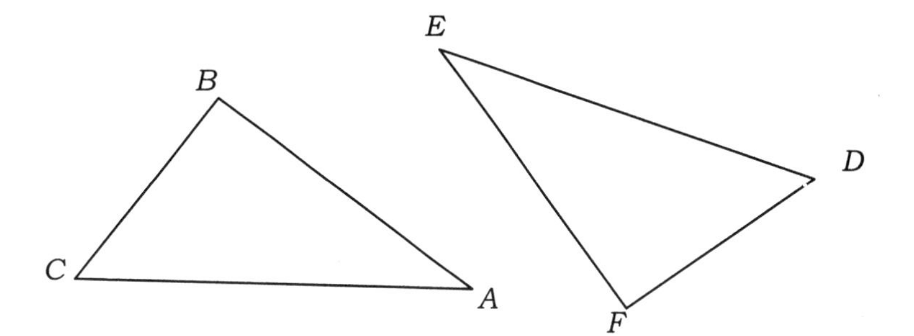

2. A teacher shows her class these two similar triangles and asks them to find the measure of angle X. A student answers, "15 degrees."

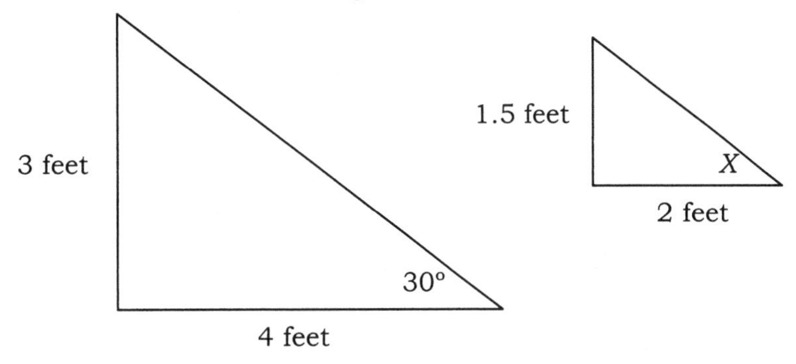

   a. Is the student correct? Why or why not?

   b. Why do you suppose the student gave this answer?

   c. What could this student do to check his answer?

3. A student says, "Angle-Side-Angle is a congruence property, so I don't see why this isn't also a similarity property." How would you respond?

4. When asked by her teacher to construct the altitude of ΔABC, using side $\overline{BC}$ as the base, Christina did the work shown at the right.

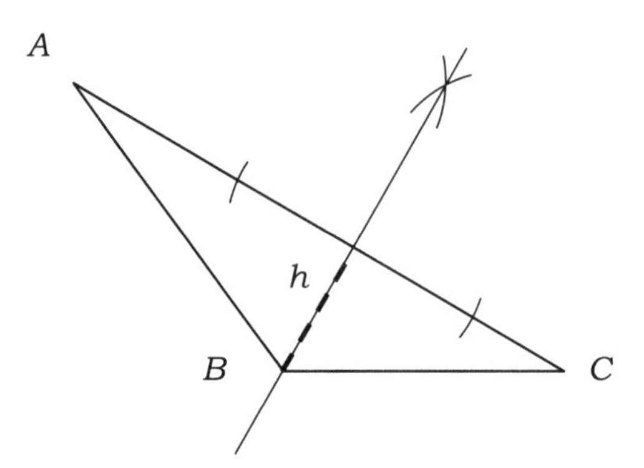

   a. Which of the seven basic constructions did she use?

      Why is this an appropriate construction for the problem?

   b. Another student in Christina's group, Paul, says, "No, you have to use vertex *A*, not *B*." What does Paul mean?

   c. How could one help Christina understand her mistake?

MENTAL MATH

For each figure below, make one cut that create two congruent halves. The cut does not have to be straight, but it must be one continuous cut.

For example:

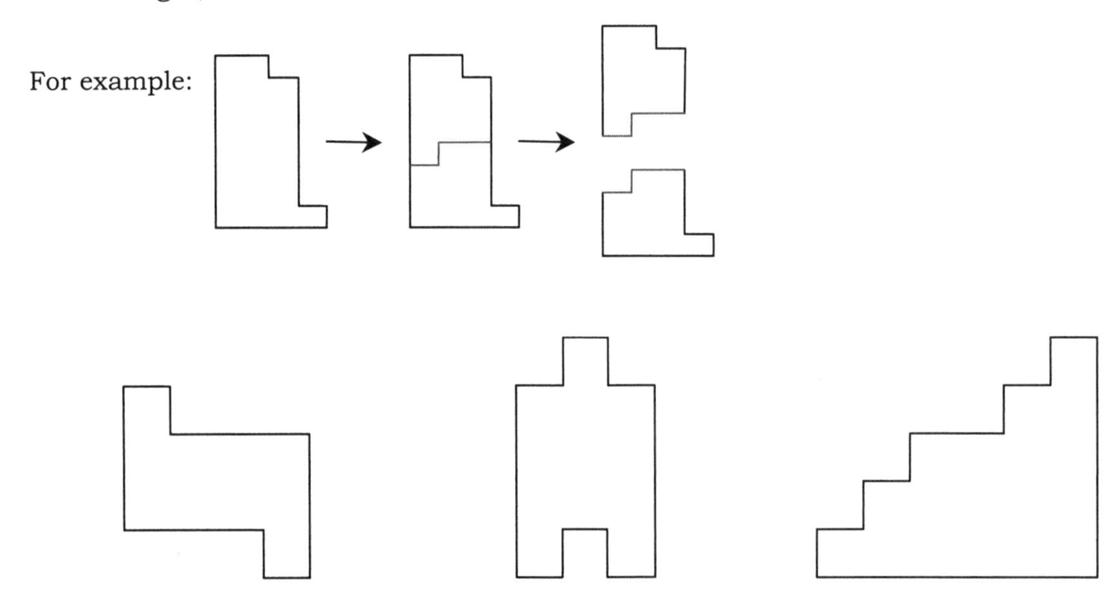

DIRECTIONS IN EDUCATION
Dealing With Diversity
Mathematicians All

American classrooms, like the American culture, are comprised of widely diverse groups of students. In efforts to meet the needs of this diverse population, educators have successively resorted to more narrow subdivisions of students within the classroom. These divisions – or tracks – have created as many problems as they have solved. For example, some students have been routinely scheduled into basic math classes where they will learn, at best, only basic math. In general, these basic skills will only prepare such students to compete against workers in the Third World for Third World wages, yet such placement decisions are frequently made by the intermediate or middle school years. Once relegated to the basic math track, students seldom have the opportunity to participate in "college prep" math courses. While research on tracking consistently shows advantages for gifted students, it is equally consistent in showing disadvantages for every other group of students in our schools.

The problem of tracking is compounded by the fact that women and minorities are more frequently referred to lower tracks in mathematics. In addition, master teachers are often given the opportunity to teach advanced tracks, thus leaving the less experienced or less skilled to teach those students least prepared to understand the concepts to be taught.

Educational trends which are currently at odds with the practice of tracking include mainstreaming of mildly or moderately handicapped youngsters into the regular classroom and the focus on cooperative learning which encourages heterogeneous grouping within the classroom. Increased demands to incorporate cultural awareness and acceptance of differences will further complicate the practice of tracking which currently predominates.

Implications for the classroom teacher:

- As schools move away from tracking in providing for the needs of individual students, classrooms become more heterogeneous. The classroom teacher must assume increased responsibility for assessing and meeting the diverse needs of individual students.

- While vertical acceleration through the sequence of courses or goals has been encouraged by the practice of tracking, enrichment in heterogeneously grouped classrooms may be more appropriately provided through extensions of learning based on a single goal – horizontal enrichment. In that way, individual student needs can be met while allowing the teacher to focus whole group instruction on a single concept.

- When handicapped students or students who speak English as a second language are placed in the mainstreamed classroom, the teacher must learn to work closely with the special education consultants and with language tutors in planning for the special needs of these students. Such students are not incapable of learning the required content; they just need different kinds of instructional assistance to achieve their goals.

- As the focus shifts to providing a common core of mathematics instruction for all students, teachers are challenged to:
  - know what comprises the core of mathematics content appropriate for students at that grade level;
  - accepts responsibility for the education of a significantly larger fraction of the population;
  - stimulate able students with the excitement and challenge of mathematics while encouraging less able students to set and meet high standards;
  - differentiate instruction by approach and speed, not by varying curricular goals or by advancement to subsequent courses.

As you think about dealing with the diverse population of a mainstreamed classroom, ask yourself:

1. Why do able students become more “remedial” when placed in a remedial program?
2. What messages are communicated to students who are placed in high level classes? In low level classes?
3. How can I establish fair and consistent grading practices which provide meaningful feedback to individual students?
4. How might students benefit from a wide range of classroom assessment techniques?
5. How can I meet the needs of students in my class who may be students acquiring English? at-risk students? gifted students?
6. How can I insure that my class is gender fair?
7. How can I create a classroom atmosphere where diversity is accepted and celebrated?

15 Geometry Using Coordinates

THEME: Using Coordinates in Geometry

HANDS-ON ACTIVITIES

Your study of geometry thus far has been without the use of coordinates; that is, you have drawn sketches of shapes, without specifying their location in the plane. In this chapter, you will investigate a two-dimensional coordinate system, called the Cartesian coordinate system. Using this coordinate system gives you a way to communicate precisely about geometric objects, such as points, lines, and polygons, because every point in the plane can be named using a pair of coordinates that are referenced to two axes. The use of a coordinate system will also help you make connections between algebra and geometry as you explore the relationship between the equation of a line and its graph in a coordinate plane.

As an elementary teacher, you will help to lay the groundwork your students will need later in high school for their further study of equations and graphs. The activities in this chapter introduce some of the elementary-level concepts related to the Cartesian coordinate system that you may teach, including the distance formula and slope. The following example illustrates how the use of a coordinate system will help your future students to expand the way they are able to reason about shapes. Consider the quadrilateral *ABCD* shown on the coordinate system below, with vertices at (-2, 4), (2, 3), (1, 0), and (-3, 1).

Using the idea that the slopes of parallel lines are equal, you can determine whether this quadrilateral is a parallelogram.

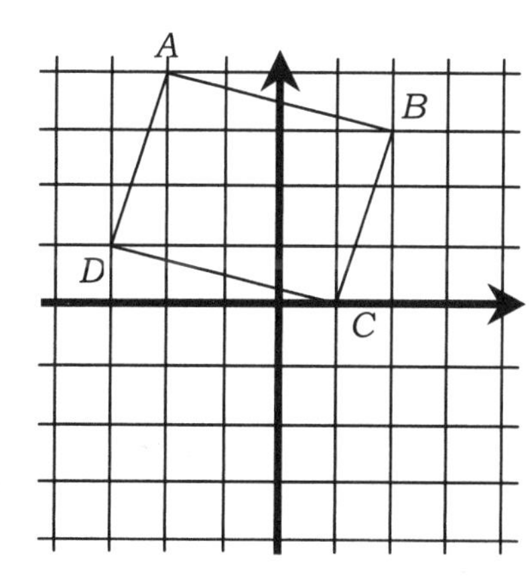

Using the idea that the product of the slopes of perpendicular lines is -1, you can determine whether this quadrilateral is a rectangle.

OBJECTIVE: **Illustrate a coordinate system on a square lattice and develop the distance formula**

You Will Need: **(eManipulative option: *Coordinate Geoboard*)**

In previous chapters, you have worked with a geoboard or a square lattice to model geometric shapes. This manipulative can also be a useful starting point for investigating the Cartesian coordinate system. Additionally, you will apply the Pythagorean theorem to develop the formula for the distance between two points in the plane.

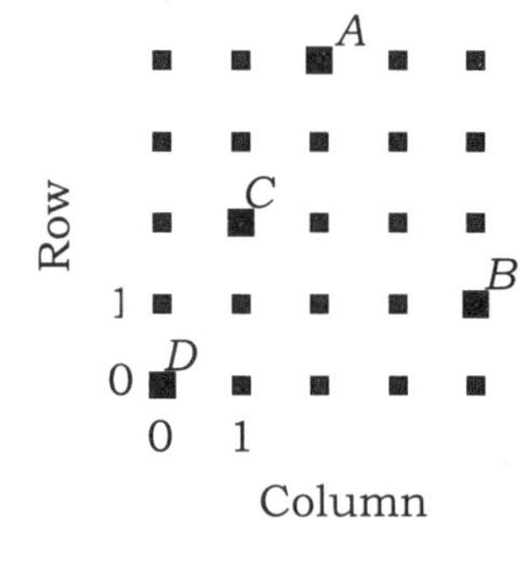

1. The lettered points in the array pictured at the right can also be identified using coordinates.

   First, continue to number the columns from left to right.

   Then, continue to number the rows beginning with the bottom row.

   A point can be identified by giving its column number and its row number, in that order; for example, point A has coordinates (2, 4).

   Find the coordinates of the following points. a. B b. C c. D

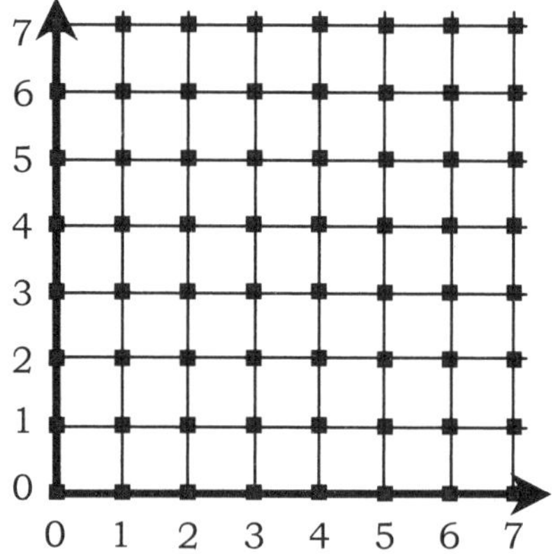

2. Connecting the points in each row and column yields the grid pattern to the right. The grid could be extended to include more points than the usual 5-by-5 geoboard as illustrated.

   a. Plot the following points on the grid at the right and label each one.

      E (1, 2), F (4, 2), G (6, 5), and H (3, 5)

   b. Connect the points you plotted in part (a) to form segments $\overline{EF}$, $\overline{FG}$, $\overline{GH}$, and $\overline{HE}$. What type of quadrilateral have you drawn?

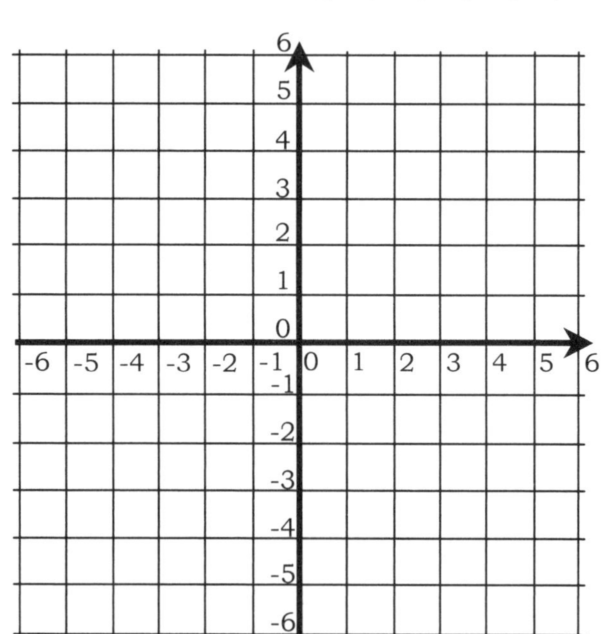

3. By extending the grid to include the negative integers (as shown at the right), you get the **Cartesian coordinate system**, named in honor of the mathematician René Descartes.

   The two darkened lines that are perpendicular to each other are called the axes (singular is axis). The horizontal axis is usually called the ***x*-axis** and the vertical axis is usually called the ***y*-axis**. They meet at a point called the **origin**, which has coordinates (0,0).

   a. Plot and label the points A (5, 2), B (3, -1), C (-2, 5), D (1, -4), E (-3, -4), and F (-6, -2)

b. The axes divide the plane into 4 regions, called **quadrants**. These are numbered I to IV, counterclockwise, beginning in the upper right-hand region. Identify the quadrant that each of the points from part (a) is in.

c. In quadrant I, both coordinates are positive. What do the coordinates of all the points in quadrant II have in common? in quadrant III? in quadrant IV?

4. In the next problem, you will develop a formula for computing the distance between two points in the plane.

   a. Plot points A (1, 5) and B (6, 1) in the coordinate plane at the right.

   b. $\overline{AB}$ is the hypotenuse of right triangle ABC. What are the coordinates of point C (there are two possibilities)?

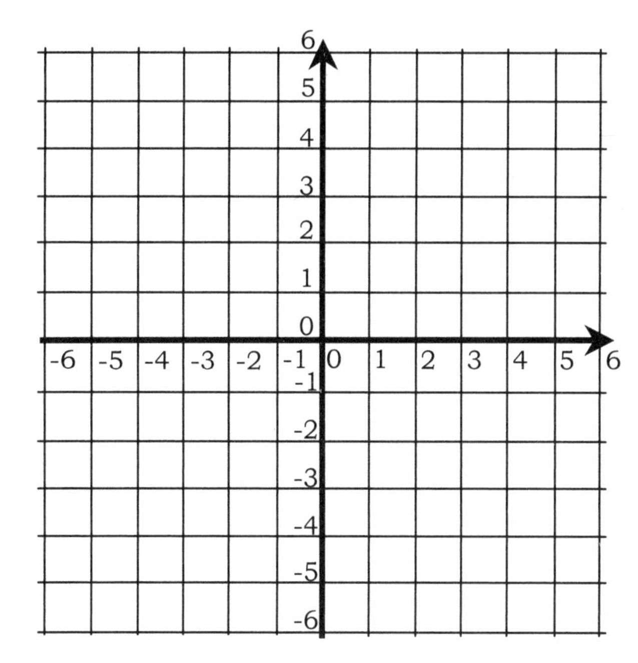

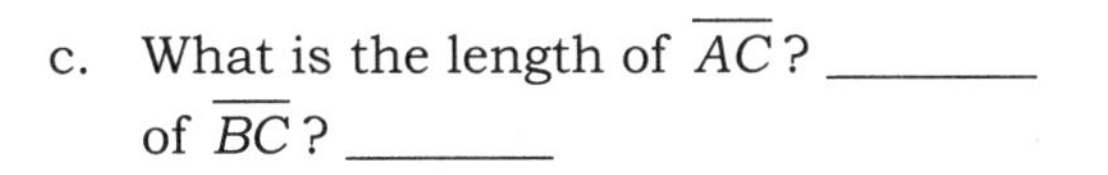

   c. What is the length of $\overline{AC}$? ________
   of $\overline{BC}$? ________

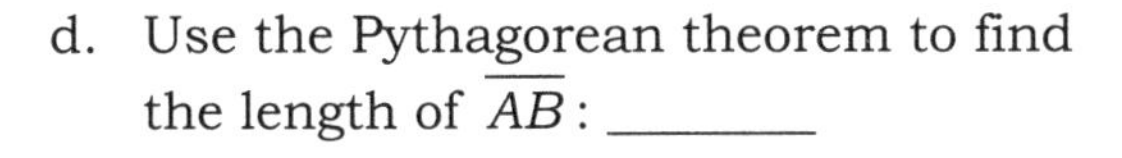

   d. Use the Pythagorean theorem to find the length of $\overline{AB}$: ________

   e. Using the same coordinate plane, repeat parts (a) – (d) to find the length of the segment with endpoints D (1, 1) and E (-4, -3). Show your process.

5. Consider the right triangle shown at the right.

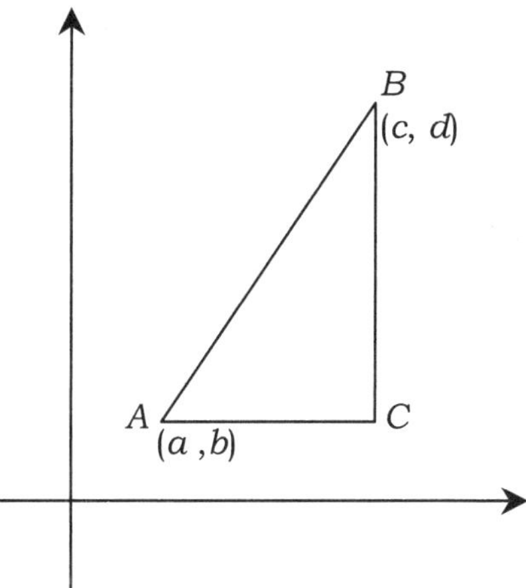

   a. What are the coordinates of point C? ________

   b. What is the distance between A and C? ______
   between B and C? ________

   c. Use the Pythagorean theorem to write an expression for the distance between A and B.

   d. The expression you wrote in part (c) is called the **distance formula**. Use your formula to compute the distance between the points D (-2, 6) and E (1, -5).

6. Consider the array from problem #1.

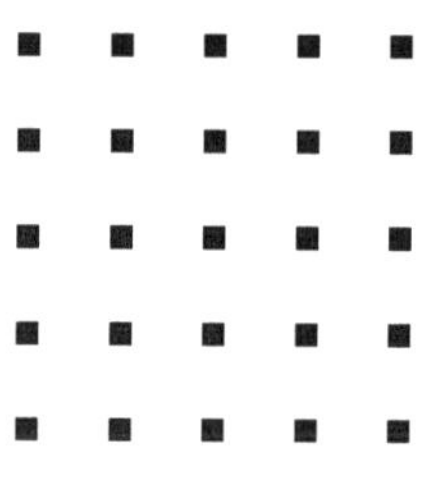

   a. On the array in #1, how many different possible distances are there between pairs of points? _______ Explain how you obtained your answer.

   b. All 'vertical' and 'horizontal' distances are whole number lengths. Are any of the diagonal lengths whole numbers? _______ How do you know?

   c. Which of the possible distances are rational numbers and which are irrational?

OBJECTIVE: **Investigate slope of a line**

You Will Need: **Geoboard (optional)**
(eManipulative option: *Coordinate Geoboard*)

Slope is a number that indicates the steepness of a line. In this activity, you will use geoboard line segments to investigate the concept of slope. Also, you will develop the slope formula using a coordinate plane.

1. The **slope** of any non-vertical line is the ratio of the vertical change (called the rise) to the horizontal change (called the run) between two points on the line.

   Run = 2
   Rise = 3
   A B

   Consider the line segment $\overline{AB}$ shown on the array at the right. To move from point A to point B, you could go up three units and right two units. Thus the slope of $\overline{AB}$ is 3/2.

   Notice, if you move from point B to point A, you could go down 3 units, a vertical change of -3, and left 2 units, a horizontal change of -2. Computing the slope with this method, you get -3/-2, which is equal to 3/2.

   a. Find the slope the line segment from point A to point C shown at the right. Record your work by labeling the rise and run in the picture.

   A C

   Slope of $\overline{AC}$ = ________

   b. Now choose another pair of points on the segment $\overline{AC}$. What is the slope between your chosen points?

   Choose another pair of points on $\overline{AC}$ and compute the slope. What do you notice?

   This example illustrates an important idea about slope: the **slope of a line is constant**; that is, the ratio of rise to run is the same for any pair of points on the line.

   c. Compare the two line segments $\overline{AB}$ and $\overline{AC}$ shown in the example above and in part (a). Which of these is steeper?

   On the array shown at the right, draw two different non-horizontal and non-vertical line segments, $\overline{AD}$ and $\overline{AE}$, so that $\overline{AD}$ is steeper than $\overline{AB}$ and $\overline{AE}$ is not as steep as $\overline{AC}$. Compute the slope of each of your segments.

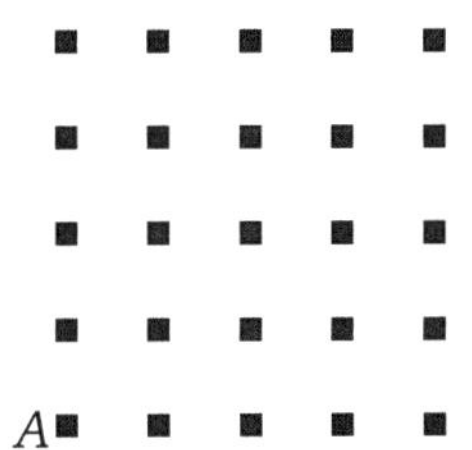

   Slope of $\overline{AD}$ = ________ Slope of $\overline{AE}$ = ________

   Write the slopes of the four line segments $\overline{AB}$, $\overline{AC}$, $\overline{AD}$, and $\overline{AE}$ in increasing order.

   What does the slope of a line have to do with its steepness?

2. Consider the line segment shown at the right.

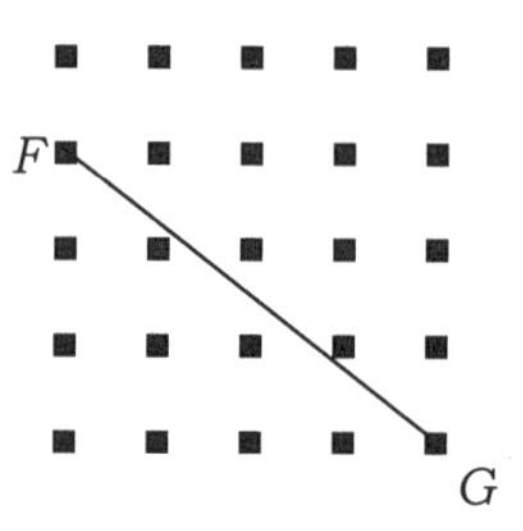

   a. Remembering that a downward movement is recorded as a negative vertical change, compute the slope from F to G. Record your work by labeling the rise and run in the picture.

      Slope of $\overline{FG}$ = ________

   b. To compute the slope from G to F, what must you remember?

   c. The four segments from problem #1 all had positive slopes and the slope of $\overline{FG}$ from part (a) is negative. What is the difference between a line segment with a positive slope and a line segment with a negative slope?

3. Horizontal and vertical lines present special cases with regard to slope.

   a. Consider the horizontal line segment $\overline{HI}$ shown at the right. Find the rise and the run from point H to point I.

      $\frac{\text{rise}}{\text{run}}$ = ____

      Since the rise between any two points on a horizontal line will always be 0, the slope of a horizontal line is _____.

   b. The picture at the right shows a vertical line segment, $\overline{JK}$. What is the rise from J to K? _____ What is the run from J to K? _____

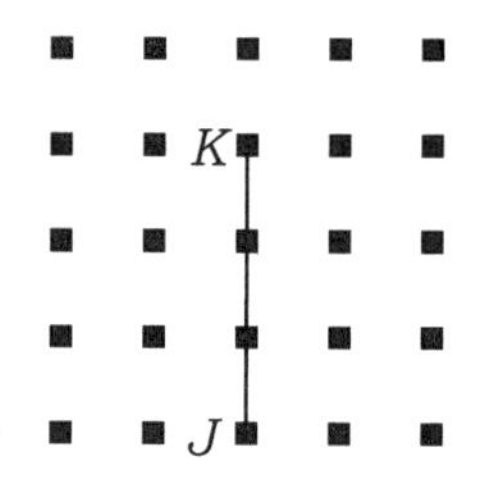

      Since the run between any two points on a vertical line will always be 0, the rise over run will be $x/0$, for some real number x. Since $x/0$ is undefined, the slope of a vertical line is said to be undefined.

      Another common way to describe a vertical line is to say that it has "no slope;" this is because there is no real number that describes the steepness of a vertical line.

4. By specifying the location of a line in a coordinate plane, you can develop a formula for computing slope, given the coordinates of two points on the line.

   a. In the coordinate plane at the right, draw the line containing the points (4, 3) and (1, 2).

   b. Label the rise and run between these two points.

      What is the slope of the line? _____

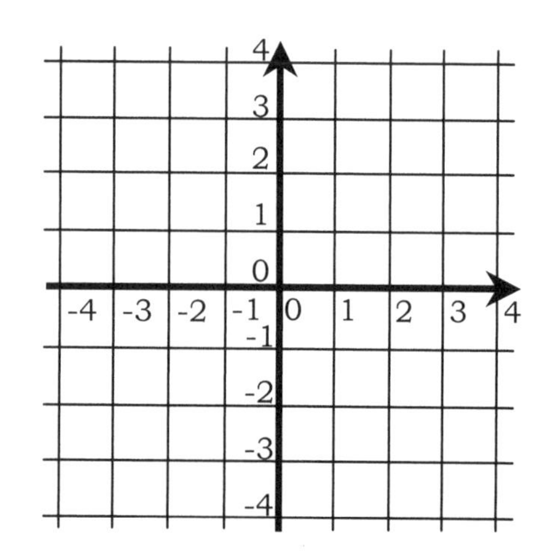

c. The line shown in the coordinate plane at the right contains the points $A\ (x_1,\ y_1)$ and $B\ (x_2,\ y_2)$.

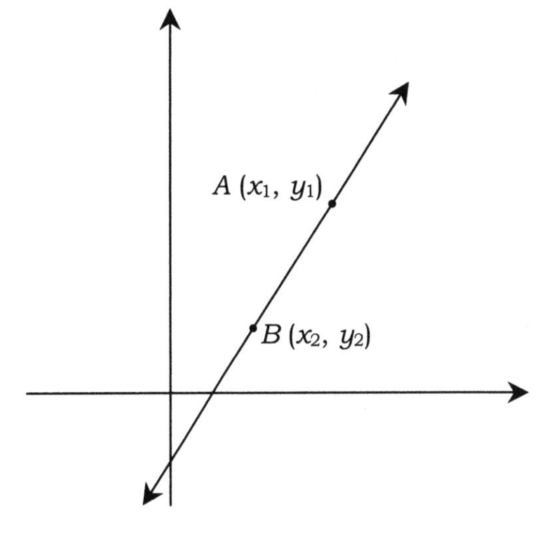

What is the rise from *A* to *B*? ___________

What is the run from *A* to *B*? ___________

What is the slope of $\overline{AB}$? ___________

d. What is the rise from *B* to *A*? ___________

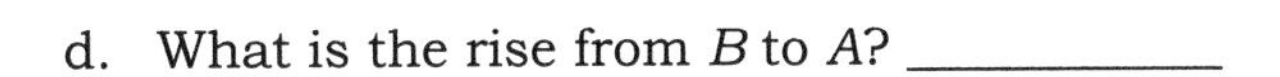

What is the run from *B* to *A*? ___________

What is the slope of $\overline{BA}$? ___________

e. Compare the answers you obtained for the slope in parts (c) and (d). What do you observe?

f. Summarize your work in parts (c) - (e):

If *A* is the point $(x_1,\ y_1)$ and *B* is the point $(x_2,\ y_2)$, then the slope of the line containing points *A* and *B* (where x_1 is not equal to x_2) is:

This result is called the **slope formula** for a line.

OBJECTIVE: **Investigate slopes of parallel and perpendicular lines**

You Will Need: **Materials Card 15.1.3 or graph paper (eManipulative option: *Coordinate Geoboard*)**

In the next activity, you will continue your investigation into the concept of slope by considering the special relationship between the slopes of parallel lines and perpendicular lines.

1. Consider line l shown on the coordinate plane at the right.

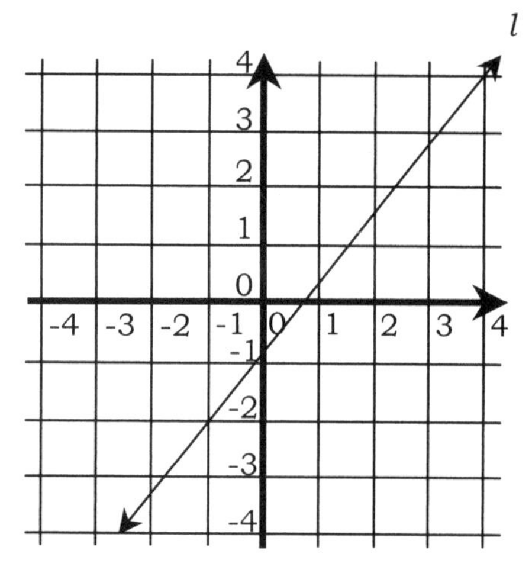

   a. Use the slope formula you developed in Activity 15.1.2 to compute the slope of line l (this may be easier if you use points with integer coordinates).

      Slope of line l = _____

   b. Sketch another line in the same coordinate plane that has the same slope as line l. Label this line m. Explain how you got your answer.

   c. Sketch two more lines, n and p, that have the same slope as l and m. What do you notice about these four lines?

2. Problem #1 illustrated that lines with equal slopes are parallel. In this problem, you will find a relationship between the slopes of perpendicular lines.

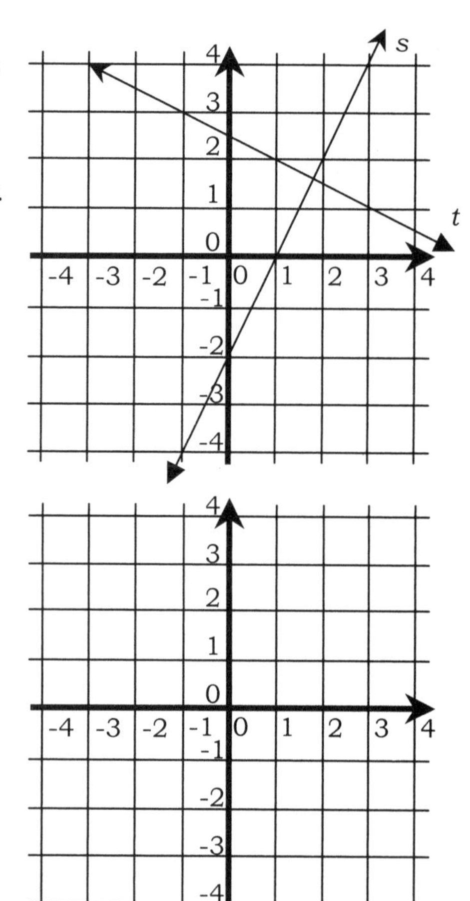

   a. Lines s and t shown at the right are perpendicular. Find the slope of each line.

      Slope of line s = _____ Slope of line t = ______

   b. What relationship do you notice between the slopes of perpendicular lines?

   c. Verify visually that your answer in part (b) is correct by sketching the graphs of two lines, one with a slope of -3/4 and one with a slope of 4/3. Do your lines look perpendicular?

   d. If the slope of a line is -3/4, what is the slope any line that is perpendicular to this line? _______

3. The relationships between the slopes of parallel lines and perpendicular lines can be useful when you are solving problems related to geometric shapes in a coordinate plane, such as the following.

   a. Give the coordinates of three points A, B, and C in the coordinate plane so that ΔABC is a right triangle *and* none of the sides of the triangle are horizontal or vertical. Use slopes to explain how you know your triangle is a right triangle.

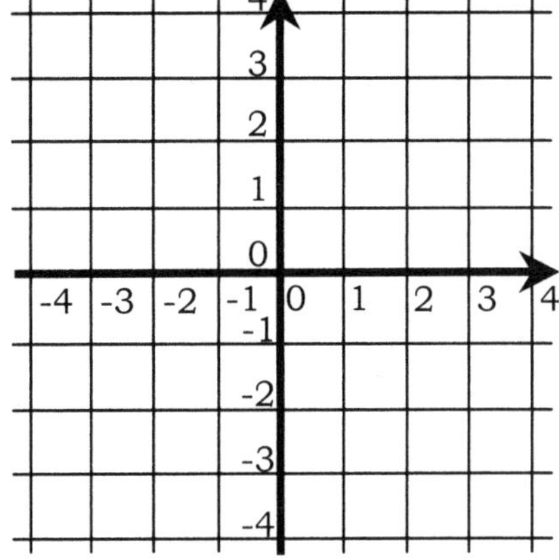

   b. The distance formula, which you developed in Activity 15.1.1, together with the converse of the Pythagorean theorem could also be used to prove that the triangle you drew in part (a) is a right triangle. Show how to do this.

4. Consider the quadrilateral shown at the right. Explain how to solve each of the following problems, using slopes.

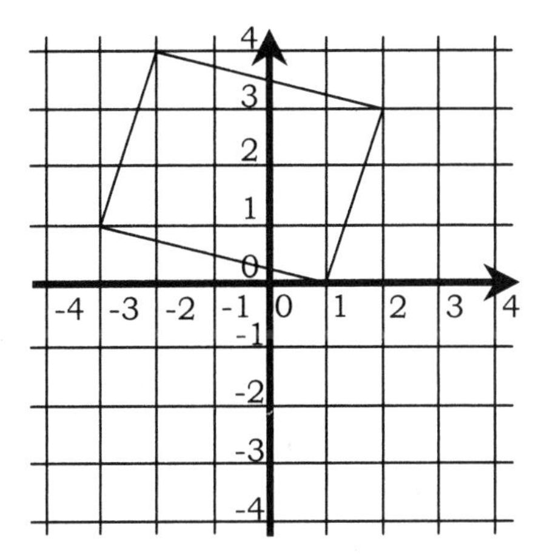

   a. Determine whether the quadrilateral is a parallelogram.

   b. Determine whether the quadrilateral is a rectangle.

5. For a class activity, a teacher creates a treasure hunt for his students. The treasure is hidden in the school playground and the students are given the following clues about its location.

   The treasure is at vertex Z of a parallelogram. The three other vertices of the parallelogram are located in the following ways: vertex W is 2 meters east and 3 meters north of the flagpole; vertex X is 5 meters east and 7 meters north of the flagpole; and vertex Y is 10 meters east and 5 meters south of the flagpole.

   a. Using the grid on Materials Card 15.1.3, create a coordinate system to represent this problem. What would be a convenient location for the flagpole?

   b. Show how to solve the problem, using your coordinate system.

   c. Did you find more than one answer? If not, look back and try to find other possible locations for the treasure.

OBJECTIVE: **Investigate equations of lines geometrically**

You Will Need: **A straightedge**
(eManipulative option: *Function Grapher*)

In this activity, you will investigate the relationship between the equation of a line and its graph. These connections show how geometry and algebra are related.

1. Consider the equation $y = 2x - 4$.

   a. To create a graph of this equation, you can find points in the form (x, y) that make the equation true. For each of the points below, the x-coordinate is given. By substituting x into the equation, find the corresponding y-coordinate.

   (0,) (2,) (4,) (1,) (3,) (-1,)

   b. On the coordinate system at the right, plot the points you found in part (a). What type of graph do you get as a result of connecting these points?

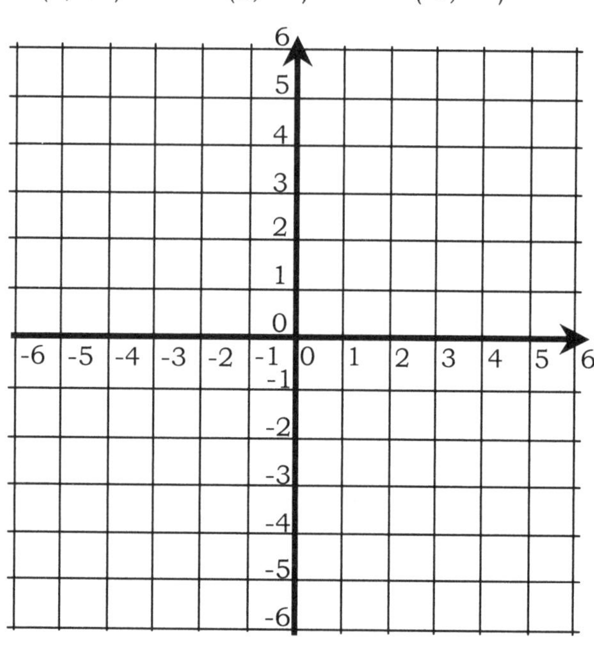

   c. What is the slope of the line you graphed in part (b)? ________

   Does this number appear in the equation? ________ Where?

   What are the coordinates of the point where the line crosses the y-axis?

   (,)

   The y-coordinate of this point is called the ***y*-intercept**. How is this point related to the equation for the line?

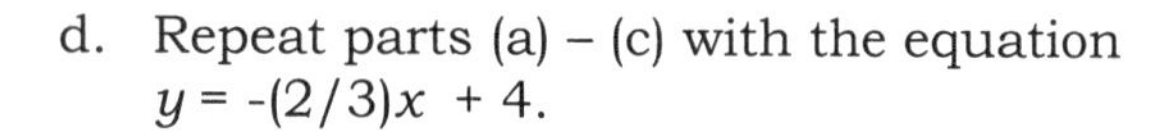

   d. Repeat parts (a) – (c) with the equation $y = -(2/3)x + 4$.

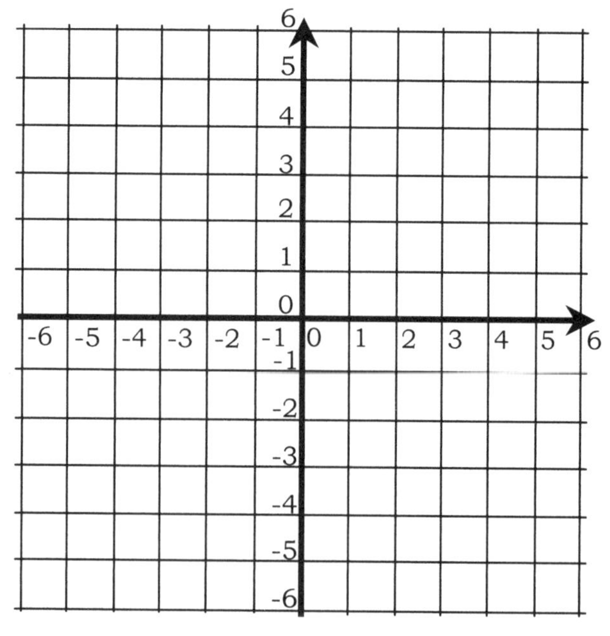

e. In general, the equation of any non-vertical line may be written in the form $y = mx + b$.

What does the value of m appear to represent?

What does the value of b appear to represent?

The equation $y = mx + b$ is the **slope-intercept form** of the equation of a line.

2. In problem #1, you plotted several points satisfying the equation of a line. Since two points are all that's needed to determine the graph of a line, the slope-intercept form can also be useful for graphing a line.

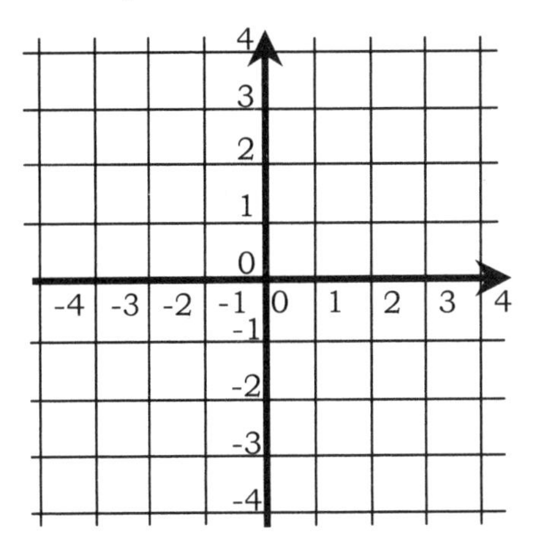

As an example, consider the equation $y = (3/4)x + 2$.

a. From the equation of the line, you know the y-intercept is: _____. Plot this point on the coordinate system at the right.

b. From the equation of the line, you know the slope is: ________. Show how to use the slope from the y-intercept to find another point on the line. Then, graph the line.

c. Check that your graph is correct by using the slope to find another point on the line.

3. Vertical lines, because their slopes are undefined, do not have equations that can be written in slope-intercept form.

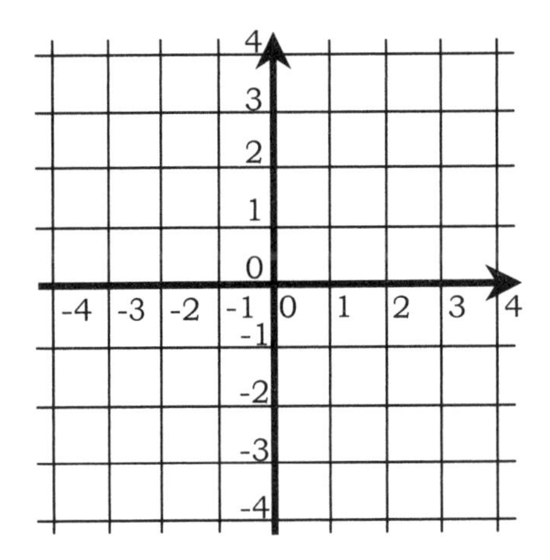

a. On the coordinate system at the right, graph the vertical line through the point (3, 2).

b. What do you observe about the x-coordinate of each point on this line?

Write an equation to express this.

c. In general, a vertical line passing through a point (a, b) can be represented by the equation: __________

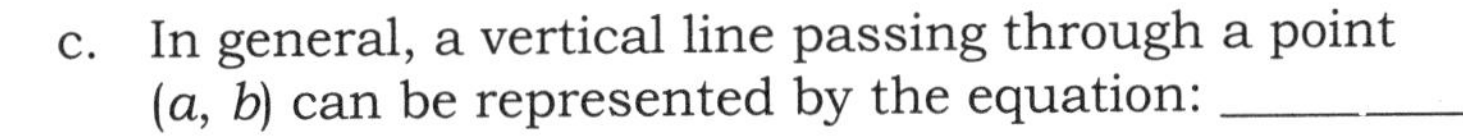

4. Horizontal lines also have equations that look unusual.

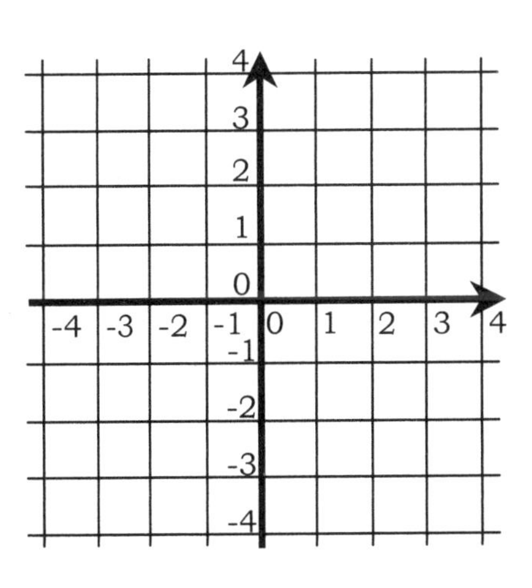

a. On the coordinate system at the right, graph the horizontal line through the point (3, 2).

b. What do you observe about the y-coordinate of each point on this line?

Write an equation to express this.

c. In general, a horizontal line passing through a point (a, b) can be represented by the equation: __________

d. Is the equation you wrote in part (c) written in slope-intercept form? ________ Explain your answer.

e. Is the equation you wrote in problem #3(c) written in slope-intercept form? Explain.

5. In problems #1 and #2, you were given the equation of a line and you graphed the equation, either by the point-plotting method or by using the slope-intercept form of the equation. It is also possible to do the reverse; namely, write the equation of a line, given its graph.

   For example, determining the equation of a line is easy if you know the slope and the y-intercept of the line – you just substitute these values into $y = mx + b$. If either or both of these values are not known, writing the equation of the line takes a few more steps.

   a. Consider the graph of the line shown at the right.

   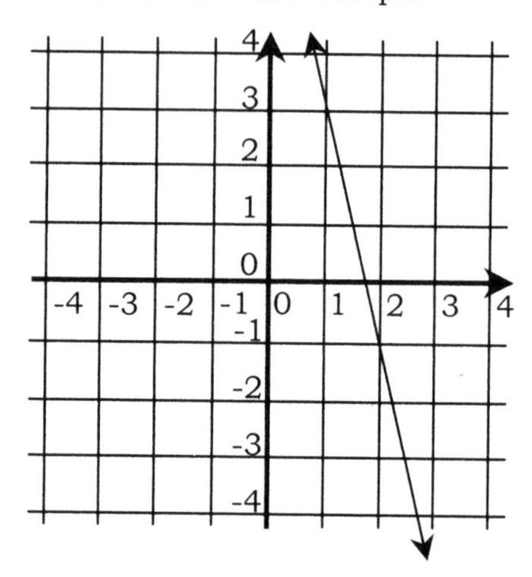

   What is the slope of the line?

   m = ________

   So far, in the equation $y = mx + b$, you have the value of m. Thus, you know $y = -4x + b$.

   b. The graph of the line does not show the y-intercept, so you must find this algebraically. One way to do this is to substitute a point from the graph into the equation $y = -4x + b$, and then solve for b. Choose one of the points on the line, say (2, -1) and find the value of b.

   b = ________

   Based on the graph of the equation, does the value of b that you found seem reasonable? ________

   c. Write the equation of the line in slope-intercept form:____________________

EXERCISE

SYMMETRY THROUGH COORDINATES

A.

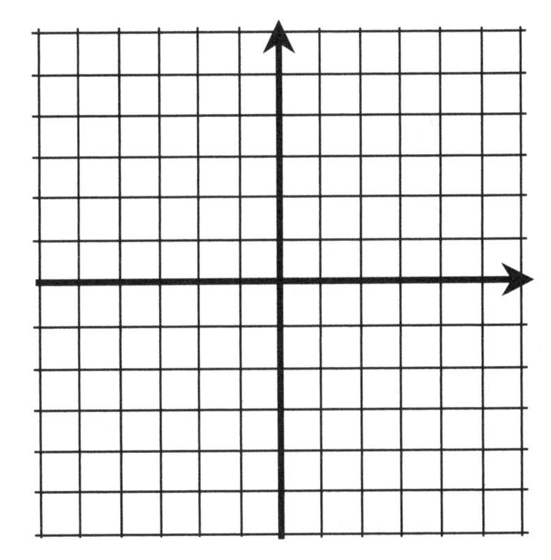

1. Graph the points and form ΔABC.

   A (3, 5) B (4, 1) C (2, -1)

2. Multiply each x-coordinate by -1 and form $\Delta A'B'C'$.

   A' (, 5) B' (, 1) C' (, -1)

3. The two triangles *together* have what kinds of symmetry?

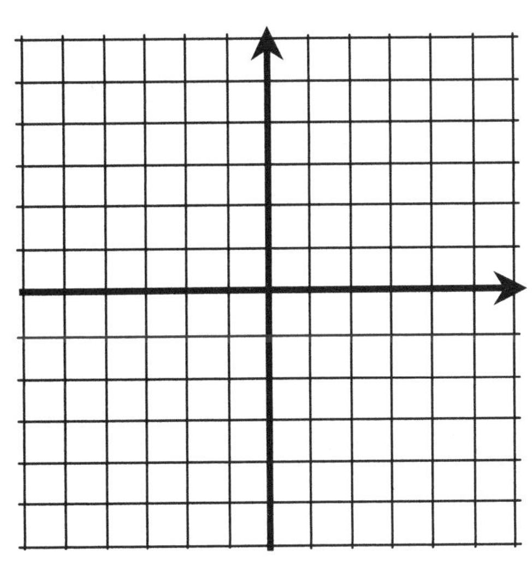

B.

1. Graph the points and form quadrilateral $ABCD$.

   A (-2, 4) B (2, 5) C (5, 1) D (-4, 0)

2. Multiply each y-coordinate by -1 and form quadrilateral $A'B'CD'$.

   A' (-2,) B' (2,) C' (5,) D' (-4,)

3. The two quadrilaterals together have what kinds of symmetry?

C.

1. Graph the following points and form segments $\overline{HI}, \overline{IJ}, \ldots, \overline{NH}$.

   H (0, 4) I (1, 2) J (3, 2) K (1, 0)
   L (-1, 0) M (-3, 2) N (-1, 2)

2. Multiply all coordinates by -1 and form $H'I'J \ldots N'$.

   H' (,) I' (,) J' (,) K' (,)
   L' (,) M' (,) N' (,)

3. The resulting figure has what kinds of symmetry?

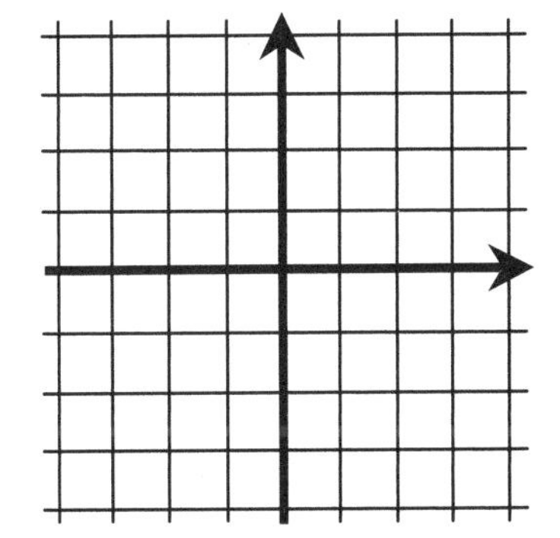

CONNECTIONS TO THE CLASSROOM

1. When asked to find the slope of the line containing the points (3, -2) and (-4, -5), two students, Hayes and Farzaneh, showed the following work. In each case, describe the student's error and explain how you could help the student.

Hayes: $\text{slope} = \dfrac{3-4}{-2-5} = \dfrac{-1}{-7} = \dfrac{1}{7}$

Farzaneh: $\text{slope} = \dfrac{3+4}{-5+2} = \dfrac{7}{-3} = -\dfrac{7}{3}$

2. A student says, "A horizontal line has no steepness, so it has no slope."

   a. Is the student's statement correct? Why or why not?

   b. How could you help this student understand the difference between "no slope" and "a slope of zero?"

3. A student graphed the lines given by the equations $y = -4x + 1$ and $y = (1/2)x - 3$ and said, "I know that the greater the slope, the steeper the line. Since 1/2 is greater than -4, the line for the equation $y = (1/2)x - 3$ should be steeper than the line for $y = -4x + 1$. In my graph, the one with a slope of -4 looks steeper. I think I must have made a mistake." How would you respond?

4. A student graphed the horizontal line and vertical line shown at the right, and wrote the equation for each line. He says, "Since the horizontal line is parallel to the x-axis, its equation is $x = 2$. Also, since the vertical line is parallel to the y-axis, its equation is $y = 3$."

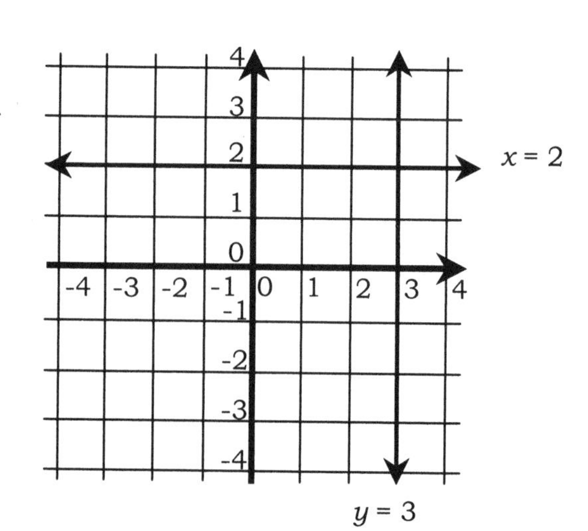

   a. Is the student correct? Why or why not?

   b. How could you help this student understand the equations for horizontal and vertical lines?

MENTAL MATH

Identify the line on the graph that matches each equation below.

1. $y = 2x + 1$
2. $x = -2$
3. $y = (-2/3)x + 1$
4. a line parallel to $y = (1/4)x + 5$
5. a line whose slope is zero
6. a line perpendicular to $y = (-2/3)x - 3$

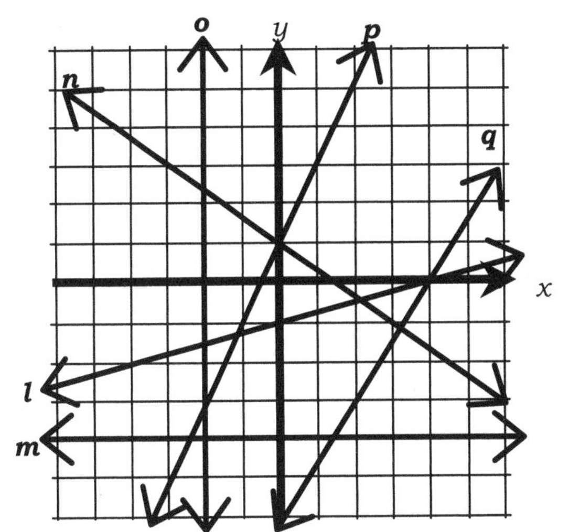

DIRECTIONS IN EDUCATION
Parental Involvement
Partnership With Parents

Parents are a precious resource for encouraging student participation and achievement in mathematics. Schools must find meaningful ways in which to involve parents in the learning process. This involvement may take various forms. Parents may participate directly in the schools in the design of curriculum or in delivery of instruction either as paid classroom assistants or as volunteers. Parents may actively participate in the linking of home and school activities by being aware of assignments and topics currently under study, by providing opportunities to link that study to the child's world, or by setting expectations for achievement in school. In addition, parents may participate indirectly by providing a mathematics-enriched environment in the home to promote learning about mathematics beyond the school setting. Parents have hopes and dreams for their children. It is the task of the teacher to help parents see the classroom as a place where those dreams are enhanced and supported. Involved parents can provide for continuity of experiences as children move from classroom to classroom, from school to school, or from town to town.

Parental involvement in the schools is enhanced by:

- Clear statements about the nature of volunteer work asked of the parent.
- Keeping open lines of two-way communication.

- Allowing parents to work with children rather than giving them mundane paperwork tasks such as cutting or correcting.

- Sharing news of student successes as well as areas requiring further attention.

- Providing resources which parents can use at home to enrich and extend classroom learning.

- Recognizing and rewarding parental involvement either informally or formally. Personal thank-you notes, certificates, or a recognition ceremony at school are ways to demonstrate that parents are appreciated in your classroom.

- Pointing out to parents negative attitudes or stereotypes about mathematics which may be unconsciously transmitted to children.

- Providing information to parents which will enable them to feel more knowledgeable as they work with their students. Parents, like teachers, do not wish to seem uninformed in front of their children.

- Sharing information about the use of instructional techniques which may be unfamiliar to parents. It is always easier to support something when it is better understood.

Advantages of parental involvement:

- Misconceptions about schools which may appear in the media can be corrected.

- Parental self-esteem is enhanced, thus contributing to the enhancement of student self-esteem.

- Consistent expectations at school and home encourage student understanding of and compliance with those expectations.

- Parents can better anticipate and support future growth and development of mathematical understanding.

- Positive attitudes toward mathematics can be linked with pleasant experiences within the family through games and other social activities which promote mathematical understanding.

- Parental support of schools will be communicated in their places of work and in the community at large.

- Teachers form strong partnerships with parents which foster positive home/school relationships and reduce confrontational situations.

As you think about involving parents in your classroom, ask yourself:

1. What do parents consider to be "meaningful" involvement in their child's classroom?

2. How can I clarify for parents the opportunities for involvement in my classroom?

3. How can I encourage more parental involvement in my classroom?

16 Geometry Using Transformations

THEME: **Using Geometric Transformations**

HANDS-ON ACTIVITIES

This final chapter introduces geometric transformations and their properties. Formally, a transformation is a function that maps each point in the plane to a unique point, called its image. Informally, you can think of the image of a geometric figure as the result of moving or changing the original figure. Transformations that move a shape without changing its size and shape are called isometries; these include slides, turns, and rotations. Transformations that shrink or enlarge a shape in the plane without changing its shape are called size transformations. Transformations are related to many of the concepts from the previous four chapters, including reflection and rotation symmetry, tessellations, and congruence and similarity of triangles. The examples below show how transformations may be related to tessellations.

By sliding a copy of this parallelogram, you can obtain the figure shown at the right.

By sliding copies of this new figure horizontally, you obtain the tessellation shown at the right.

By sliding a copy of this parallelogram and then reflecting the copy over a vertical line, you obtain the figure shown at the right.

By sliding copies of this new figure horizontally, you obtain the tessellation shown at the right.

Transformations will allow you to give more mathematically precise meaning to these concepts, as well as new ways to view shapes in the plane and their relationships to one another. The hands on activities in this chapter will involve many of the techniques for introducing the ideas of transformations in the elementary classroom, including tracings and paper-folding.

OBJECTIVE: **Investigate isometries and their properties**

You Will Need: **Tracing paper (eManipulative options: *Reflection, Translation,* and *Rotation Transformations*)**

A geometric transformation involves moving or changing a given shape in the plane. In the first activity, you will investigate three types of transformations – translations, rotations, and reflections; informally, these are called slides, turns, and flips, respectively. These three transformations are called "isometries" because they involve moving a geometric object in the plane, while preserving its size and shape.

1. A **translation** is a transformation associated with a sliding motion of a specified distance and direction, without turning. The distance and direction may be indicated with an arrow, called a **directed line segment**. Follow the steps given in parts (a) – (d) to translate the given point P.

   a. The arrow shown at the right indicates a translation. Extend the arrow in both directions by drawing a dashed line $\overleftrightarrow{AB}$.

   b. Lay a piece of tracing paper on the figure in part (a) and trace the dashed line, point A (the "tail" of the arrow), and the point P you want to translate.

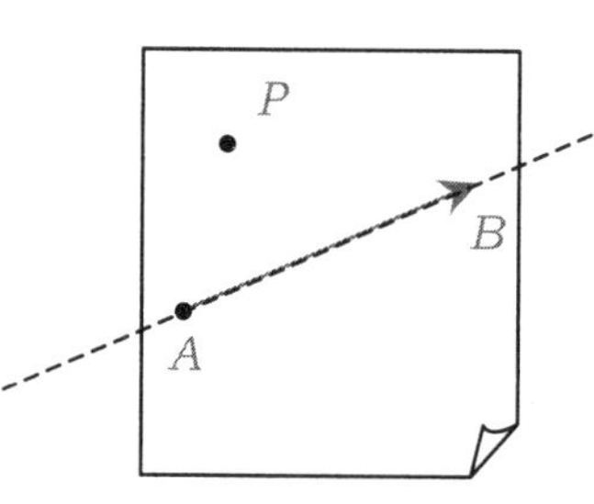

   c. Slide the tracing paper along the arrow until the traced point A is at the head of the arrow (on top of point B). Be sure that the dashed line is still on top of itself.

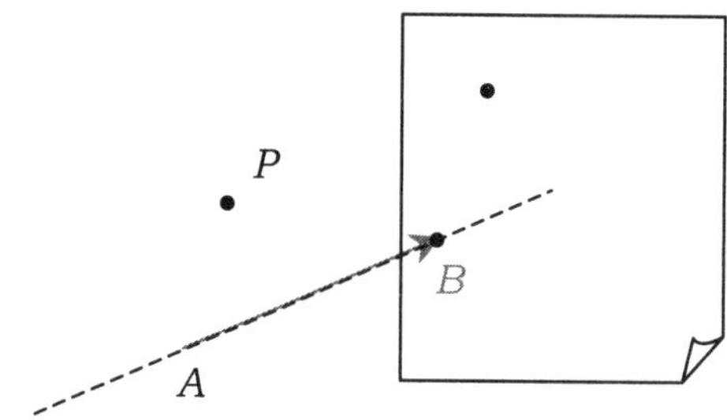

   d. With your pencil, press an indentation through the tracing paper at point P. Remove the tracing paper and, at the mark, label the point P'. The point P' is the **translation image** of point P.

2. Finding images of transformations using tracings provides a good visual model. Imagine the tracing paper is a copy of the plane. As you move the tracing paper, not only does point P correspond to its image P', but each point in the plane corresponds to a unique point.

   a. Using tracing paper, find the images of points X, Y, and Z under the translation indicated by the directed line segment $\overrightarrow{AB}$. Mark these points X', Y', and Z', respectively.

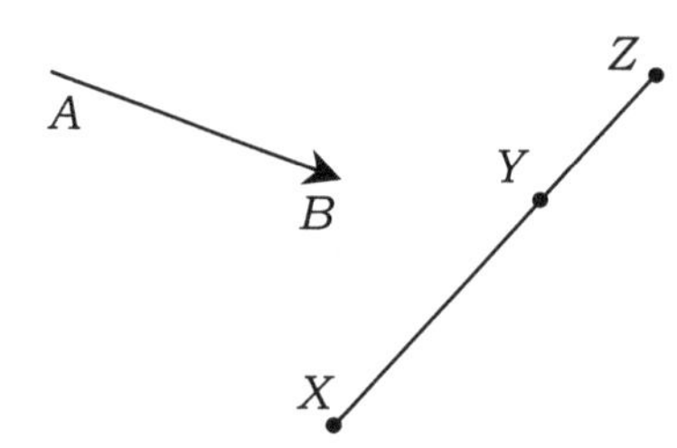

b. Draw the segment with endpoints X' and Z'. How does the image compare to $\overline{XZ}$?

c. Is Y' on $\overline{X'Z'}$?________

d. Choose the *most appropriate* way to complete the following statement:
The translation image of a segment is a

A. segment.
B. segment congruent to the original segment.
C. segment parallel to the original segment.
D. segment congruent and parallel to the original segment.

3. For the translation indicated by $\overrightarrow{AB}$ at the right, use tracing paper to find the images of the given angle, triangle, and ray.

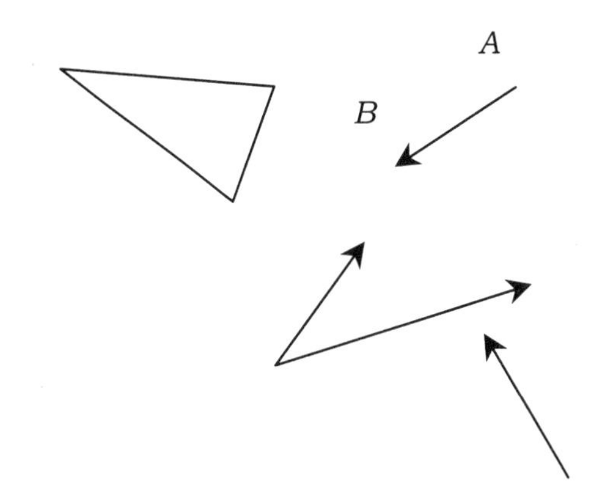

a. How do the measures of the angle and its image compare?

b. How do the triangle and its image compare?

c. What is the relationship between the ray and its image?

4. A **rotation** is a transformation associated with a turning motion through a specified **directed angle** around a fixed point, called the **center**. The center can be the vertex of the directed angle. Follow the steps given next to rotate point *P*.

a. The given angle and point *C* determine the rotation. Here the angle is drawn with its vertex at the center *C*.

Lay your tracing paper on top of the figure and trace the initial side of the angle, $\overrightarrow{CA}$, and point *P*.

b. Fix the center *C* with your pencil point and turn the tracing paper until the traced initial side of the angle coincides with the terminal side of the original angle, $\overrightarrow{CB}$.

c. With your pencil, press an indentation through the tracing paper to mark the **rotation image** of *P*. Remove the tracing paper and mark the image point P'.

d. Using dashed lines, draw the line segments $\overline{PC}$ and $\overline{P'C}$. How does the measure of $\angle PCP'$ compare to the original directed angle with vertex *C*?

e. How does the length of $\overline{PC}$ compare to the length of $\overline{P'C}$?

5. Using tracing paper, find the images of the given figures under the rotations indicated.

   a. How does the image of the line segment compare to the original line segment?

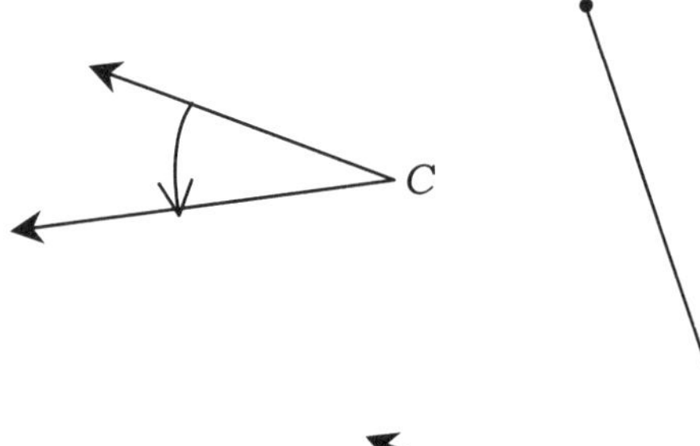

   b. How does the measure of the image of the angle compare to the original angle?

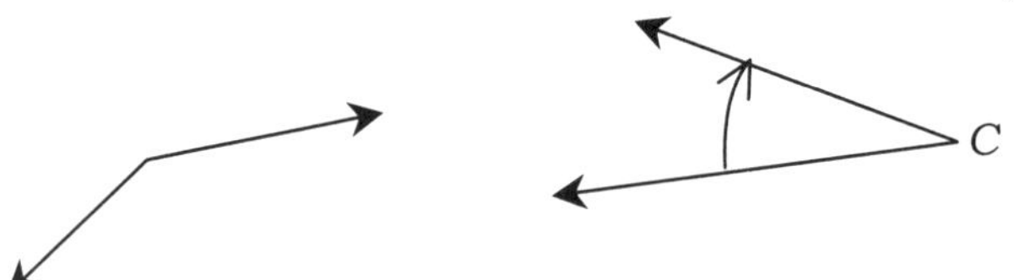

   c. How do the triangle and its image compare, with regard to shape and size?

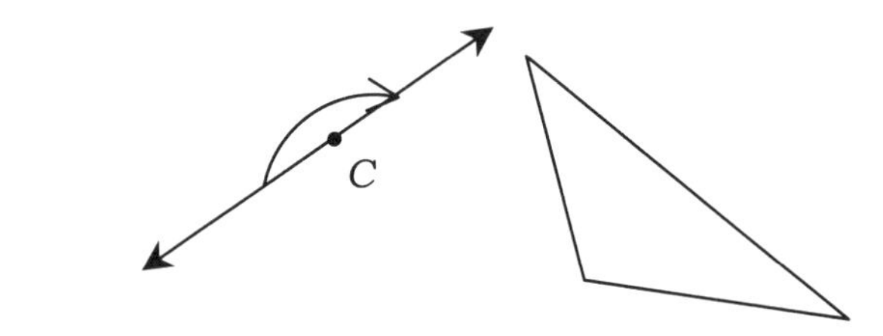

6. A **reflection** is a transformation that gives a "mirror" image across a line – the line is called the **reflection line**. Follow the steps given next to find the reflection image of point *P*, which is not on the reflection line, and the point *Q*, which is on the reflection line.

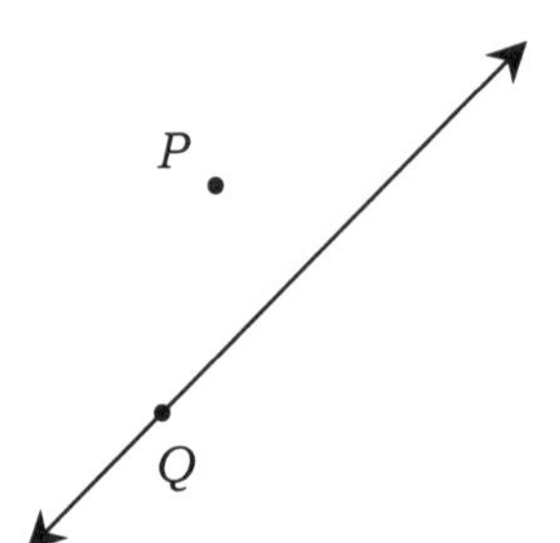

   a. The given line determines a reflection. Mark a reference point, *R*, on the line. Lay your tracing paper on top of the figure and trace the reflection line and the points *P*, *Q*, and *R*.

   b. Turn over the tracing paper so that it is face down. Be sure that the copy of the reflection line and *R* coincide with the original line and *R*, respectively.

   c. Using your pencil, make an indentation through the tracing paper to mark the reflection images of points *P* and *Q*. Remove the tracing paper and mark these image point *P*' and *Q*'.

   d. What do you notice about points *Q* and *Q*'?

   e. Using a dashed line, draw the segment $\overline{PP'}$. What appears to be the relationship between the reflection line and $\overline{PP'}$.

7. Using tracing paper, find the reflection images of the given figures under the reflection in line *l*.

a.

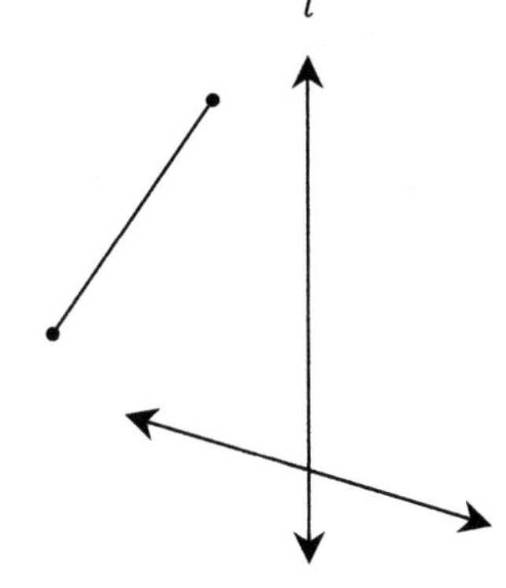

b.

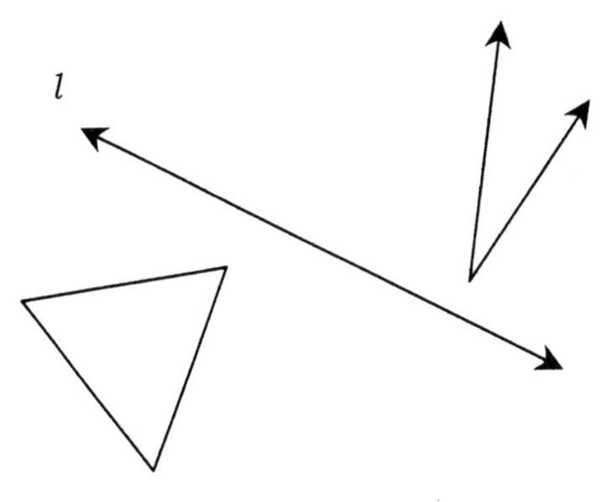

8. A **fixed point** is a point that corresponds to itself (i.e. it is its own image) after a transformation.

   a. Are there any fixed points under translation? _______ Explain your answer.

   b. Are there any fixed points under rotation? _______ Explain your answer.

   c. Are there any fixed points under reflection? _______ Explain your answer.

OBJECTIVE: **Investigate combinations of translations and reflections**

You Will Need: **Tracing paper (eManipulative option: *Compositions of Transformations*)**

In the next activity, you will explore transformations that are combinations of slides and flips. You will also discover when reversing the order in which you perform two transformations will produce the same results.

1. A **glide reflection** is a transformation that combines a translation and a reflection where the reflection line is chosen to be parallel to the translation direction. Follow the steps given next to find the glide-reflection image of point *P*.

   a. The given line *l* and the directed line segment $\overrightarrow{AB}$ determine a glide reflection. Notice that *l* is parallel to $\overrightarrow{AB}$.

   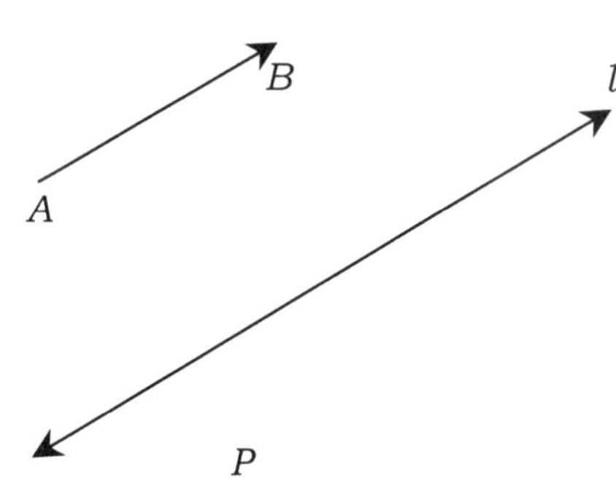

   Using your tracing paper and the techniques from Activity 16.1.1, first translate point *P* determined by the directed line segment $\overrightarrow{AB}$. Mark the translation image as *P'*. Then, reflect *P'* over line *l* and mark the reflection image as *P''*.

   b. Repeat the glide reflection in part (a), except do the reflection over line *l* first, followed by the translation. What do you notice?

2. Perform each of the following glide reflections that are determined by the directed line segment $\overrightarrow{AB}$ and the line *l*.

   a.

   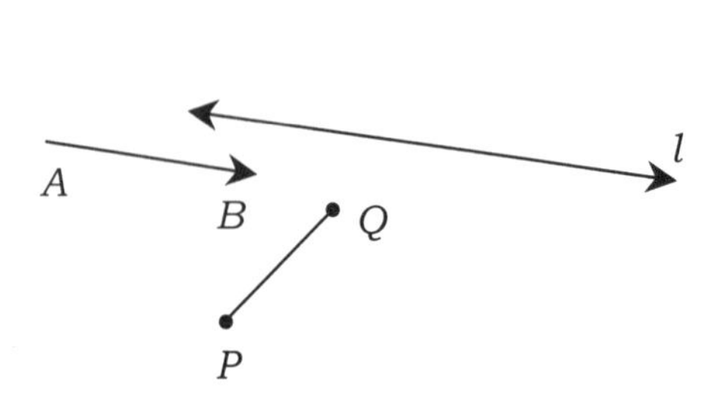

   b.

   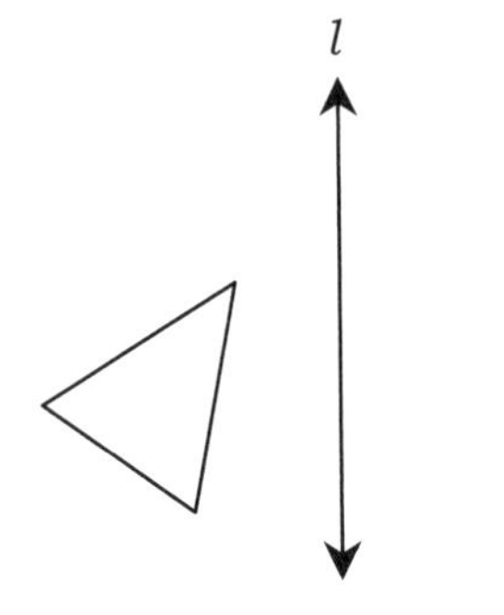

   c. Try each of the glide reflections in parts (a) and (b) again, except reverse the order of the translation and reflection. Do you obtain the same results when the order is reversed?

3. Next consider a transformation that is a combination of a translation and reflection that is *not* a glide reflection; that is, where the translation is *not* parallel to the reflection line.

   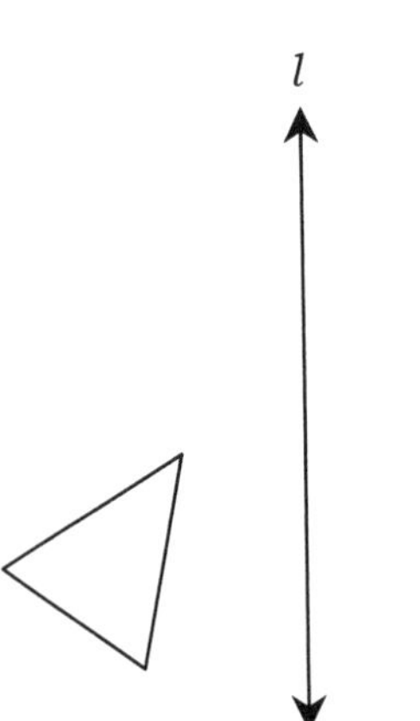

   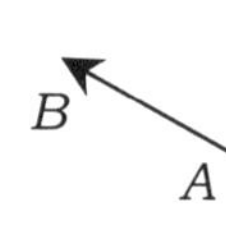

   a. Use your tracing paper to find the image of the triangle under the translation, followed by reflection over line *l*.

b. Now use your tracing paper to find the image of the triangle in part (a) by reflecting and then translating.

c. What do you notice about your answers in parts (a) and (b)?

4. Problems #1 – 3 illustrated that, when a translation is parallel to the reflection line, the order in which you do a combination of a translation and a reflection is not important.

In this problem, you will investigate combinations of two reflections. For each part, use your tracing paper to first reflect the triangle over line *l*, then over line *m*. Then, reverse the order – reflect over line *m*, followed by reflection over line *l*.

a.

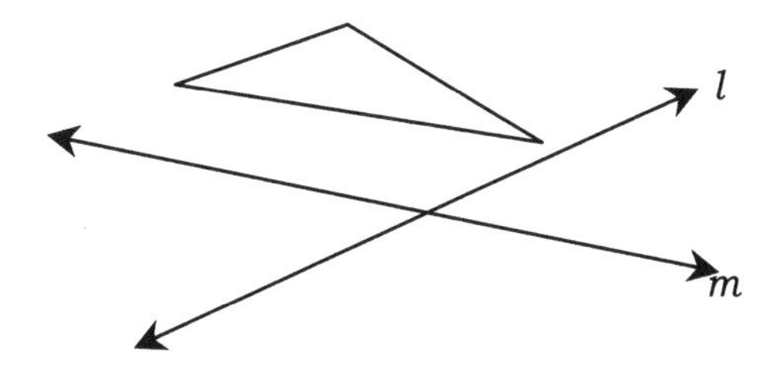

b. lines *l* and *m* are parallel:

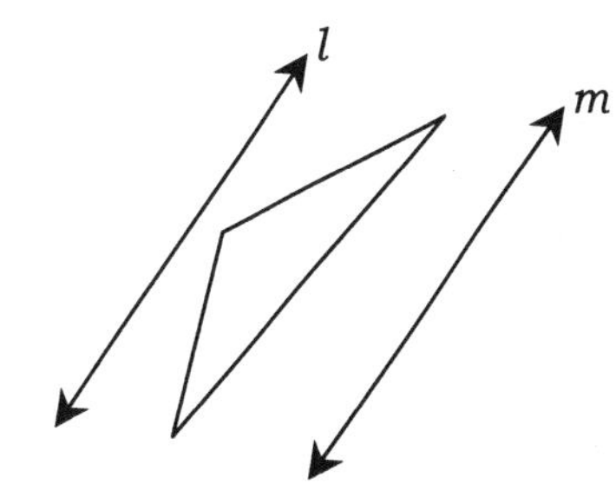

c. lines *l* and *m* are parallel:

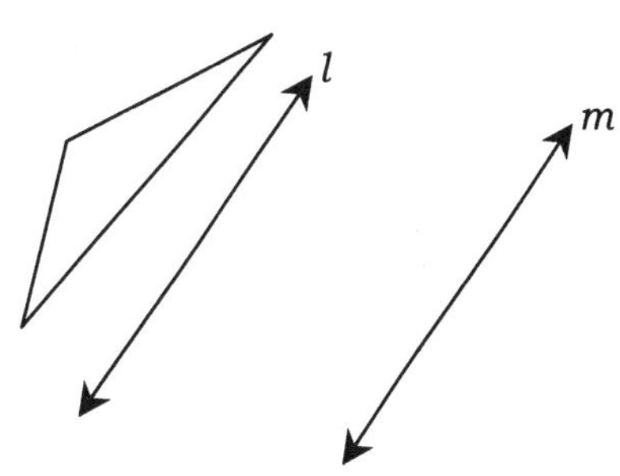

d. lines *l* and *m* are perpendicular:

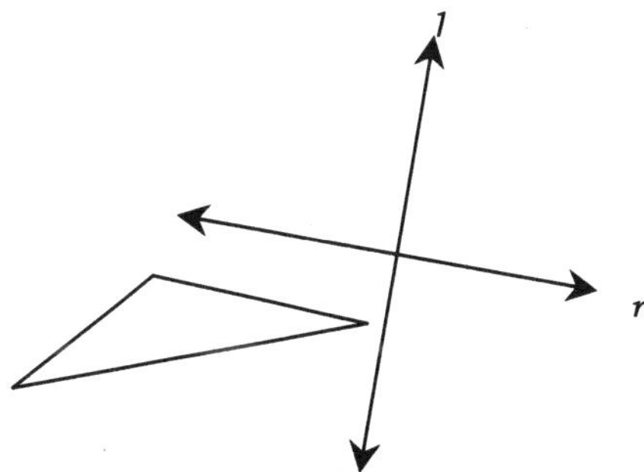

e. In which of parts (a) – (d) do you obtain the same image when you reverse the order of the two reflections? ________Try to state a generalization based on what you found in parts (a) – (d).

OBJECTIVE: **Investigate size transformations and their properties**

You Will Need: **Ruler or compass and straightedge (eManipulative option: *Dialation Transformations*)**

In the first two activities, you investigated isometries, which are transformations that involve moving geometric objects in the plane without changing the size or shape of the object. Another type of transformation, called a size transformation, is one that may produce a change in the size of an object, but not its shape.

1. A **size transformation** is a transformation that 'stretches' or 'shrinks' the plane away from or toward a specific point. Follow the steps given next to find the size transformation image of point P.

   a. The **center** point O and a **scale factor** k determine a size transformation. Here, let $k = 2$. Draw ray $\overrightarrow{OP}$.

   b. Measure $\overline{OP}$ with your ruler or with a compass. Mark point P' on the ray $\overrightarrow{OP}$, a distance of $2OP$ from point O. The point P' is the size transformation image of point P with center point O and a scale factor of 2.

2. Using the technique from problem #1, find the size transformations images of points P, R, and S under the size transformation where point O is the center and the scale factor k is 3. Label these points P', R', and S', respectively.

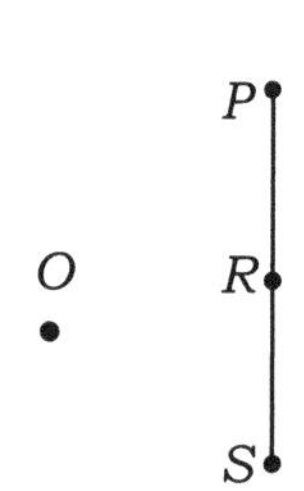

   a. Does the segment $\overline{P'S'}$ contain R'?

   b. How do the lengths of $\overline{PS}$ and $\overline{P'S'}$ compare?

   c. How else are $\overline{PS}$ and $\overline{P'S'}$ related?

3. Using the technique described in problem #2, find the size transformation image of ΔABC where O is the center and $k = 1/2$. Label the image points of the vertices as A', B', and C', respectively.

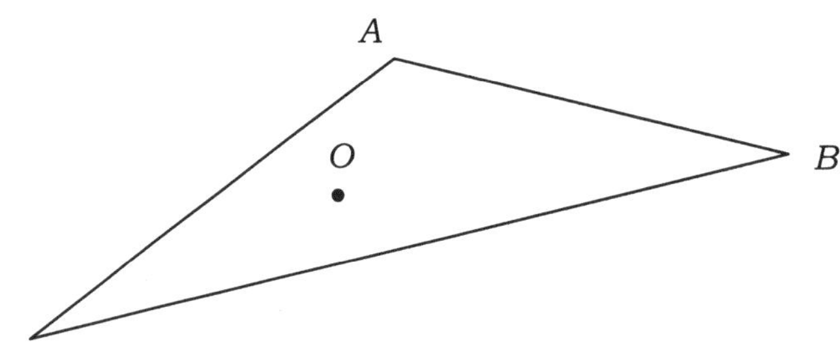

   a. Is the resulting figure also a triangle?

   b. How do the measures of the angles of $\Delta A'B'C'$ compare to the measures of the angles of ΔABC?

   c. How do the lengths of the corresponding sides of $\Delta A'B'C'$ and ΔABC compare?

   d. How do the sizes and shapes of the two triangles compare?

OBJECTIVE: **Find isometries that map one triangle onto another**

You Will Need: **Tracing paper, ruler or compass and straightedge**

In Chapter 12, you used tracings and paper-folding techniques to investigate symmetries of polygons. These methods are also useful for finding an isometry that maps one shape onto another.

1. ΔABC and a reflection image are shown at the right. One method for finding the reflection line is to use a tracing.

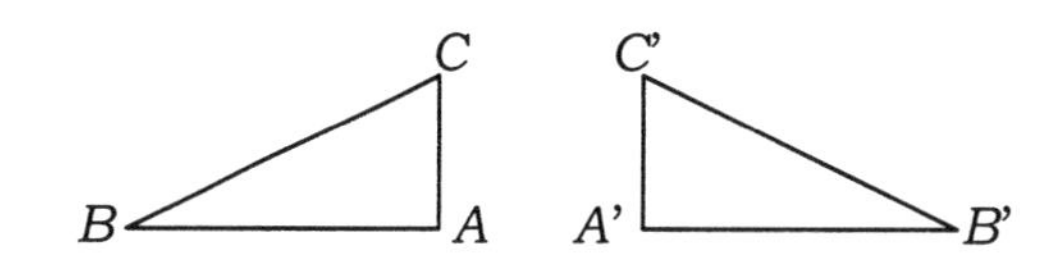

   a. Trace both triangles, and then fold the tracing paper so the reflection line is on the crease. Sketch the reflection line at the right.

   b. Explain how you decided where to fold the paper.

   If you trace ΔABC from A to B to C, you travel in a clockwise direction, so this triangle is said to have **clockwise orientation**. Tracing $\Delta A'B'C'$ from A' to B' to C', you go in a counterclockwise direction, so this triangle is said to have **counterclockwise orientation**.

   This example illustrates that a figure and its reflection image have *opposite* orientations.

2. Consider the congruent triangles ΔRST and $\Delta R'S'T'$ shown at the right.

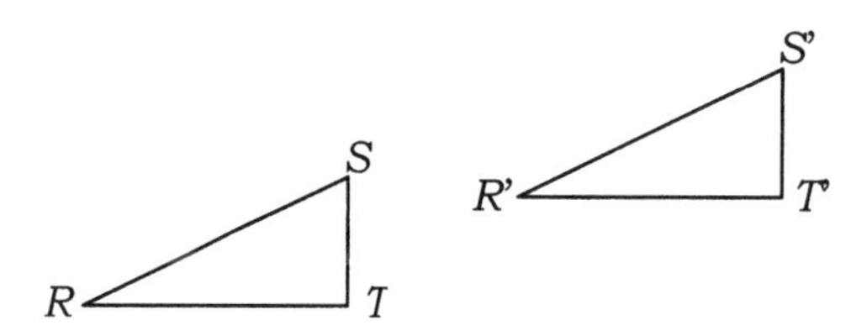

   a. Can you visualize an isometry that maps ΔRST onto $\Delta R'S'T'$? What type is it?

   b. Using dashed lines, draw the directed line segments $\overrightarrow{RR'}$, $\overrightarrow{SS'}$, and $\overrightarrow{TT'}$ with your ruler. What relationships do you notice between these three segments?

   c. Using tracing techniques from Activity 16.1.1, find the translation image of ΔRST according to the directed line segment $\overrightarrow{RR'}$. Do you get $\Delta R'S'T'$?________

   d. Do the triangles in part (a) have the same orientation or opposite orientations?

3. From Activity 16.1.1, recall that a translation image of a line segment is congruent and parallel to the original segment.

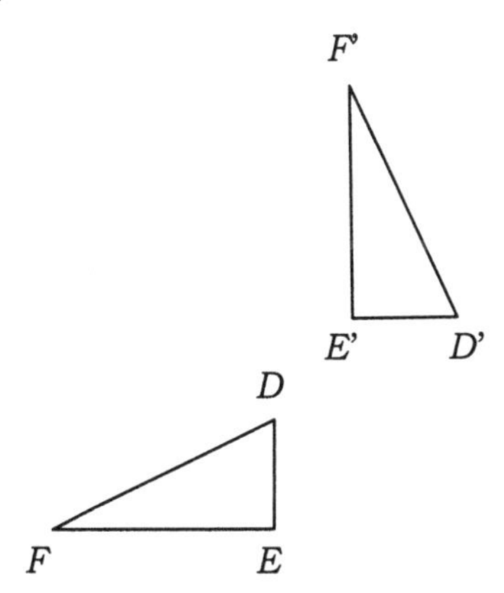

   Consider the congruent triangles ΔDEF and $\Delta D'E'F'$ shown at the right. Because, for example, $\overline{DE}$ is not parallel to $\overline{D'E'}$, you can see that $\Delta D'E'F'$ could not be the translation image of ΔDEF.

a. Although there is no translation that maps ΔDEF onto $\Delta D'E'F'$, there is an isometry that will. Can you visualize what type it is?

b. Follow the steps given next to find the center point, O, of the rotation that maps ΔDEF onto $\Delta D'E'F'$.

Use your straightedge to draw the segments $\overline{DD'}$ and $\overline{EE'}$.

Using either a tracing and paper-folding techniques or your compass and straightedge, construct the perpendicular bisectors of $\overline{DD'}$ and $\overline{EE'}$.

The intersection point of these two bisectors is the center point O and the angle $\angle DOD'$ is the angle of rotation.

c. Using a tracing and techniques from Activity 16.1.1, find the rotation image of ΔDEF, using the center point O and the angle $\angle DOD'$ as the angle of rotation. Do you get $\Delta D'E'F'$? ________

d. Do the triangles in part (a) have the same orientation or opposite orientations?

4. A theorem in geometry states: two triangles are congruent if and only if there is an isometry that maps one triangle onto the other.

The following pairs of triangles are congruent and, in each case, there is either a reflection, a translation, or a rotation that maps one of the triangles onto the other. Keeping in mind what you discovered about the orientations of a figure and its image under these three types of transformations, find a transformation that maps one triangle in each pair onto the other. In the case of a reflection, draw the reflection line; in the case of a translation, indicate the directed line segment; and in the case of a rotation, indicate the center point and angle of rotation.

a.

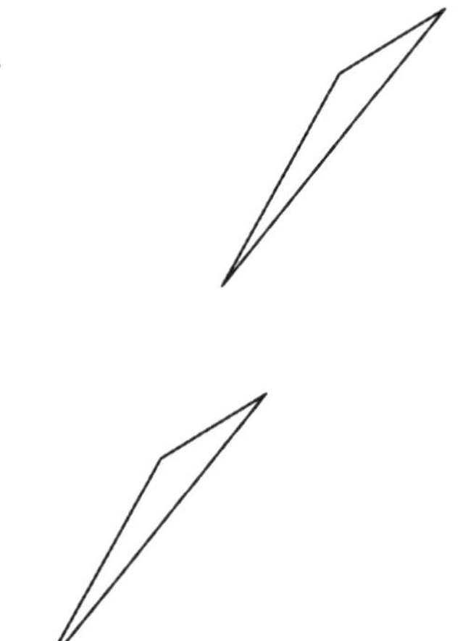

b.

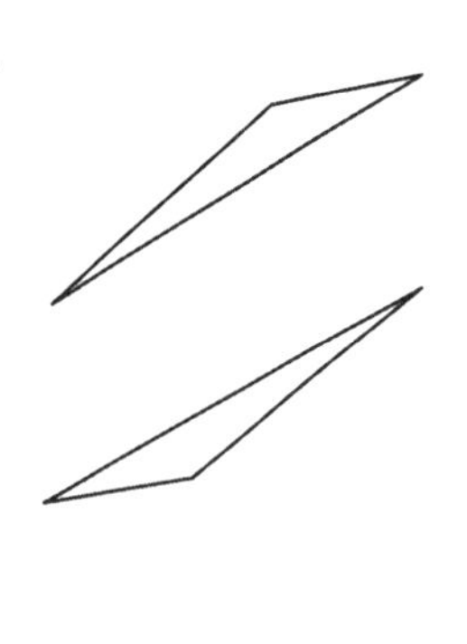

c.

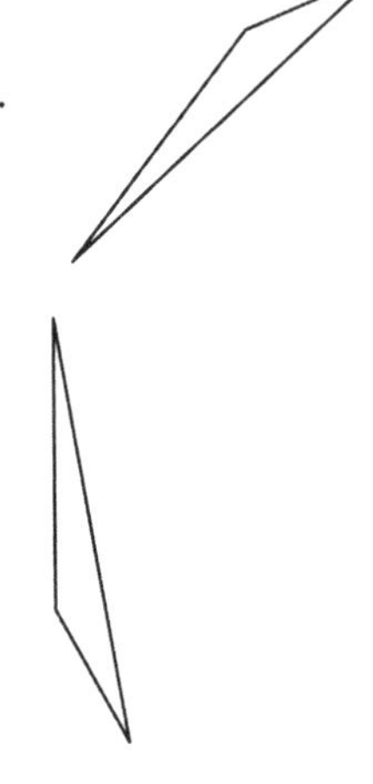

OBJECTIVE: **Find a similitude that maps one triangle onto another**

You Will Need: **Tracing paper, ruler or compass and straightedge**

In the previous activity, you investigated the connections between isometries and the concept of congruent triangles. In the next activity, you will explore similitudes, which are combinations of a size transformation and an isometry, and their connection to similar triangles.

1. Consider the triangle ΔABC and a size-transformation image ΔAB^*C^* shown at the right. The points B^* and C^* are the images of points B and C, respectively, under the size transformation. Notice that the size transformation in this case is a shrinking of the points in the plane toward vertex A, the center point of the size transformation.

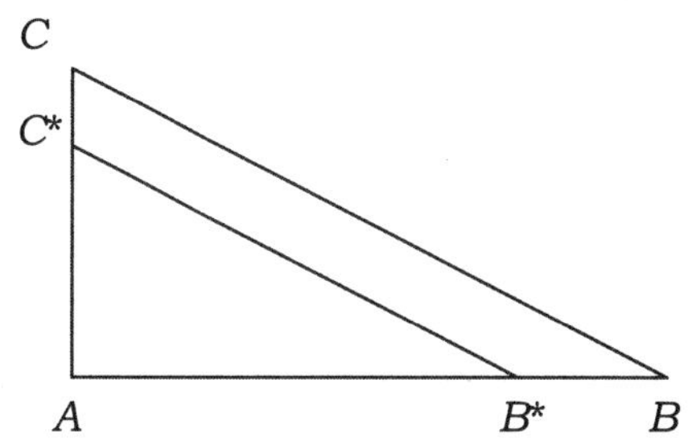

   a. To determine the scale factor k for the size transformation that maps ΔABC onto ΔAB^*C^*, use your ruler to measure the three pairs of corresponding sides of these two triangles. Record your measurements and explain how to use these to determine k.

      The scale factor k = ________

   b. What is the image of vertex A under this size transformation? ________

2. A **similitude** is a combination of a size transformation and an isometry.

   As an example, consider the triangles ΔABC and $\Delta A'B'C'$ shown at the right. First, the size transformation from problem #1(a) was applied to obtain ΔAB^*C^*. Then an isometry was applied to ΔAB^*C^* to obtain $\Delta A'B'C'$.

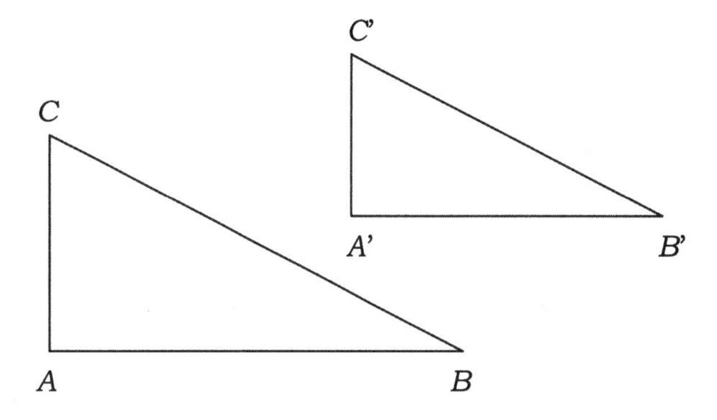

   a. Can you visualize the type of transformation that maps ΔAB^*C^* onto $\Delta A'B'C'$?

   b. To determine the isometry for a similitude, it can be helpful to first sketch the size-transformation image, as shown by the dashed segments at the right.

      Using techniques from Activity 16.2.1, determine the translation that maps ΔAB^*C^* onto $\Delta A'B'C'$.

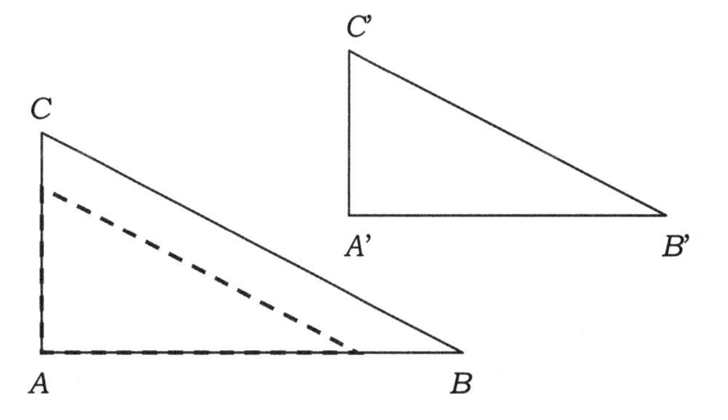

3. A theorem in geometry states: two triangles are similar if and only if there is a similitude that maps one triangle onto the other.

   The following pairs of triangles are similar. In each case, there is a similitude that maps one of the triangles onto the other.

   For each pair, determine a center point and scale factor for the size transformation (a convenient choice for the center is one of the vertices of the triangle). Then find the isometry. If the isometry is a reflection, draw the reflection line; if it is a translation, indicate the directed line segment; and if it is a rotation, indicate the center point and angle of rotation.

   a. 

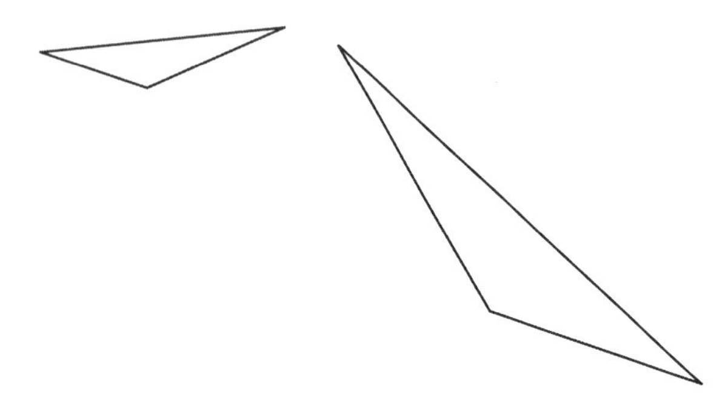

   b.

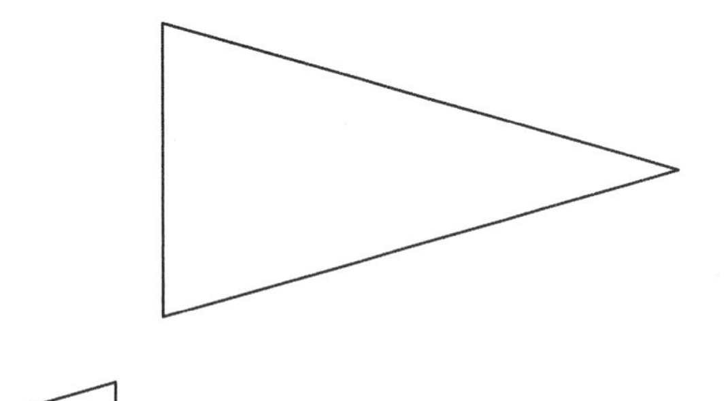

   c.

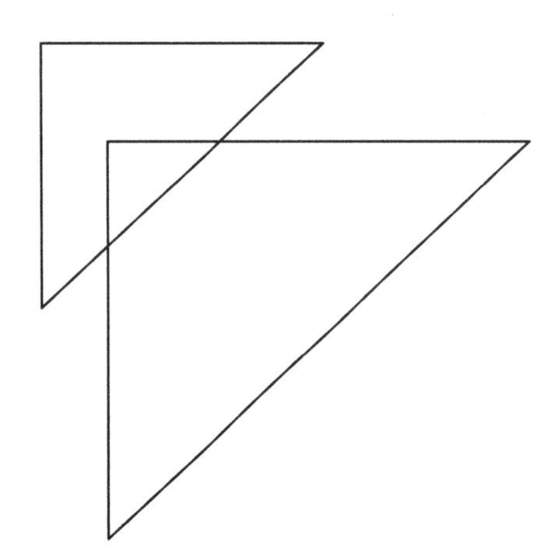

4. Consider the similar triangles ΔABC and $\Delta A'B'C'$ shown at the right.

   a. Can a similitude that takes ΔABC onto $\Delta A'B'C'$ be the combination of a size transformation followed by a translation? Why or why not?

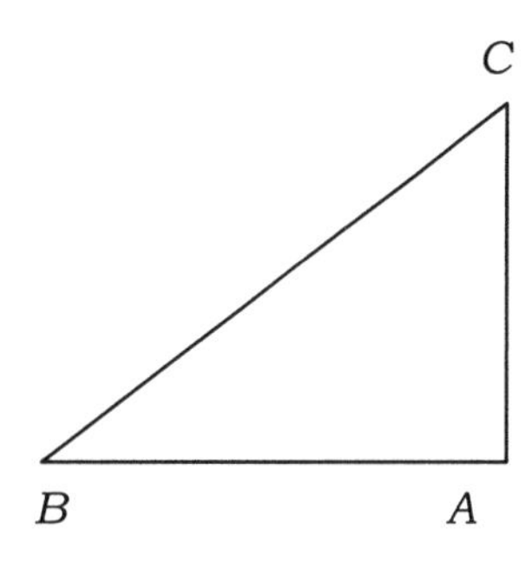

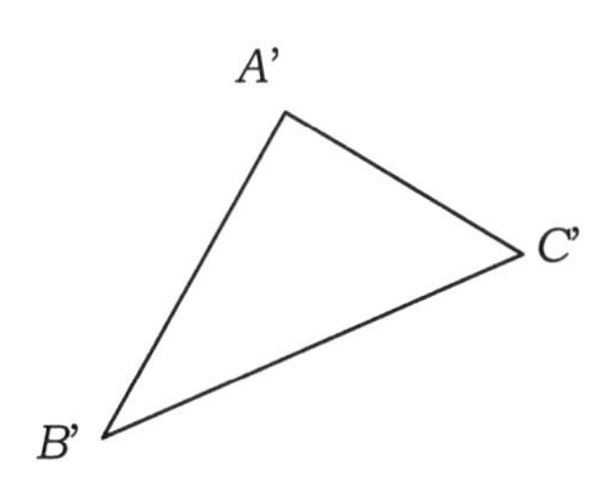

   b. Can a similitude that maps ΔABC onto $\Delta A'B'C'$ be the combination of a size transformation followed by a rotation? Why or why not?

   c. Follow the steps given next to find the similitude that maps ΔABC onto $\Delta A'B'C'$.

      A. First, find ΔAB^*C^*, the size transformation image of ΔABC, using point A as the center point (as in problem #1). Draw this triangle and label the images of B and C as B^* and C^*, respectively.

      B. Is there a reflection line that will map ΔAB^*C^* onto $\Delta A'B'C'$? Why or why not?

      C. Use your straightedge to draw the line segments $\overline{AA'}$, $\overline{B^*B'}$, and $\overline{C^*C'}$. With your ruler or compass and straightedge, locate the midpoints of these three segments. Label these midpoints as P, Q, and R. Using your straightedge, connect the points P, Q, and R to form the line $\overleftrightarrow{PR}$.

      D. Reflect ΔAB^*C^* over $\overleftrightarrow{PR}$ and label the reflection image $\Delta A''B''C''$.

      E. Finally, what isometry appears to one that will map $\Delta A''B''C''$ onto $\Delta A'B'C'$?

      F. Summarize what you have done above. That is, describe the similitude you constructed as a size transformation followed by an isometry. Be specific in describing both the size transformation and the type of isometry.

EXERCISE

Fancy "Freezing"

The one-dimensional strip patterns given below are called frieze patterns. Imagine that they extend indefinitely to the left and right. Each one has translation symmetry; that is, sliding it to the right or left a certain distance maps it onto itself. Identify the other symmetries of these frieze patterns, if they exist. Note: no two of these patterns are equivalent, i.e. have the same list of symmetries.

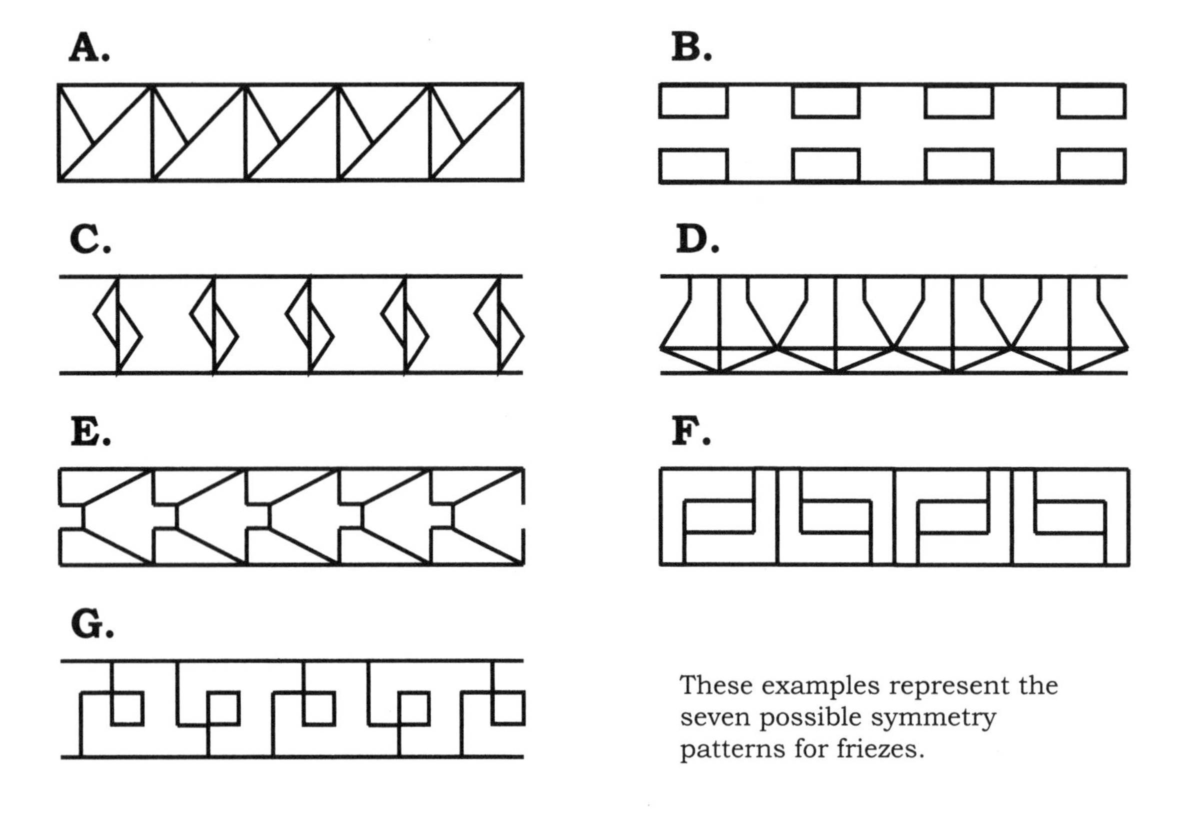

These examples represent the seven possible symmetry patterns for friezes.

CONNECTIONS TO THE CLASSROOM

1. A student wants to rotate the triangle shown on the coordinate grid at the right, using the origin as the center of rotation and a 90° clockwise rotation.

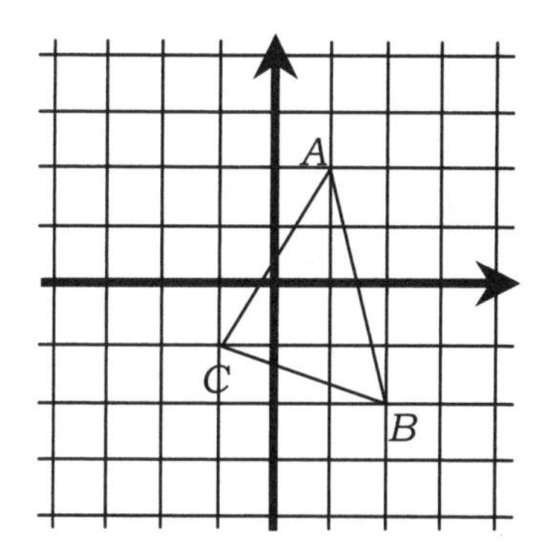

She says, "I have to go up 2 and right 1 from the origin to get to vertex *A*, so I have to go down 1 and right 2 from the origin to get to *A*'."

a. Find *A*', using the student's directions. Is her method correct? ________ How can you tell?

b. How does the concept of slope relate to the student's method?

c. Using a similar method, find the rotation images of points *B* and *C*.

d. If the student wanted to rotate the triangle 90° in a counterclockwise direction, how would her method be different?

2. A student wants to use only his compass and straightedge to make a triangle that is 4 times as large as the one shown at the right, while keeping the same shape. He says, "I know that I need to do a size transformation, but I don't know where the center point should be."

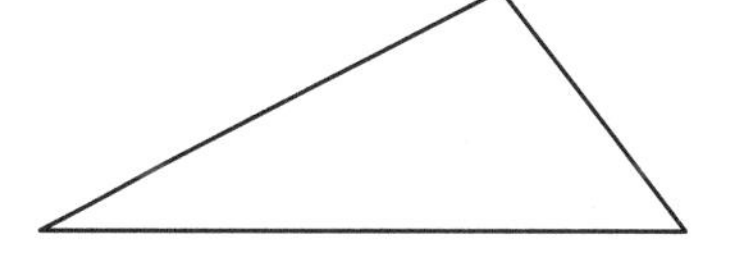

a. How could you help this student?

b. Write a description of the student's steps, after he chooses a center point.

3. Two students are discussing the figures shown at the right.

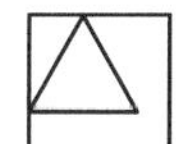

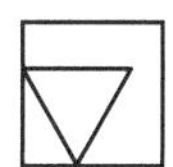

a. Sherry says, "The transformation that maps the figure on the left to the figure on the right is a combination of two reflections." Is Sherry correct? ________ If so, sketch the two lines of reflection. If not, explain why she is incorrect.

b. Is there a glide reflection that maps the figure on the left to the figure on the right in part (a)? ________ If so, show the reflection line and the translation direction and distance.

MENTAL MATH

Given below are patterns that can be folded together to make a cube. Match each of the patterns on the left with its respective cube on the right.

A. 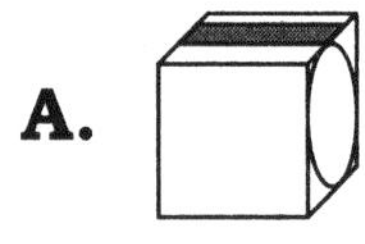

1. 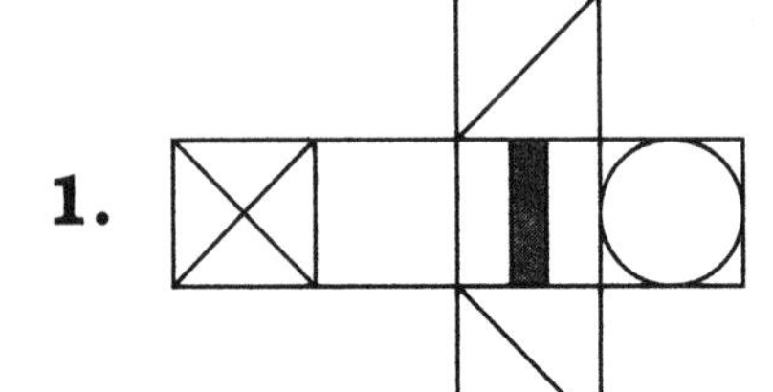

2. 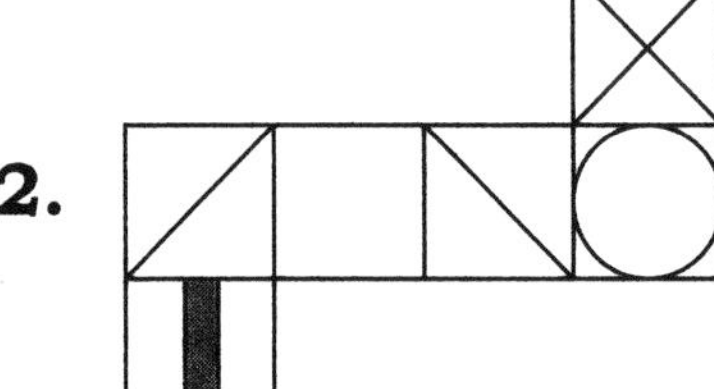

B. 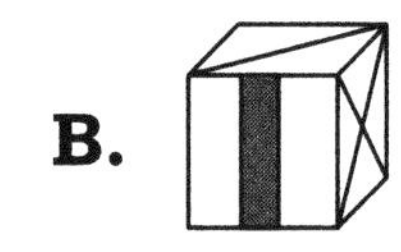

3. 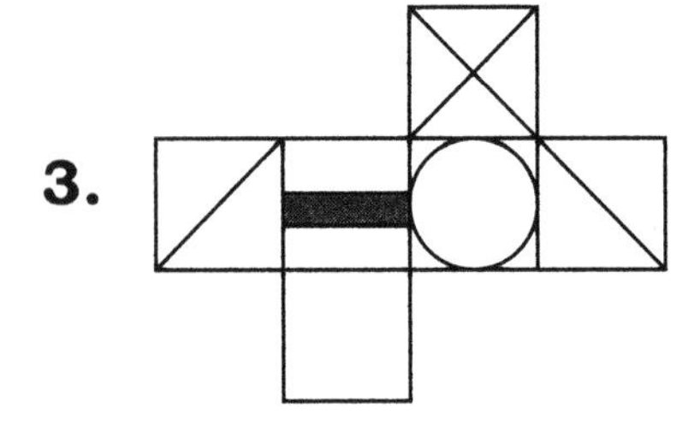

4. 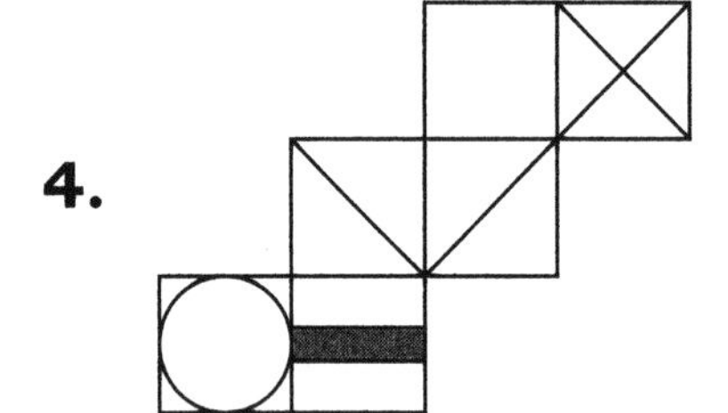

C. 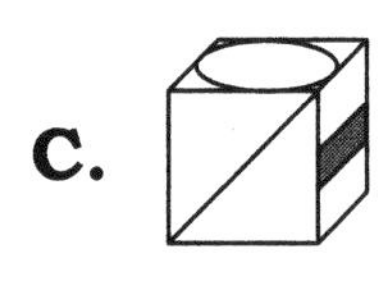

D. 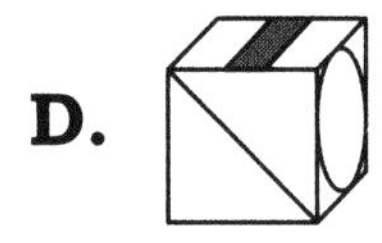

DIRECTIONS IN EDUCATION
Professional Growth
The Teacher As Lifelong Learner

Establishing a pattern of continuous professional growth and lifelong learning may be the single most important challenge faced by teachers. Futurists have stated that most of the jobs which are held by the students of today have not even been thought of yet. As technology continues to expand and to change the way we manage the world, mathematicians must continue to expand. Once math teachers taught their students to calculate cords of wood, to compute the number of pecks in a bushel, and to do long division such as eight digits into ten digits. Today calculators and computers exist which can convert units of measure into equivalent units, which can solve equations and symbolically graph those solutions, and which can perform multiple long division problems in the time needed to complete just one using a paper-and-pencil algorithm. The mathematical skills which were once considered basic for students and teachers are now seen as outdated or as limited in scope. Continuous professional growth can assist teachers in moving ahead to the basics which will be required of students as they move into the world of work.

Components of professional growth for keeping teaching skills current:

- **Opportunities to keep content area knowledge and skills up-to-date.** College courses, workshops, and specific subject-related professional conferences all provide opportunities to enhance content expertise.

- **Opportunities to keep instructional methodology current and exciting.** In addition to college courses, workshops, and conferences, many districts offer opportunities to enhance instructional techniques. It is essential that teachers take advantage of these opportunities.

- **Opportunities to know and understand rationale behind change in instructional techniques and/or content.** Journals such as *Teaching Children Mathematics*, *Mathematics Teaching in the Middle School*, and *Mathematics Teacher* or more general journals such as *Educational Leadership* should be a part of the classroom teacher's regular professional reading. These journals provide a research base and the opportunity to hear from recognized experts in the field.

- **Opportunities to see and experience the use of instructional materials in the classroom.** As new materials are developed – manipulatives, supplemental printed materials, software or hardware, or text-related materials – teachers need to explore options for the use of these materials in the classroom. Workshops and conferences provide excellent opportunities to discover new instructional materials.

- **Opportunities to interact with peers in a professional growth environment.** The classroom can isolate teachers from interactions with their peers. Collegial support of peers in your building or district will eliminate some of the feelings of loneliness – of being out on a limb – that teachers experience when they implement changes in their classroom.

- **Understanding that change takes time.** As teachers implement change, they experience the need for more information, the need for more time, and the need for support in their efforts. Teachers need to give themselves time to make a change in their classroom instruction without becoming frustrated by early problems. Self-analysis of the nature of change needed, the problems encountered and potential solutions to those problems, and of the personal feelings encountered during the change will be helpful. In addition, teachers should seek out peers or supervisors who can provide support for making changes in the classroom.

- **Setting both short-term and long-term goals for professional growth.** Many teachers follow patterns throughout their teaching career which were established in the first three years of teaching. Beginning teachers must project themselves into the mid-years of their career and see how dull a career can be if one does the same things in the same way for thirty years or more. In addition to personal dissatisfaction, the loss of self-esteem which occurs when fellow teachers no loner regard the teachers as current and knowledgeable in the field will make membership in the teaching profession far less rewarding than it can be. Beginning teachers should enter the profession with short-term goals to explore the various professional organizations and to participate in some professional growth opportunity in each of the first three years of teaching. At the end of the three year period, the teacher should establish long-term goals which must then be continually revised to meet changing needs.

As you think about lifelong professional growth, ask yourself:

1. Can I name three professional journals which would be of use to me in my teaching?

2. Do I know how to obtain memberships in professional organizations?

3. What professional conferences or conventions are being held in my geographic area in the next year? NOTE: *Educational Leadership*, the journal of the Association for Supervision and Curriculum Development, *Education Week*, and state education departments usually provide information about such conferences.

SOLUTIONS

CHAPTER 1

Hands-on Activities

Activity 1.1.1

1. Answers will vary: "sum": result of adding two or more numbers together; "square number": result of multiplying a number by itself; ex: 4,9,16,25...; other important information: the two numbers must be less than or equal to 125, the two numbers are consecutive.
2. Answers will vary: example clues: book has a lot of pages, there are many square numbers; make a guess for a square number – the guess gives an answer if it can be written as the sum of two consecutive whole numbers, or make a guess for two consecutive numbers – the guess gives an answer if their sum is a square number.
3. Answers will vary: any two consecutive numbers that add to a square number less than (124 + 125) = 249
4. First find all square numbers from 0 to 249: 4, 9, 16, 25, 36, 49, 64, 81, 100, 121, 144, 169, 192, 225. Two consecutive numbers consist of one even and one odd, so the sum of two consecutive numbers is always odd. All odd square #'s are possible sums: 9, 25, 49, 81, 121, 169, 225; possible solutions are 4&5, 12&13, 24&25, 40&41, 60&61, 84&85, 112&113
5. Possible numbers of adult tickets sold are 2, 5, 8, 11, and 14

Activity 1.1.2

1. Answers will vary, most will say: 20 ÷ 5 = 4 and 4 × 10 = 40, or 10 ÷ 5 = 2 and 2 × 20 = 40, so answer is 40 minutes
2. The missing information is the cuts made; 20 minutes ÷ 4 cuts = 5 minutes/cut
   5 minutes/cut × 9 cuts = 45 minutes - different from the previous answer of 40 minutes.

Activity 1.1.3

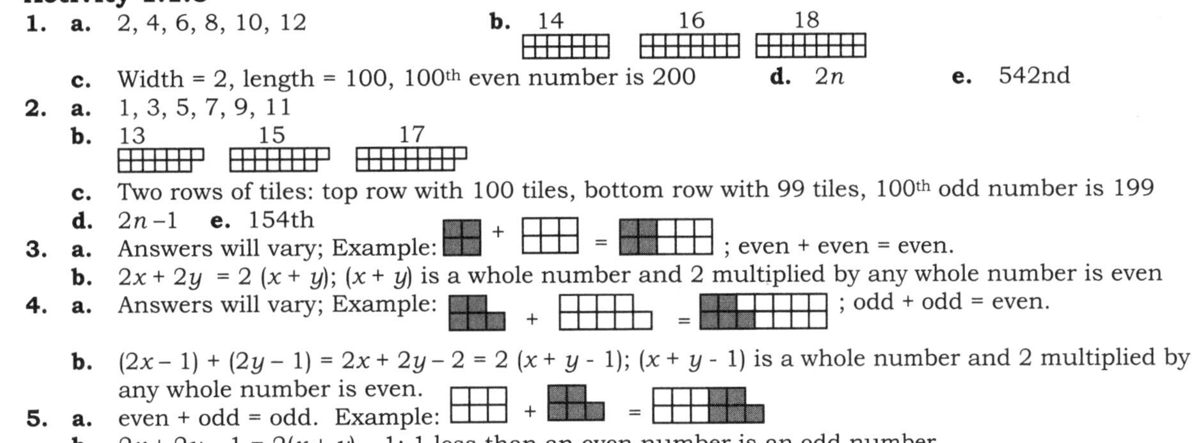

1. **a.** 2, 4, 6, 8, 10, 12 **b.** 14 16 18
   c. Width = 2, length = 100, 100th even number is 200 **d.** $2n$ **e.** 542nd
2. **a.** 1, 3, 5, 7, 9, 11
   b. 13 15 17
   c. Two rows of tiles: top row with 100 tiles, bottom row with 99 tiles, 100th odd number is 199
   d. $2n-1$ **e.** 154th
3. **a.** Answers will vary; Example: + = ; even + even = even.
   b. $2x + 2y = 2(x + y)$; $(x + y)$ is a whole number and 2 multiplied by any whole number is even
4. **a.** Answers will vary; Example: + = ; odd + odd = even.
   b. $(2x - 1) + (2y - 1) = 2x + 2y - 2 = 2(x + y - 1)$; $(x + y - 1)$ is a whole number and 2 multiplied by any whole number is even.
5. **a.** even + odd = odd. Example: + =
   b. $2x + 2y - 1 = 2(x + y) - 1$; 1 less than an even number is an odd number.

Activity 1.2.1

1. Answers will vary
2. 7, 15, 31
3. Answers will vary: 2 times the previous number of moves + 1, ex: (2 × 7) + 1 = 15;
4. 31 + 32 = 63; $2^3 - 1$, $2^4 - 1$, $2^5 - 1$; the number of disks is the exponent of 2; $2^n - 1$
5. $2^6 - 1 = 63$
6. Yes; $2^1 - 1 = 1$ move for 1 disk and $2^2 - 1 = 3$ moves for 2 disks

Activity 1.2.2

1. a. ex:  ; 3 per person; 6 total;

4×3 counts each handshake twice - once for each of the two people shaking hands.

b. Drawings should show 1, 3, and 10 handshakes total; table: shakes per person: 1, 2, 3, 4, total shakes: 1, 3, 6, 10

c. Answers will vary; examples: number of shakes per person is one less than the number of people, total # of shakes is $(n - 1)$ + previous total; 15 handshakes

2. 3, 2, 1, 0; $4 + 3 + 2 + 1 + 0 = 10$

3. a.

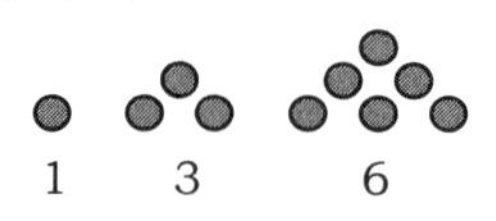 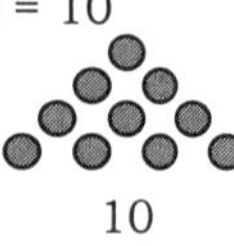 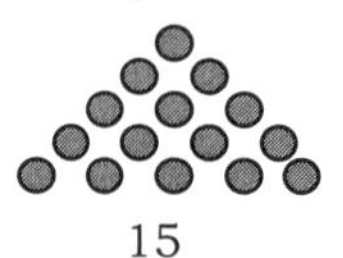

1 3 6 10 15

b. All whole numbers 1 through 30

4. a. 31; 15; $(15 \times 31) = 465$; 465

b. If n is even, sum of each pair × number of pairs $(n + 1) \times (n \div 2)$. What if n is odd?

c. $(n \times (n + 1)) \div 2$

d. The number of handshakes for n people is the $(n - 1)^{st}$ triangular number = $(n - 1) \times (n) \div 2$ (not the nth triangular number, since a person does not shake his/her own hand.)

5. a. $(100 \times 101) \div 2 = 5{,}050$

b. $(99 \times 100) \div 2 = 4{,}950$

6. Make a table, draw a picture, look for a pattern

Exercise

Pascal's Triangle: I-6, E-10, N-15, C-21
Numerical Sequences: O-13
Fibonacci Sequences: K-13, W-21
Triangular Numbers: N-15
NICE WORK!

Connections to the Classroom

1. a. Draw a Picture would be a better strategy

b. Ask the student if he/she can visualize the field.

2. a. The Guess and Test strategy is appropriate in this situation. Guess values for x, substitute into the equation, and solve for y – your guess gives an answer when y is a whole number.

b. There is a unique solution since you have a second equation, $(x + y) = 20$. You can now solve for one variable in terms of the other to obtain an equation in one variable.

c. You would substitute values for both x and y that have a sum of 20.

Mental Math

The first cut through the cake would be horizontal as shown:

Cut the doughnut as shown:

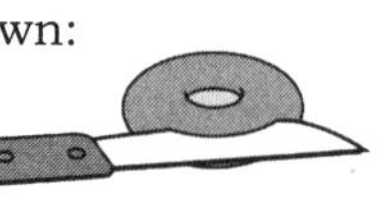 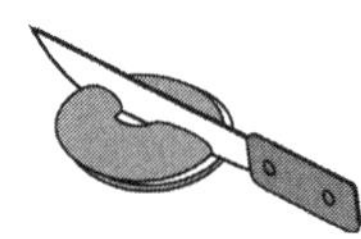 

CHAPTER 2

Hands-on Activities

Activity 2.1.1

1. a. 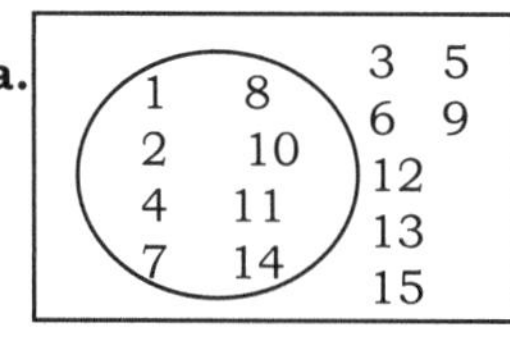

b. square buttons

c. A = {4, 11, 14}

$\overline{A}$ = all square buttons and all non-black round buttons = {1, 2, 3, 5, 6, 7, 8, 9, 10, 12, 13, 15}

2. a. These sets have no buttons in common.

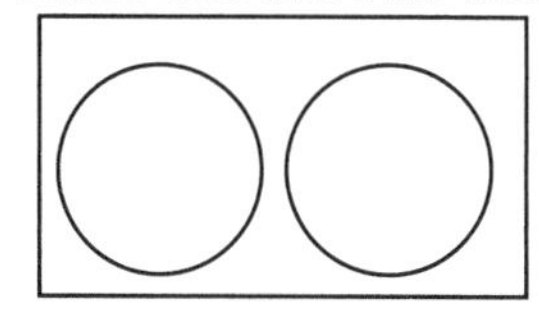

b. These sets have some buttons in common.

c. All the buttons in B are also in A.

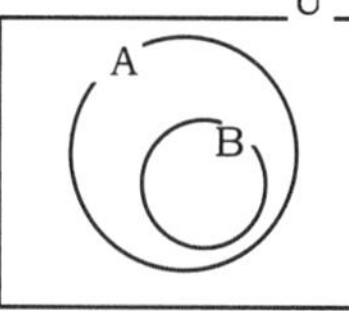

3. **a.** the set of buttons that are round <u>and</u> have two holes
   b. the set of buttons that are round <u>or</u> have two holes
   c. the set of round buttons that do not have two holes
   d. the set of buttons that have two holes that are not round

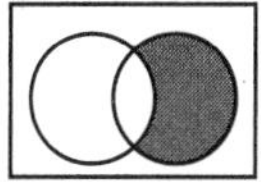

4. Answers will vary; **a.** **b.** **c.**

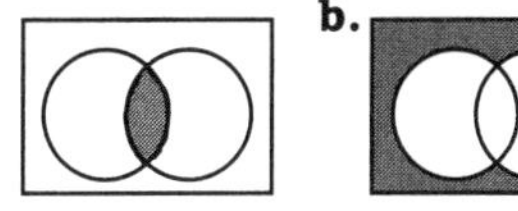

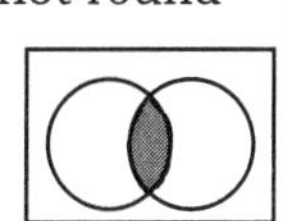

Activity 2.1.2

1. **a.**

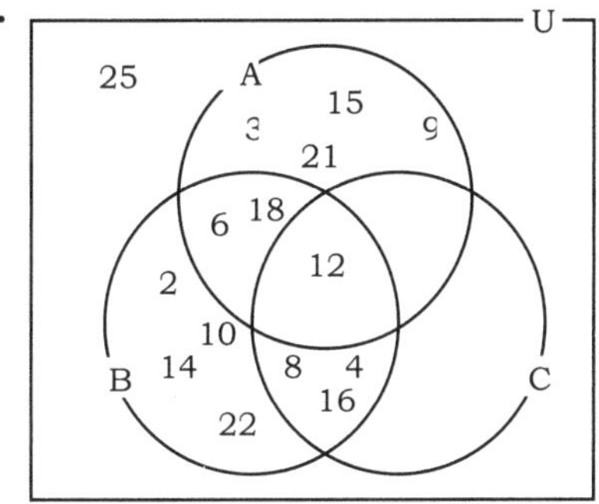

b. No, all multiples of 4 are also multiples of 2 since 4 is a multiple of 2 (i.e. *C* is a subset of *B*).

2. **a.** Yes, all multiples of 8 are also multiples of 4 and 2
   b. $\overline{A}$ is the set of all odd whole numbers

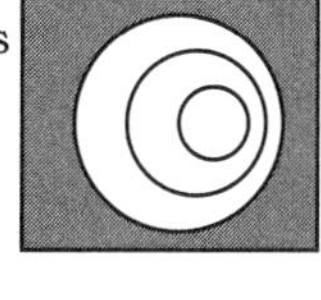

   c.

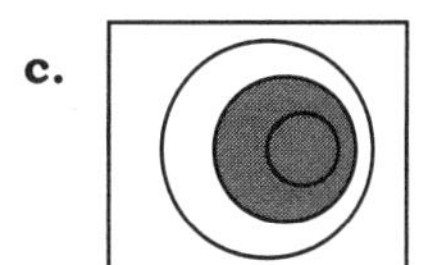

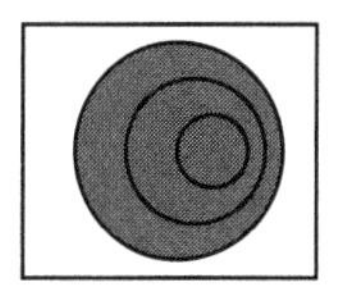

3. **a.** $\overline{A \cap B}$ is the set of all whole numbers that are not multiples of 6
   b. same as (a).
   c. $\overline{A \cup B}$ is the set of all whole numbers which are not a multiple of 2 or 3 or both; Ex: 1, 5, 7, 11, 13

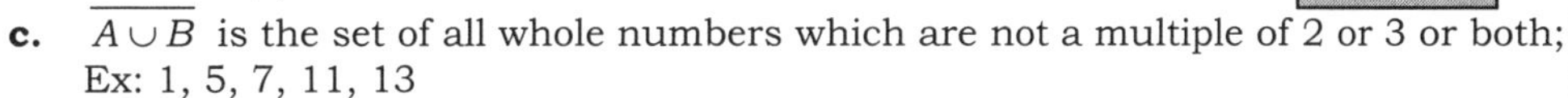

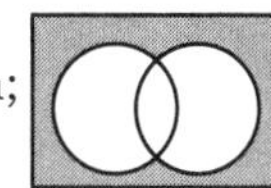

   d. same as (c)
   e. $\overline{A \cap B} = \overline{A} \cup \overline{B}$ and $\overline{A \cup B} = \overline{A} \cap \overline{B}$

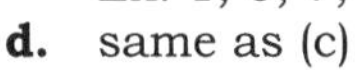

Activity 2.3.1

1. Block = 3 flats = 9 longs = 27 units; Flat = 3 longs = 9 units; Long = 9 units
2. **a.** Answers will vary **b.** Answers will vary
   c. 0, 1 or 2; every group of 3 like pieces can be exchanged for one of the next largest piece.
3. **a.** 1120_{three} **b.** 112_{three} **c.** 10_{three} **d.** 100_{three}
4. **a.** 1000_3, 100_3, 10_3, 1_3 **b.** 27, 9, 3, 1
5. **a.** $2 \times 100_{three} + 0 \times 10_{three} + 1 \times 1_{three}$ **b.** $2 \times 1000_{three} + 0 \times 100_{three} + 1 \times 10_{three} + 0 \times 1_{three}$
6. **a.**

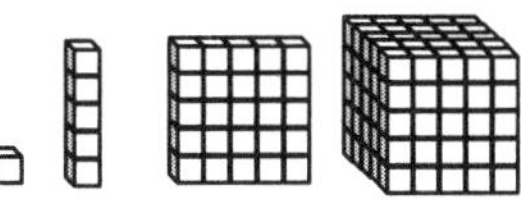

   b. 1_5, 10_5, 100_5, 1000_5 **c.** 1, 5, 25, 125

Activity 2.3.2

1. **a.** 27 + 9 + 9 + 1 = 46 **b.** 27 + 27 + 9 + 3 + 3 + 2 = 71

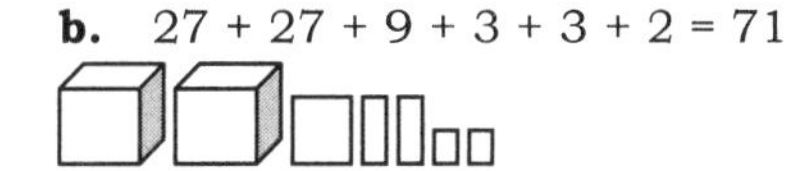

   c. 27 + 9 + 9 + 3 + 3 + 1 =52

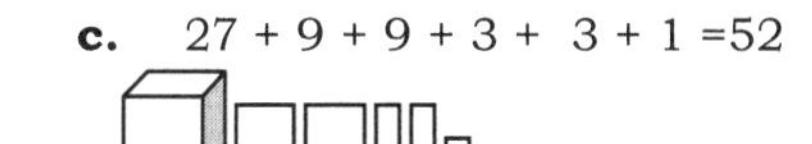

2. **a.** 125 + 25 + 25 + 5 + 5 + 5 + 4 = 194

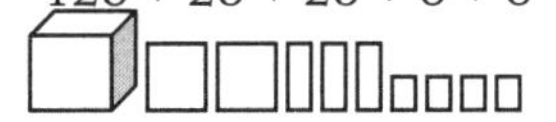

   b. 125 + 125+ 25 + 25 + 25 + 5 + 2 = 332

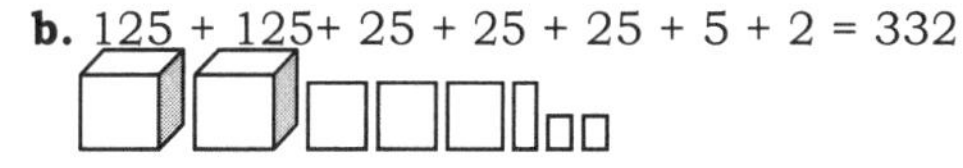

   c. 125 + 25 + 25 + 25 + 5 + 5 + 5 + 5 + 5 + 1 = 221

3. No **a.** Write the numeral in expanded notation with the place values written in base-ten.
   b. You need to know the place values of the given base and how to write them using base ten numerals.
4. 112_{three}
5. **a.**

   b.

   c.

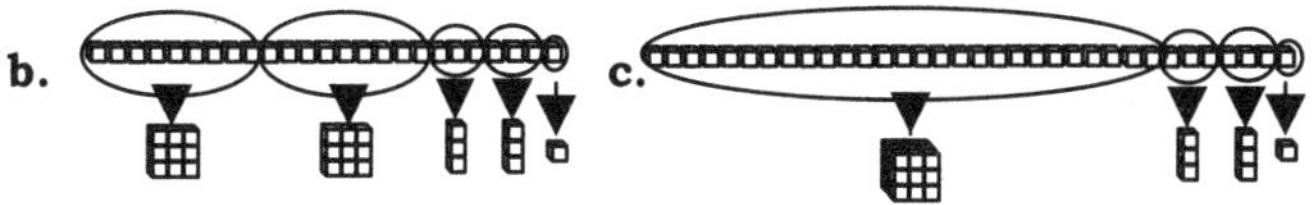

6. **a.** 124_{five} **b.** 303_{five} **c.** 1111_{five}

7. To convert from base 10 to base n, start with the highest power of n that will divide into the given base-10 number - the quotient is the digit in the highest place value of the new base-n number. Next, divide the remainder by the next smaller power of n – this new quotient is the next digit of the new base-n number. Continue until the remainder is less than the base you are converting to – the final remainder is the units digit.

Activity 2.4.1

1. **a.** Domain: people in my family; codomain: their birth months
   b. Yes, every element of the domain corresponds to exactly one element of the codomain.
2. **a.** Yes, every element of the domain corresponds to exactly one element of the codomain.
   b. No, 0 doesn't correspond to any element of the codomain. **c.** Yes; same reason as (a).
   d. No, for example, 6 corresponds to both 2 and 3. **e.** Yes; same reason as (a).
   f. Yes; same reason as (a)
3. **b.** Each element in the domain corresponds to the quotient of 10 and that number (under this correspondence, why doesn't 0 correspond to any element in the codomain?)
   c. Each animal in the domain corresponds to its newborn name.
   d. Each number in the domain corresponds to its prime factors.
   e. Each number in the domain corresponds to 6
   f. Each letter in the domain corresponds to its number order in the alphabet.
4. People may see different relationships between the elements or find different ways of describing the same relationship.
5. There should be exactly one arrow pointing from every element in the domain.
6. This is impossible when there is an infinite number of elements in the domain.
   a. Answers will vary; ex: Every whole number corresponds to its triple.
   b. Answers will vary; ex: Every whole number n would have an arrow pointing to $3n$.

Exercise

1. 158 **2.** 56 **3.** 34 **4.** 31 **5.** 31 **6.** 56 **7.** 98 **8.** 56 **9.** 56
10. 98 **11.** 84 **12.** 34 **13.** 69 **14.** 31 **BASES ARE A BLAST!**

Connections to the Classroom

1. **a.** Since there are 25 students, all the numbers in the Venn diagram should add up to 25.
   b. In the diagram, change 10 to 9 (10 minus the one student already counted in the center of the diagram), 3 to 2, 2 to 1, 15 to 3 [15 minus the (9 + 2 + 1) students already counted in Read for Fun], 6 to 2, and 17 to 6; check your work by making sure the sum of all the numbers in the diagram is 25.
   c. The number of students who play video games and watch TV but don't read for fun; play video games – (watch no TV $\cup$ read for fun); because it had the largest number.
   d. The student counted many classmates more than once; the 1 in the center and the 1 outside the three sets are correct since they were the most inclusive and exclusive numbers.
2. The student is reasoning that the names for numbers in the twenties follow the same pattern as numbers in the teens; some adults may understand her, but they will recognize that her counting is nonstandard. A teacher could have this student write out the numbers she is saying so she can see that she needs to regroup.
3. **a.** The student is forgetting the role of zero as a place-holder; show the base blocks for 21 and 201 and ask the student to explain why 201 has a 0 in it, but 21 does not.
   b. When thinking of twenty-one, the student writes the twenty 20 followed by the one 1; show the student 2 longs and ask him to say that number. Then show 2 longs and 1 unit – you still have the twenty from the two longs so this is "twenty-one" or 21. With 201, you have 2 in the "flats" place, so this is 2 flats, 0 longs and 1 unit. Ask the student to say this number.

Mental Math

a. 124_{five}, 130_{five}, 131_{five} **b.** 1344_{five}, 1400_{five}, 1401_{five}; Total = 2023_{five}

CHAPTER 3

Hands-on Activities

Activity 3.1.1

1. **a.** Y **b.** D **c.** E
2. **a.** O + D **b.** O + P **c** O + P

3. **a.** O + Y **b.** O + Y **c.** K **d.** K **e.** E **f.** E
Sums are the same, order doesn't matter.

4. **a.** and **b.**

+	W	R	G	P	Y
W	R	G	P	Y	D
R	G	P	Y	D	K
G	P	Y	D	K	N
P	Y	D	K	N	E
Y	D	K	N	E	O

c. for any two whole numbers a and b, $a + b = b + a$

5. **a.** G + Y = N **b.** R + Y; W + K; they both equal N **c.** D + G = E and P + Y = E they are equal
d. for any three whole numbers a, b, and c, $(a + b) + c = a + (b + c)$

Activity 3.1.2

1. **a.** 15 – 6 = 9  **b.** 8 – 5 = 3 crackers cheese

c. 7 + ? = 12; 12 – 7 = 5  **d.** ? + 8 = 17; 17 – 8 = 9

2. Answers will vary
3. Problem 1(a) because you find the result of taking away a certain amount from a given quantity.
4. Problems 1(c) and (d) because you find the number that must be added to a given number to obtain a certain sum.
5. $8 = 5 + c$, c pieces of cheese would have to be added if all 8 crackers were to have cheese.
6. Word problems will vary; diagrams are as follows
Take Away: Missing Addend: 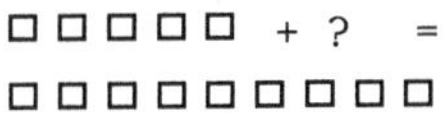 + ? = Comparison

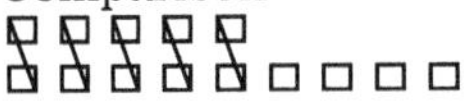

Activity 3.1.3

1. **a.** G **b.** Y; Y; N **c.** O – Y = Y; (O + R) – K = Y; (O + P) – D = N
d. R + <u>D</u> = N; D + <u>W</u> = K; Y + <u>D</u> = O + W
2. **a.** G **b.** Y **c.** G **d.** R
3. **a.** O – Y = Y; O – Y = Y; Y + Y = O; Y + Y =O
b. K – P = G; K – G = P; P + G = K; G + P = K
c. D – P = R; D – R = P; P + R = D; R + P = D

Activity 3.2.1

1. 4 + 4 + 4; $n + n + \ldots + n$, m times
2. 0 + 0 + 0 + 0 + 0 = 0; 5 × 0 = 0
3. 1 by 24; 2 by 12; 3 by 8; 4 by 6
4. 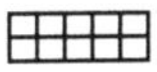
5. 3 **a.**  **b.** Instead of 3 sets of 4, as in #1, there is 4 sets of 3.
c. There are 4 rows and 3 columns, instead of 3 rows and 4 columns. 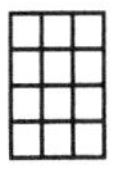
d. $a \times b = b \times a$

6.  $4 \times (2 \times 3) = 24$;  $(4 \times 2) \times 3 = 24$;
Associative property: $a \times (b \times c) = (a \times b) \times c$
7. 10 + 6 = 16; 2 rows and 8 columns; 8 is the sum of 5 and 3; $2 \times (5 + 3) = 2 \times 8 = 16$; Distributive property: $a \times (b + c) = (a \times b) + (a \times c)$

Activity 3.2.2

1. **a.** count # of groups of 4
   $24 \div 4 = 6$

   b. count # of groups of 3
   $24 \div 3 = 8$

   c. count # of groups of 5
   $24 \div 5 = 4$ remainder 4

2. **a.** Make 4 groups and count the # in ea. group.
   $24 \div 4 = 6$

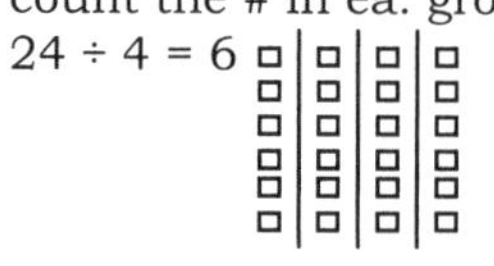

   b. Make 3 groups and count the # in ea. group.
   $24 \div 3 = 8$

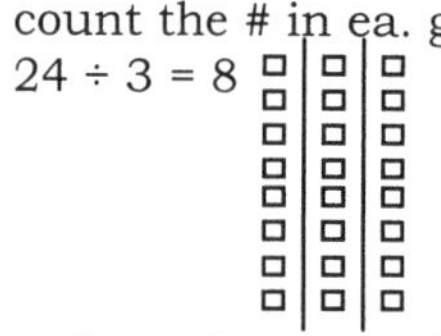

   c. Make 5 groups and count the # in ea. group.
   $24 \div 5 = 4$ remainder 4

3. Both methods produce the same quotient. In method #1, the divisor tells you the number to put in ea. group and you count the # of groups. In method #2, the divisor tells you the number of groups and you count the number in ea. group.
4. Measurement: make columns of 5 squares until all squares are used and count the # of columns; Partitive: divide the squares into 5 rows and count the # in each row.
5. **a.** Answers will vary

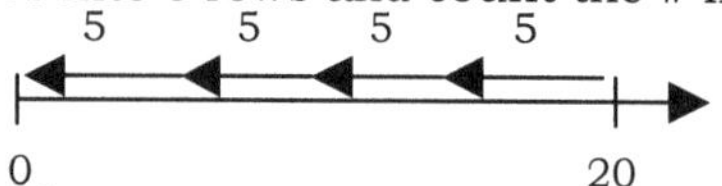

   b. Keep taking away 5 until you reach 0.
6. **a.** They are inverse operations.
   b. What # multiplied by 4 equals 24? $4 \times 6 = 24$, so $24 \div 4 = 6$
   What # multiplied by 4 equals 0? $4 \times 0 = 0$ so $0 \div 4 = 0$
   What # multiplied by 0 equals 24? This is impossible or "undefined."
   0 can be divided by any number, but no number can be divided by zero.
7. Answers will vary

Activity 3.3.1

1. $a \times a \times \ldots \times a$, m times
2. $3 \times 3 \times 3 \times 3$; 3×3; $3 \times 3 \times 3 \times 3 \times 3 \times 3$; 3^6; $a^m \times a^n = a^{m+n}$
3. **a.** 5^9 **b.** 2^{11} **c.** Because 2 and 8 are different bases.
4. 5^4; 5^4; Subtract exponents when dividing two numbers with the same base; $a^m \div a^n = a^{m-n}$
5. **a.** 7^6 **b.** $3^1 = 3$
6. $2 \times 2 \times 2$; $5 \times 5 \times 5$; $2 \times 2 \times 2 \times 5 \times 5 \times 5$; $(2 \times 5)^3$; commutative; $a^m \times b^m = (a \times b)^m$
7. $a \times a \times a$; $5^4 \times 5^4 \times 5^4$; $5^{4+4+4} = 5^{12}$; $(a^m)^n = a^{m \times n}$
8. 625, 125, 25, 5; Divide by 5 at ea. step; $5^0 = 1$; $a^0 = 1$; 1; 0

Exercise

1. M **2.** T **3-4.** M, C **5.** A **6.** E **7.** S **8-9.** A, H **10.** I **11.** T
MATHEMATICS

Connections to the Classroom

1. **a.** She has the correct answer and likely the correct thinking.
   b. No, 1 + 2 doesn't equal 3 + 3, and 3 + 3 doesn't equal 6 + 4.
   c. 1 + 2 = 3, 3 + 3 = 6, 6 + 4 = 10
2. **a.** Take away **b.** Missing addend **c.** Comparison with take away
3. No, $2 \times (3 \times 4) = 2 \times 12 = 24$; $(2 \times 3) \times (2 \times 4) = 6 \times 8 = 48$; 24 is not equal to 48.
4. No; for example, $(2 + 3)^2 = 5^2 = 25$, but $2^2 + 3^2 = 4 + 9 = 13$ and 25 is not equal to 13.

Mental Math

3, 5, 1, 6, 9, 2   14, 16, 11, 15, 18, 17, 13

CHAPTER 4

Hands-on Activities

Activity 4.1.1

1. Answers will vary

2. **a.** Because the sum is a multiple of 10.
   b. No, they are not easy to multiply mentally; ex: 17 × 10 =1700
   c. commutative and associative
3. **a.** 25 × 101 = 25 × (100 + 1) = (25 × 100) + (25 × 1) = 2500 + 25 = 2525; distributive
   b. 19 × 21 – 19 × 11 = 19 × (21 – 11) = 19 × 10 = 190; distributive
4. **a.** 5 × 36 × 2 = 5 × 2 × 36 = (5 × 2) × 36 = 10 × 36 = 360
   b. 82 + 15 + 18 + 41 + 55 = (82 + 18) + (15 + 55) + 41 = 100 + (70 + 41) = 100 + 111 = 211
5. **a.** The distance on the number line between 47 and 18 is the same as the distance between 49 and 20.
   b. The method is correct. It *is* compensation because you compensate for subtracting 2 from 27, by subtracting 2 from the difference of 75 and 25.
   c. Yes, this is compensation. Not recommended because the new problem is not any easier to do mentally than the original.
6. **a.** Subtract 3 from 38, 130 + 35 = 165 **b.** Multiply 4 by 2, 7 × 8 = 56
   c. Divide 18 by 2, 72 ÷ 9 = 8
7. **a.** ex: (163 + 4) - (46 + 4) = 167 – 50 = 117 **b.** ex: (57 + 3) + (24 - 3) = 60 + 21 = 81
   c. ex: (28 ÷ 4) × (25 × 4) = 7 × 100 = 700 **d.** ex: (1700 × 2) ÷ (50 × 2) = 3400 ÷100 = 34

Activity 4.1.2

1. For the low estimate, both numbers were rounded down to the nearest hundred. For the high estimate, both numbers were rounded up to the nearest hundred.
2. Answers will vary.
3. **a.** 1000 **b.** 440 **c.** 1200 **d.** 1300 – 1245 > 1245 – 1200
4. **a.** ex: 620 – 300 – 130 = 320 – 130 = 190 **b.** ex: 50 × 30 = 1500 **c.** ex: 500 ÷ 50 = 10
5. **a.** 70 × 100 = 7000 **b.** 60 × 20 = 1200 **c.** 250 ÷ 50 = 5 or 200 ÷ 40 = 5
   The two numbers are easy to compute with mentally (e.g. one may end in 0).
6. **a.** ex: 15 + 60 + 25 = 100 **b.** ex: 3 × 8 = $24 **c.** Round up to the nearest dollar when you want to make sure you have enough money to make a purchase.
   d. The scientist should overestimate in case the lava goes farther than expected.
   e. You will be able to buy 5 dozen because ($1.20 - $.25) × 5 = $.95 × 5 = $4.75. If there was no budget, an estimate would be appropriate. Exact calculation was necessary in this situation.

Activity 4.1.3

Answers will vary

Activity 4.1.4

1. Answers will vary
2. **a.** Answers will vary **b.** From left to right **c.** Multiplication before addition and subtraction
   d. Answers will vary
   e. 12 ÷ 3 + 3 × 5 = 19
   12 ÷ (3 + 3) × 5 = 10
   (12 ÷ 3 + 3) × 5 = 35

   25 + 20 ÷ 5 – 2 × 4 = 21
   (25 + 20) ÷ 5 – 2 × 4 = 1
   (25 + 20) ÷ (5 – 2) × 4 = 60
   ((25 + 20) ÷ 5 – 2) × 4 = 28

   $3 + 2^2 + 12 \div 2 \times 3 = 25$
   $3 + 2^2 + 12 \div (2 \times 3) = 9$
   $3 + (2^2 + 12 \div 2) \times 3 = 33$

   f. Answer will vary for various calculators
3. Error message. Division by zero and square roots of negative numbers are not defined for the set of real numbers
4. **a. & b.** Answers will vary **c.**

```
                9898 9898
           ×    1919 1919
                1899 4262        (9898 × 1919)
           1899 4262 0000        (98980000 × 1919)
           1899 4262 0000        (9898 × 19190000)
      1899 4262 0000 0000        (98980000 × 19190000)
      1899 8061 0423 4262
```

Activity 4.1.5

1. **a.** – 1, - 60, ÷ 100, + 6000 **b.** – 8, ÷ 10, × 10000, + 200, ÷ 10 **c.** Answers will vary
5. **a.** 42 + 963 = 1005 (or exchange 6 and 4, 2 and 3); 64 × 932 = 59,648; 6432 × 9 = 57,888; 964 – 23 = 941
   b. 236 + 49 = 285 (or exchange 3 and 4, 6 and 9); 2346 – 9 = 2337; 2 × 3469 = 6938; 369 × 24 = 8856

Activity 4.2.1

1. **a.** $2 \times 5 = 10$ **b.** $4 \times 4 = 16$
2. **a.** 13 by 24 **b.** 13×24 **c.** 312, the product of 13×24
3. **a.** 10×20, 10×4, 3×20, 3×4

b.
$$\begin{array}{rl} & 24 \\ \times & 13 \\ \hline & 12 = 4 \times 3 \\ & 60 = 3 \times 20 \\ & 40 = 4 \times 10 \\ & \underline{200} = 10 \times 20 \\ & 312 \end{array}$$
The product is the sum of the four smaller rectangles.

c.
$$\begin{array}{r} 24 \\ \times 13 \\ \hline 72 \\ \underline{240} \\ 312 \end{array}$$

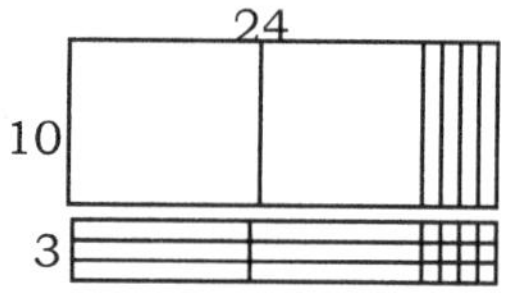

4. **a.**
$$\begin{array}{rl} & 52 \\ \times & 14 \\ \hline & 208 = (4 \times 2) + (4 \times 50) \\ & \underline{520} = (10 \times 50) + (10 \times 2) \\ & 728 \end{array}$$

b.
$$\begin{array}{rl} & 26 \\ \times & 34 \\ \hline & 104 = (4 \times 6) + (4 \times 20) \\ & \underline{780} = (30 \times 6) + (30 \times 20) \\ & 884 \end{array}$$

c.
$$\begin{array}{rl} & 17 \\ \times & 24 \\ \hline & 68 = (4 \times 7) + (4 \times 10) \\ & \underline{340} = (20 \times 7) + (20 \times 10) \\ & 408 \end{array}$$

Activity 4.2.2

1. **a.** Thought process: "2 goes into 2 one time; 2 goes into 3 one time, with a remainder of 1." **b.** 1,1,1; 1,1,1

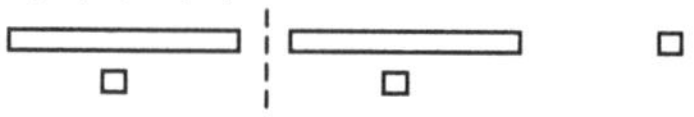

c. First, divide longs into two groups, then divide units into two groups. You have 1 long and 1 unit (or 11) in each group with one left over

d. Thought process: "3 does not go into 2, so find how many times 3 goes into 23. It's 7 with a remainder of 2."

e.

f. In the algorithm, 3 does not go into 2 since 3 > 2, so you must think about dividing a larger number (23) by 3. In the base piece model, you have the extra step of breaking the longs down into units before you can divide.

2. 100, 400, 253; 60, 240, 13; 3, 12, 1; In the standard algorithm, you do not write out the zeros at the end of each number (e.g. you write 4 in the hundreds place, instead of 400).
3. **a.** It represents 3 flats; written in this position because 3 flats is 300.

b. It represents the 1 unused flat.

c. Break the unused flat into 10 longs.

d. This is represented by regrouping the 10 longs with the original 7 longs.

$$\begin{array}{r} 2\ 1\ 7 \\ 3\overline{)6\ 5\ 3} \\ \underline{6\ 0\ 0} \\ 5\ 3 \\ \underline{3\ 0} \\ 2\ 3 \\ \underline{2\ 1} \\ 2 \end{array}$$

e.
$$\begin{array}{r} 1\ 5 \\ 3\overline{)4\ 7\ 6} \\ \underline{3} \\ 1\ 7 \\ \underline{1\ 5} \\ 2\ 6 \end{array}$$

$$\begin{array}{r} 1\ 5\ 8 \\ 3\overline{)4\ 7\ 6} \\ \underline{3} \\ 1\ 7 \\ \underline{1\ 5} \\ 2\ 6 \\ \underline{2\ 4} \\ 2 \end{array}$$

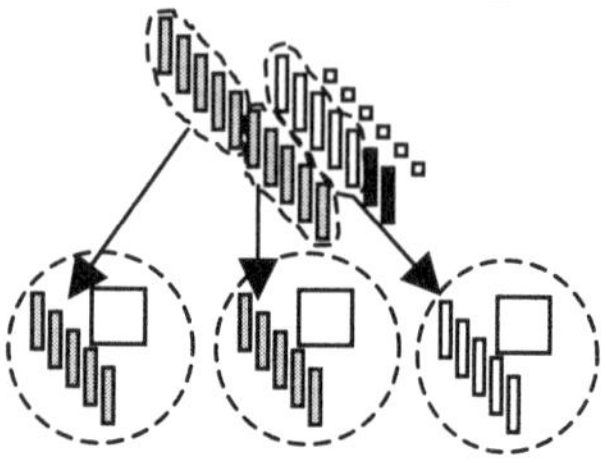

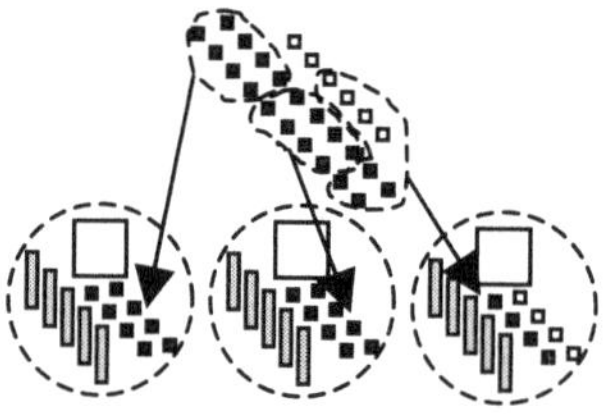

Activity 4.3.1

1. **a.** 100_{three}, 10_{three}, 1_{three} **b.** 9, 3, 1 **c.** 2, because when you get 3 chips in a column, you exchange for a chip in the next column to the left.
2. 212_3, 1012_3
3. **a.**

100_3	10_3	1
●	● ●	

b.

10_3	1
●	●●

c.

1000_3	100_3	10_3	1
●			● ●

4. a. 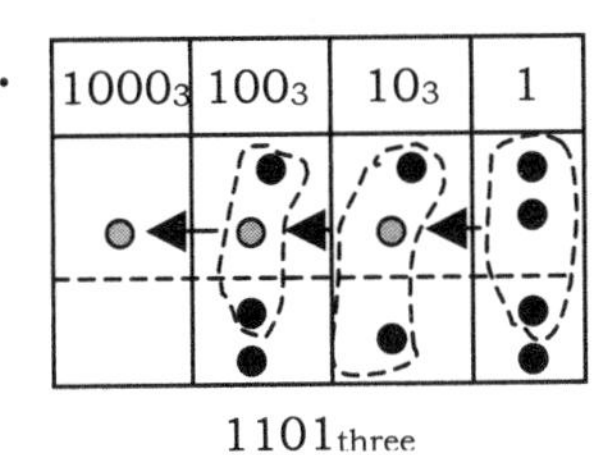

First trade three unit chips for one 10_3 chip. Next trade three 10_3 chips for one 100_3 chip. Lastly, three 100_3 chips for one 1000_3 chip.

b. 1101_{three} First trade three 100_3 chips for one 1000_3 chip. Then three unit chips for one 10_3 chip. Lastly, three 10_3 chips for one 100_3 chip.

c. It's easier to work from right to left. The example in (b) illustrates that working from left to right may require you to work from left to right and then to the left again.

d. 1101_3 **e.** 1120_3

5. a. 110_3; 122_{three} 12_{three}

b. With the Take Away approach, you represent only the minuend, and then you take away the subtrahend from these chips. With the comparison approach, you represent both numbers and compare.

6. a. One chip from the 10_3 column and two from the 1_3 column.
b. There are not enough unit chips to take away two pieces.
c. Trade one 10_3 chip for three unit chips and trade one 100_3 for three **d.** 122_3

7. a. 11120_3 **b.** 222_3

Activity 4.3.2

1. c. They have the same number of chips, but 121_{three} is shifted one place to the left.
d. Put one zero on to the end of the number; put two zeros on the end of the number.

2. 211_3

3. 2110_3; associative and commutative

4. a. Distributive; $102_{three} + (2_{three} \times 102_{three}) \times 10_{three} = 102_{three} + (211_{three}) \times 10_{three}$
$= 102_{three} + 2110_{three} = 2212_{three}$;
b. You first multiply the units digit of 21_{three} by 102_{three}, and then the 10_3 digit of 21_{three} by 102_{three}.

5. a. 2102_3 **b.** 10101_3 **c.** 11221_3 **d.** 2210_3

Activity 4.3.3

1. a. 1211_3; 1197 **b.** You still add from right to left, but regrouping is done at the end, instead of column by column.

2. a. 665; 2736 **b.** 133,416

3. Digits with the same place value line up on the same diagonal.

Exercises

(There may be more than one correct answer.)
3 × 5 – 6 = 9; 6 + 3 × 5 = 21; 5 + 6 + 9 – 3 = 17; 5 × 6 – 3 – 7 = 20; 5 × 6 ÷ 3 + 9 = 19
3 × 6 ÷ 9 = 2; 3 × 7 – 6 = 15; (3 + 7) × 6 = 60; 3 × 5 + 6 – 9 = 12; (3 + 5) × (9-7) = 16; 3 × 6 + 5 × 7 = 53

Connections to the Classroom

1. a. The student first added the tens digits, then the units digits, then added these two sums together.
b. (35 + 41) = (30 + 5) + (40 + 1) = (30 + 40) + (5 + 1) = 70 + 6 = 76

2. When you type in 6,289,214 divided by 92,365 on your calculator, what does the whole number part of the answer tell you? What does the decimal part of the answer tell you? What do you get when you multiply the decimal part by the divisor? What do you think this number means? How could you check your answer?

3. Incorrect. In the order of operations, which comes first, division or multiplication?

4. a. 324; 4014 **b.** Ask the student do these problems using a chip abacus.

5. a. 1200 ÷ 4 = 300; 6300 ÷ 7 = 900; 4800 ÷ 6 = 800
b. Forgetting to use zeros as holders; likely not thinking about place value.
c. e.g. for 1229 ÷ 4, have the student start with 1 block, 2 flats, 2 longs and 9 units and divide these into four equally-sized groups.

Mental Math

17, 8, 14, 6, 2, 75

CHAPTER 5

Hand-on Activities

Activity 5.1.1

1. a. 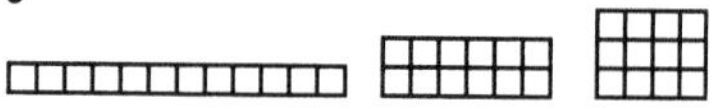 b. 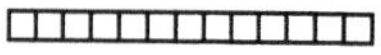
   c. When you multiply the dimensions, you get the number of squares in the rectangle.
   d. Factors of 12: 1, 2, 3, 4, 6, 12; factors of 13: 1, 13
   e. 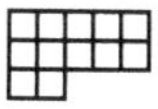 f. 1, 2, 3, 4, 6, 12 g. They are the same; we think of factors in terms of multiplication and divisors in terms of division.
2. a. 1, 7 b. 1, 3, 9 c. 1, 19 d. 1, 2, 4, 5, 10, 20 e. 1, 5, 25
   f. 1, 2, 3, 4, 6, 9, 12, 18, 36
3. 7, 19; 9, 20, 25, 36; 1 has one factor, not exactly two or more than two.
4. a. 9, 25, 36 b. It is a square.
   c. A square has the same dimension for the length and width instead of having a two different numbers for its dimensions.
5. a. 1, 2, 4, 8, 16, 32, 64
   b. All factors > 8 were already listed when you found its partner that is < 8.
   c. $8 = \sqrt{64}$
   d. 1, 2, 3, 5, 6, 10, 15, 25, 30, 50, 75, 150; when you get to 13, you can stop checking for factors.

Activity 5.1.2

1. a. 2, 3 b. Only the prime numbers remain in the sieve.
   c. 2, 3, 5, 7, 11, 13, 17, 19, 23, 29, 31, 37, 41, 43, 47, 53, 59, 61, 67, 71, 73, 79, 83, 89, 97
2. a. They have the same prime factors, but listed in a different order.
   b. $14 = 2 \times 7$; $16 = 2^4$; $150 = 2 \times 3 \times 5^2$; $252 = 2^2 \times 3^2 \times 7$
3. $2^0 \times 3^0 \times 5^0 = 1$, $2^0 \times 3^0 \times 5^1 = 5$, $2^0 \times 3^0 \times 5^2 = 25$, $2^0 \times 3^1 \times 5^0 = 3$, $2^0 \times 3^1 \times 5^1 = 15$, $2^0 \times 3^1 \times 5^2 = 75$, $2^1 \times 3^0 \times 5^0 = 2$, $2^1 \times 3^0 \times 5^1 = 10$, $2^1 \times 3^0 \times 5^2 = 50$, $2^1 \times 3^1 \times 5^0 = 6$, $2^1 \times 3^1 \times 5^1 = 30$, $2^1 \times 3^1 \times 5^2 = 150$
4. a. The units digit must be even.
   b. $1 + 2 + 7 = 10$ and 10 is not divisible by 3, so 127 is not divisible by 3; 127 does not end in 5 or 0, so 127 is not divisible by 5; $127 \div 7$ is not a whole number so 127 is not divisible by 7; $127 \div 11$ is not a whole number so 127 is not divisible by 11.
   c. All composite numbers are made of prime factors; she only needs to check for prime factors (e.g. since 2 and 3 are not factors of 127, 6 is not a factor of 127).
   d. The square root of 127 is approximately 11.2 and 11 is the greatest prime number < 11.2.
   e. 127 is prime

Activity 5.2.1

1. Red = 2; Green (light) = 3; Purple = 4; Yellow = 5; Dark green = 6; black = 7; brown = 8; blue = 9; Orange = 10
2. a. $2 \times 3 = 6$, 2 is a factor of 6
   b.

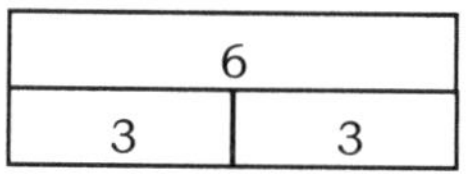

6	
3	3

   c. sixteen 1 strips, eight 2 strips, four 4 strips, two 8 strips, and one 16 strip; twenty-four 1 strips, twelve 2 strips, eight 3 strips, six 4 strips, three 8 strips, two 12 strips, and one 24 strip.
   d. 1, 4, 8 e. 8
3. a. 8 b. 1
4. a; a is the largest number that is a factor of a, so it is the largest factor of both a and b.
5. a. 4, 8, 12, 16, 20, 24, 28, 32
   b. 6, 12, 18, 24, 30, 36
   c. 12, 24 d. 12
6. a. 16 b. 36
7. b; b is the smallest multiple of b, so it is the smallest multiple of both a and b.
8. $\text{LCM}(a, b) = a \times b$ when a and b have no common factors.

Activity 5.2.2

1. a. $84 = 2 \times 2 \times 3 \times 7$; $90 = 2 \times 3 \times 3 \times 5$; common prime factors: 2, 3
b. The product of the common prime factors is also a common factor. **c.** 6
2. a. 9 **b.** 1 **c.** 105
3. 3×5
4. a. 378 **b.** 420 **c.** 1575

Exercise

GCF(10, 35) = 5; GCF(21, 36) = 3; GCF(24, 64) = 8; GCF(21, 84) = 21
GCF(150, 375) = 75; LCM(3, 9) = 9; LCM(10, 12) = 60; LCM(24, 16) = 48
LCM(15, 9) = 45; LCM(6, 7) = 42; GCF(12, 18, 30) = 6; GCF(15, 45, 90) = 15
LCM(5, 7, 15) = 105; LCM(5, 6, 15) = 30; THEY BOTH HAVE THE SAME MIDDLE NAME.

Connections to the Classroom

1. a. If 15 divides two numbers, 15 must divide their sum.
b. Yes; for example, use centimeter strips to represent two numbers with a common factor and show the common factor is also a factor of the sum.
c. Ex. 288 = 240 + 48, 24 divides 240 and 24 divides 48, so 24 divides 288.
2. a. Kathy is correct since 2 and 3 are both prime factors of 12; Al is incorrect since there is only one 3 in the prime factorization of 12.
b. She could use 10 and 2; since 720 ends in zero, 720 is divisible by 10 and since 720 is even, it is divisible by 2. Thus, 720 is divisible by 20.
3. a. When the two numbers have no common factors **b.** 40 **c.** GCF (8, 10) = 2
d. Multiply the two numbers together and divide the product by the GCF.

Mental Math

The sum is 19, a prime. The next sums are 37, a prime; 61, a prime; and 91, not a prime.

CHAPTER 6

Hands-on Activities

Activity 6.1.1

1. a. 1, 1/2, 1/3, 1/4, 1/6, 1/8 **b.** No, the pieces are not equal-sized. **c.** 1/3 **d.** 1/4, 1/6, 1/8
2. a The number of equal sized pieces that make up a whole.
b. The number of pieces shaded or considered
c. Two shaded pieces out of three equal-sized pieces.
3. a. 5/6 **b.** 3/12 **c.** 6/6 **d.** 0/4
4. a. A piece with 12 subdivisions, 7 of which are shaded.
b. A piece with 6 subdivisions, 4 of which are shaded.
c. A piece with 4 subdivisions, 1 of which is shaded.
d. A piece with 12 subdivisions, 9 of which are shaded.
5. a. Divide the strip into 9 equal-sized pieces and shade 1 of the pieces.
b. Divide the strip into 5 equal-sized pieces and shade 4 of the pieces.
6. Fold over shaded section and see how many folds it takes to reach the end of the strip.
7. a. Draw two whole fraction strips, each divided into 4 equal-sized pieces. Shade all 4 pieces on one strip and 1 of the pieces on the other strip, for a total of 5 shaded pieces.
b. $1\frac{1}{4}$, 1 full strip and 1/4 of another strip shaded.
c. Addition of the whole number part and the fraction part. **d.** $2\frac{3}{5}$; $3\frac{2}{3}$
e. The reason you divide 13 by 5 is to express 13/5 as 10/5 + 3/5. Thus, 13/5 is the mixed number $2\frac{3}{5}$.

Activity 6.1.2

1. a. 1/2 **b.** W = 1/10; R = 1/5; E = 9/10
2. a. P **b.** W **c.** R **d.** O **e.** D

3. a. 1/4 b. 1/5 c. 1/10
4. a. R b. P c. P d. G e. The size of the whole is different.
5. a. 1/3 b. 1/4 c. 2/5 d. 2/3 e. The size of the whole is different.

Activity 6.1.3

1. 1/6; a. 2/4; 3/6; 4/8 b. 4/6; 2/8; 6/6; 6/8; 10/12 c. there is no 1/12 piece
2. a. 1/4 b. 2/8 c. Yes, the total size of the shaded pieces is the same.
   d. Both the numerator and denominator doubled; when you divide each piece of a fraction strip by 2, the total number of pieces is multiplied by 2.
   e. 2/4; 6/12; the numerator and denominator have been multiplied by 3.
   f. Multiply both *a* and *b* by the same number: $(a \times c)/(b \times c) = a/b$
3. 0/2 = 0/3 = 0/4 = 0/6 = 0/12; 1/2 = 2/4 = 3/6 = 6/12; 2/2 = 3/3 = 4/4 = 6/6 = 12/12; 1/3 = 2/6 = 4/12; 2/3 = 4/6 = 8/12; 1/4 = 3/12; 3/4 = 9/12; 1/6 = 2/12; 5/6 = 10/12

Activity 6.1.4

1. a. 3/5 < 4/5; 3/5 is less than 4/5 b. The one with the larger numerator.
   c. 9/15 < 7/10
2. a. When you divide a fraction strip into 5 pieces and shade 4 of them, you have a larger shaded area than when you divide a fraction strip into 7 pieces and shade4 of them, since each of the 7 pieces is smaller than the 5 pieces.
   b. > c. The one with the smaller denominator.
3. a. Half of 7 is 3.5, so 4/7 is greater than 1/2.
   b. It is easy to mentally compute half of the denominator of a fraction and compare this to the numerator.
4. a. c/d b. 7/13 > 5/11 since 7/13 > 1/2 and 5/11 < 1/2.
   c. 3/8 is 1/8 less than 4/8 = 1/2 and 4/10 is 1/10 less than 1/2 = 5/10. Since 1/8 > 1/10, 3/8 is farther below 1/2 than 4/10. Thus, 3/8 < 4/10;
   d. > ; <

Activity 6.2.1

1. b. 3/6
2. a. 5/8 b. 6/6
3. Add the numerators and keep the same denominator.
4. a. The denominators are different. b. 1/6; 3; 2; 5; 5/6
5. Different denominators indicate the wholes are divided into different sized pieces. We need to convert all pieces to be the same size.
6. a. LCD = 35; 41/35 b. LCD = 90; 35/90
   c. Same answer; if you simplify, you get 210/540 = 35/90
   d. Common denominator is easy to find, using this method. Disadvantages may be: larger numbers to work with and the need for more steps when simplifying your answer.
7. a. $2 + 3/5 = 2/1 + 3/5 = (2 \times 5)/(1 \times 5) + 3/5 = 10/5 + 3/5$
   b. 10/5 + 3/5 = 13/5; keep the same denominator and add the numerators.
   c. Multiplying 5 times 2 is equivalent to representing the whole number part as a fraction with the same denominator as 3/5. Then 3 may be added to 10 to form 13/5.

Activity 6.2.2

1. a. 2/8; subtract numerators and keep the same denominator.
   b. Take away 2 whole fraction strips and 1/5 of another fraction strip. $2\frac{2}{5}$
2. a. Divide one of the 4 whole fraction strips into 5 pieces. Then you have 7/5 to take the 3/5 away from. b. $3\frac{7}{5}$; $1\frac{4}{5}$
3. a. You took away 1 from the 21. How many 12ths is this equal to? b. $6\frac{8}{12}$
4. a. $4\frac{24}{30}$ b. $20\frac{11}{28}$
   c. 144/30; advantage: you don't have to think about regrouping; disadvantage: you have larger numbers to work with and you need to convert your answer back to a mixed number.

Activity 6.3.1

1. **a.** 1/4 + 1/4 + 1/4 + 1/4; 3/4; 3/4 **b.** 2/6; 3/2
c. Multiply the whole number by the numerator and keep the same denominator.

2. 1; 1/2; $1\frac{1}{2}$; #1(b) shows three 1/2's shaded, whereas this drawing has one-half of three shaded; they have the same shaded amount.

3. **a.** 1/6; 1/6 **b.** 2/6 **c.** 3/8; 1/6

Activity 6.3.2

1. **a.** 2 × 3 = 6; 3 × 4 = 12 **b.** 1/12; the paper is divided into 12 equal-sized pieces, 1 is shaded.
c. 1/6; Fold in thirds in one direction and fold in half in the opposite direction and shade the top.
d. 2/6; Fold in thirds in one direction, unfold once, fold in half in the opposite direction, and shade the top two squares; 8/15; fold in fifths in one direction, unfold 3 times so 4/5 are showing, fold in thirds in the opposite direction, unfold once so 2/3 are showing, and shade the top.

2. **a.** 15; when multiplying fractions, you multiply the denominators: 3 × 5.
b. When multiplying fractions, you multiply the numerators: 2 × 4.
c. The number of shaded pieces represents the product of the numerators, the total number of pieces you divide the paper into is the product of the denominators

Activity 6.3.3

1. **a.** Count the number of groups.
b. How many Green strips are in a blue? 3; How many Red strips are in a browN? 4; How many Yellow strips are in an Orange? 2

2. **a.** 2 **b.** 3/4; you have 3 of the 4 eggs needed for a whole batch; $2\frac{3}{4}$. **c.** $2\frac{1}{3}$; $2\frac{2}{4}$; $2\frac{2}{5}$

3. **a.** 3/4 **b.** How much of a Y is R? 2/5; How much of a D is G? 3/6; How much of an O and R is P? 4/12

4. Ask "how many?" when the divisor (denominator) is less than the dividend (numerator); ask "how much" when the divisor is greater then the dividend.

5. How many 1/4's are in 5/6? $3\frac{1}{3}$; How many 1/6's are in 3/4? $4\frac{1}{2}$; How much of 5/6 is 2/3? $\frac{8}{10}$;
How much of 2/3 is 5/12? $\frac{5}{8}$; How many 1/3's are in 5/4? $3\frac{3}{4}$; How much of 7/4 is 1/2? $\frac{2}{7}$

Exercise

Round Holes: First row: 3/4, 1/12; Second row: 1/4; Third row: 11/12, 1/2, 5/6; Fourth row: 7/12
Square Holes: First row: 3/4, 1/2; Second row: 3/2, 3; Third row: 1/4, 1/3; Fourth row: 6

Connections to the Classroom

1. No, 3/4 is not equal to 5/6.
2. **a.** You could ask: is 1/3 equal to 11/13? **b.** Divide numerator and denominator by the same #.
3. The units are different sizes; they should be equal.
4. Incorrect, ask the student to show the shaded fraction strips for these fractions.
5. **a.** Incorrect, to compare two fractions using their numerators, they must have the same denominator.
b. Half of 20 is 10, so 13/20 > 1/2, and half of 32 is 16, so 17/32 is also > 1/2. Since 13/20 is 3/20 greater than 1/2 and 17/32 is only 1/32 greater than 1/2, 13/20 > 17/32.
6. Jamal is correct - cross-multiplication works for determining whether two fractions are equal because it is really a shortcut for finding a common denominator between two fractions. Alex is incorrect; you could have him use paper-folding (Activity 6.3.2) to develop the idea that, when you multiply fractions, you multiply the numerators together and the denominators together.
7. **a.** Incorrect, the units are different sizes. **b.** Answers will vary.
8. **a.** Correct answer, but inefficient method - you don't need common denominators to multiply.
b. Simplify the fractions first: 3/6 × 10/15 = (3 ÷ 3)/(6 ÷ 3) × (10 ÷ 5)/(15 ÷ 5) = 1/2 × 2/3 = 2/6.

Mental Math

Answers will vary

CHAPTER 7

Hands-on Activities

Activity 7.1.1

1. 1000, 100, 10; chart: 1000, 100, 10, 1
2. 10, 1, 1/10; chart: 10, 1, 1/10, 1/100
3. **a.** 10 times larger **b.** 1/10 as large
c. & d. 100, 10, 1, decimal point marker, 1/10, 1/100, 1/1000
4. These representations all have the same arrangement of chips. The decimal point is in a different position for each one.
5. 3.2104 = 3(1) + 2(1/10) + 1(1/100) + 0(1/1000) + 4(1/10000)
32.104 = 3(10) + 2(1) + 1(1/10) + 0(1/100) + 4(1/1000)
6. **a.** 5.12 **b.** 12 **c.** Five and twelve hundredths
7. Twenty-one and four hundred twelve thousandths; three hundred ten and forty five hundredths
8. **a.** 32.13 **b.** 500.021 **c.** 19.010 **d.** 34.11

Activity 7.1.2

1. **a.** 0.1 **b.** 0.01 **c.** 1000; 0.001
2. **a.** 0.25 **b.** 0.8 **c.** 0.08
3. **a.** 0.6; 0.60
b. They have the same area shaded, but the #s of subdivisions of the unit are different; 0.6 = 0.60.
4. $<$; 0.4 has four full columns shaded, 0.36 has only 3 full columns and 6 small squares shaded.
5. **a.** 0.709 = 7(1/10) + 0(1/100) + 9(1/1000); 0.71 = 7(1/10) + 1(1/100); 0.7 = 7(1/10)
b. $0.7 < 0.709 < 0.71$
c. Both methods allow you to compare decimals using the idea of place value. The decimal squares give a picture of the place values as columns and small squares and you can compare the total shaded areas. The expanded notation method shows the place values written numerically so you can compare the digits place-by-place.
d. The decimal square for 0.71 has 7 full columns (or seven 1/10 pieces) and 1 square (or one 1/100 piece) shaded, while 0.7 has only 7 full columns (or seven 1/10 pieces) shaded.
e. $0.700 < 0.709 < 0.710$; Answers will vary.

Activity 7.1.3

1. **a.** Exchange for 1 long; 1 flat; 1 block **b.** 20 small cubes; 2 longs **c.** 5 flats
2. 7 small cubes → 70 sm. cubes = 7 longs; 5 longs → 50 longs = 5 flats; 3 flats → 30 flats = 3 blocks; move the decimal point (and column labels) 1 place to the right.
3. The decimal point and (column labels) will move 2 places to the right; 2 sm. cubes → 200 sm. cubes = 2 flats, 7 longs → 700 longs = 7 blocks, 1 flat → 100 flats = 1 long block (made up of 10 blocks); shortcut: move the decimal point 1 place to the right when multiplying by 10, 2 places to the right when multiplying by 100, and 3 places right when multiplying by 1000.
4. **a.** 1 long **b.** 1 small cube
5. **a.** 2 blocks = 20 flats → 2 flats0; 3 longs = 30 sm. cubes → 3 sm. cubes; 6 flats = 60 longs → 6 longs; move the decimal point (and column labels) 1 place to the left.
b. Move the decimal point (and column labels) 2 places place to the left; shortcut: move the decimal point 1 place to the left when dividing by 10, 2 places to the left when dividing by 100, and 3 places left when dividing by 1000.
6. It is multiplied by 10. It is multiplied by 100. It is divided by 10. It is divided by 100.
7. Move 1 place to the right. Move 3 places to the left. Move 1 place to the left. Move 4 places to the right.

Activity 7.2.1

3. The algorithms for adding and subtracting decimals are similar to those for whole numbers in that you line up digits with the same place value. Since you need to line up the decimal points, decimals numbers may not line up on the left, as whole numbers do.

Activity 7.2.2

1. long: width = 0.1, area = 0.1; small cube: length and width = 0.1, area = .01
2. **a.** A: 0.2 × 0.6 = 0.12; B: 0.5 × 0.4 = 0.20; C: 0.3 × 0.5= 0.15; D: 0.7 × 0.1= 0.07; E: 0.3 × 0.3 = 0.09; F: 0.3 × 0.6 = 0.18; G: 0.2 × 0.9 = 0.18; H: 0.1 × 0.1 = 0.01
b. Tenths multiplied by tenths gives you hundredths.

3. a. 1.2 by 2.3 **b.** 1.2 × 2.3 **c.** 2 flats + 7 longs + 6 small squares = 2.76

4. 1 × 2 is shown by the 2 flats, 0.2 × 2 is shown by the 4 horizontal longs, 1 × 0.3 is shown by the 3 vertical longs, 0.2 × 0.3 is shown by the 6 small squares.

5. a. 12/10, 23/10 **b.** 276/100
- **c.** Hundredths means there are 2 decimal places to the right of the decimal point.
- **d.** Count the number decimal places to the right of the decimal point in each number and add these. This sum is the number of decimal places to the right of the decimal point in the product.
- **e.** The reason you add the number of decimal places is because when you multiply 2 powers of 10, the number of zeros in the product is the sum of the number of zeros in the factors (e.g. 1<u>0</u> × 1<u>0</u> = 1<u>00</u>, 1<u>0</u> × 1<u>00</u> = 1<u>000</u>, etc.). If a fraction has a denominator that is a power of 10, the number of zeros is the same as the number of places to the right of the decimal point, when the fraction is written as a decimal.

6. 3,754,433/100,000; there are 5 places to the right of the decimal point since there are five zeros in the denominator.

Activity 7.2.3

1. a. 75 ÷ 30 **b.** 0.75 ÷ 0.30 **c.** 2; 1/2 **d.** 2.5
- **e.** In the usual algorithm, you move the decimal point to the right in the divisor to make it a whole number, and then move the decimal point in the dividend the same number of places to the right. Parts (a)–(c) show that these decimal point moves create an equivalent division problem.

2. a. Dividing by a whole number is easier to do than dividing by a decimal number.
- **b.** 10^2; 518.4 ÷ 216
- **c.** 203.415 ÷ 18.7789 = (203.415 × 10^4) ÷ (18.7789 × 10^4) = 2,034,150 ÷ 187,789

Activity 7.3.1

1. a. 2:3 **b.** 4:9 **c.** 3:9 **d.** 4:5

2. Part-to-whole: b & c; part-to-part: a & d

3. a. part-to-whole **b.** part-to-part **c.** part-to-part

4. a. 2; 1/2; 1:2 **b.** 1:3; 1:6

5. a. hexagon : trapezoid **b.** hexagon : rhombus **c.** 1:6; no, these are each the smallest and largest shapes, respectively.

6. a. 1:4 **b.** 1:6 **c.** 6:4; It takes 6 new triangles to cover the trapezoid and 4 new triangles to cover the rhombus.
- **d.** 3:2; It takes 3 original triangles to cover the trapezoid and 2 original triangles to cover the rhombus.
- **e.** 1.5:1; It takes 1.5 rhombuses to cover the trapezoid.
- **f.** 6 × 2 = 4 × 3

Activity 7.3.2

1. Answers will vary

2. a. (i) and (iv)
- **b.** The ratios are set up in a consistent way according to the units associated with the numbers.
- **c.** The units in the numerators and denominators do not correspond. **d.** 4.125 = $4\frac{1}{8}$ tsp.

3. a. No, 3/8 is not equal to 2/7. **b.** 5 1/3 cups **c.** Answers will vary

Activity 7.4.1

1. a. Shade 10 small squares; shade 45 small squares; shade 1.5 small squares.
- **b.** Shade 2 whole unit squares and 50 small squares out of another whole unit.

2. a. 4; 37%; multiply 37% squares by the value of 4 per 1% square to get a value of 148 **b.** 18.5

3. a. 0.25 **b.** 60%;

4. a. 7 **b.** 700

5. a. 1% square has a value of 0.15; 93.3% **b.** 1% square has a value of 970; $2,910
- **c.** 1% square has a value of 330.43; $33,043

Activity 7.4.2

1. 75/100; 75%; 0.75

2. *n*/100; **a.** 87/100 **b.** 321/100 **c.** ½ /100 = 1/200

3. a. 0.87; 19% = 19/100 = 0.19 **b.** Move decimal point 2 places left.

4. a. 5/100, 5%; 70/100, 70%; 135/100, 135% **b.** Move decimal point 2 places to the right.

5. a. 0.05 = 5%; 0.1 = 10%; 0.2 = 20%; 0.25 = 25%; 0.5 = 50%; 0.75 = 75%
   b. $0.\bar{3} = 33.\bar{3}\%$ or $33\frac{1}{3}\%$; $0.\bar{6} = 66.\bar{6}\%$ or $66\frac{2}{3}\%$

Activity 7.4.3

1. a. Shade ten 1% squares b. 5 c. Move decimal 1 place left d. 2.5
   e. 5 is half of 10. f. 10% + 5% = 5 + 2.5 = 7.5 g. $3.69
2. a. 1/2 × 1047= 523.5 b. 10% of 25 = 2.5, 2.5 × 2 = 5
   c. 1/4 × 80 = 20
   d. 50% of 60 = 30, 25% of 60 = 1/2 × 30 = 15, 75% of 60 = 30 + 15 = 45
3. 50% is the same as 50 shaded squares out of 100, so 50% = 5/100 = 1/2.
4. 50% = 1/2 (so you can divided by 2), 20% = 1/5 (so you can divide by 5), 25% = 1/4 (so you can divide by 4), and 75% = 50% + 25% = 1/2 + 1/4 (so you can divide the number by 2 and by 4 and add the two quotients).

Activity 7.4.4

1. a. Associative b. 21.875%; $833\frac{1}{3}$
2. a. 75% b. The whole c. 90 = 75/100 × x; $120
3. a. 105% b. The whole c. 315 is 105% of what number? d. 315 = 105/100 × x; 300
4. a. $87.50 b. $70.00
   c. No, 50% of $125 is $62.50; 20% was taken off the sale price, not the original price d. 56%

Exercise

A. 20 B. 15 C. 19 D. 36 E. 14 F. 38 G. 30 H. 37 I. 17 J. 55 K. 8 L. 11
Each straight line of four circles totals 100.

Connections to the Classroom

1. a. Incorrect; for example, in "thirty two and thirteen hundredths, the student begins writing the 13 in the hundredths place (the 13 should end in the hundredths place). To help, you could ask the student to ask the student to tell you the place value of the last digit in each number he/she wrote.
   b. Incorrect, the student forgot that "and" separates the whole number part and the decimal part; "five hundred twenty-one thousandths."
2. Ask the student to draw and shade decimal squares for 0.20 and 0.2 so he/she can see that 0.20 is 20 shaded parts out of 100 and 0.2 is 2 shaded parts out of 10, but both decimal squares have the same amount of shaded area.
3. Incorrect; to compare 0.3 and 0.25 using whole numbers, the student should think of 0.3 as 0.30. Then he/she would see that 30 > 25 so 0.3 > 0.25.
4. Does not always work.
5. a. e.g. 0.8 is close to 1, so 0.8 + 0.5 should be greater than 1. The student's answer of 0.13 is less than 1.
   b. The student places the decimal point to the left of all the digits in the answer (perhaps because the decimal point appears to the left of the digits in each of the addends).
   c. The student understands that adding decimals is like adding whole numbers, but does not understand where to place the decimal point in the answer.
   d. It may help to have the student do these problems using a chip abacus.
6. Incorrect; the 30% of the unknown original price is not the same as 30% of the sale price.

Mental Math

3 lbs. for $1.67; 4 packs for $1.19; 30 oz. for $3.00

CHAPTER 8

Hands-on Activities

Activity 8.1.1

1. a. Counting numbers **b.** Whole numbers
2. 5; -6
3. Answers will vary.
4. a. 2 red chips and 2 black chips cancel out, leaving 5 black chips with a value of 5.
b. Answers will vary **c.** An infinite number of ways.

Activity 8.1.2

1. a. 3 + 4 = 7
b. -5 + (-2) = -7 **c.** -5 + 2 = -3 **d.** 5 + (-2) = 3
2. a. When both are positive or when the positive integer is greater than the absolute value of the negative integer; that is, there are more black chips than red chips.
b. When both are negative or the absolute value of the negative integer is greater than the positive integer; that is, there are more red chips than black chips.
c. Add the absolute values of the numbers and keep the same positive or negative sign
d. Subtract the integer with the smaller absolute value from the integer with the greater absolute value, and keep the sign of the integer with the greater absolute value.
3. a. 1 **b.** -14 **c.** -2

Activity 8.1.3

1. a. -4 **b.** Five black; eight black; -3 **c.** three red; five black; -8
2. a. -4 **b.** -2 **c.** 7 **d.** -7
3. a. They were taken away. **b.** They were added to the original 3 red chips.
c. Just add four red chips.
4. c. Opposite

Activity 8.2.1

1. a. 4 times put in 2 black chips; 8 **b.** 3 times put in 4 red chips; -12
c. 2 times put in 3 red chips; -6
2. a. 6 red chips **b.** -6 **c.** -6
3. The number of chips you start with may differ from the following examples:
a. From a set with 8 black and 8 red chips, 4 times take out 2 black chips; -8
b. From a set with 4 black and 4 red chips, 2 times take out 2 red chips; 4
c. From a set with 8 black and 8 red chips, 4 times take out 2 red chips; 8
4. a. negative **b.** negative **c.** positive
5. a. 5 × -1, 5 × -2, 5 × -3; decrease the second integer in the product by 1; subtract 5 at each step.
b. -5 × -1, -5 × -2, -5 × -3; decrease the second integer in the product by 1; add 5 at each step.

Activity 8.2.2

1. a. There are 2 red chips in each group, so the answer is -2.
b. When you divide 8 black chips into 2 equal-sized groups, there are 4 black chips in each group, so 8 ÷ 2 = 4; when you divide 10 red chips into 5 equal- sized groups, there are 2 red chips in each group, so -10 ÷ 5 = -2.
2. a. Count the number of groups, 2; No, it doesn't make sense to divide chips into a negative number of groups.
b. When you divide 8 red chips into groups of 2 red chips each, there are 4 groups, so -8 ÷ (-2) = 4; when you divide 12 red chips into groups of 4 red chips each, there are 3 groups, so -12 ÷ (-4) = 3.
3. a. You can't divide chips into a negative number of groups and you can't divide black chips into groups of 3 red chips each.
b. n = -2; To check using multiplication, show -3 × 2 = -6: from a set with 6 red and 6 black chips, 3 times take out 2 black chips. 6 red chips remain. **c.** -4; 1; -3; 2

Exercise

U = 35 T = -23 H = 6 A = -75 F = 20 E = -17 V = 26 L = 12 S = 30 W = -18
D = -13 O = -30 I = 19 G = 5 C = -115 N = 13 Y = -11
He was having cycle-logical difficulties

Connections to the Classroom

1. Negative numbers are not used to describe numbers of things. They are used to describe quantities that are less than zero, such as...
2. **a.** Incorrect; "larger" refers to the larger of the absolute values of the two numbers.
   b. You could ask the student to state the rule in terms of red and black chips to help them see that "larger" means the integer with the greater number of chips.
3. A negative and a negative make a positive with multiplication, not addition.
4. **a.** No, $-x$ means "the opposite of x."
   b. Yes; but if one of x or y is negative, then the product is negative.

Mental Math

41 degrees

CHAPTER 9

Hands-on Activities

Activity 9.2.1

1. 1 square unit **a.** 2 square units **b.** 5 square units **c.** 2 square units
2. **a.** 1, 1, and 2 sq. units **b.** 1, 4, and 5 sq. units
   c. The sum of the areas of the squares on the sides of a right triangle is equal to the area of the square on the hypotenuse.
   d. 5 sq. units; Yes, the 4 triangles can be arranged to form one of the squares and the square at the center covers the other square.
   e. Parts (a) – (d) give you a way to visualize the relationship described by the equation: $a^2 + b^2 = c^2$.
3. **a.** $x, y, x + y, (x + y)^2$ **b.** z, z^2 **c.** $1/2(xy)$ **d.** $4((1/2)xy) = 2xy$
   e. $S + A_2 = A_1$; $x^2 + y^2 = z^2$
   f. The sum of the areas of the squares of the side lengths of a right triangle is equal to the area of the square of the length of the hypotenuse.
4. **a.** Yes, $6^2 + 8^2 = 10^2$ **b.** Yes, $5^2 + 12^2 = 13^2$ **c.** Yes, $15^2 + 112^2 = 113^2$
   d. No, $7^2 + 25^2 \neq 26^2$

Activity 9.2.2

1. **a.** 9 sq. units **b.** 3 units
2. $\sqrt{5}$; **a.** $x^2 = 2^2 + 3^2$; $x = \sqrt{13}$; $y^2 = \sqrt{18}^2 + \sqrt{2}^2$; $y = \sqrt{20}$
   b.

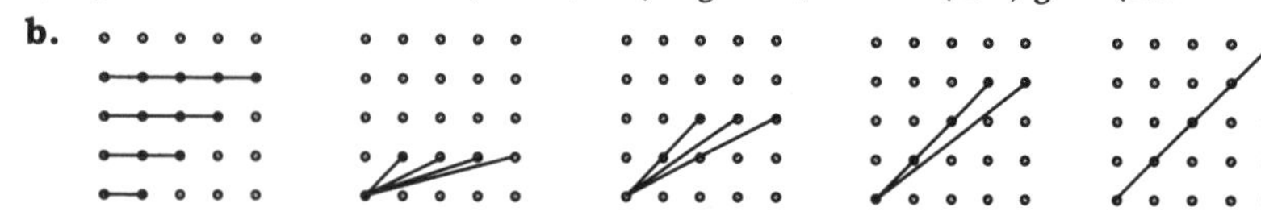

   d. & e. 1, $\sqrt{2}$, 2, $\sqrt{5}$, $\sqrt{8}$, 3, $\sqrt{10}$, $\sqrt{13}$, 4, $\sqrt{17}$, $\sqrt{18}$, $\sqrt{20}$, 5, $\sqrt{32}$

Activity 9.2.3

1. **a.** Replace each box on the original balance scale with three chips. Since both sides of the balance scale now have 14 chips, this shows $x = 3$ is a solution to the equation.
   b. $3x + 5 = 14$, $3(3) + 5 = 14$, $9 + 5 = 14$, $14 = 14$
2.

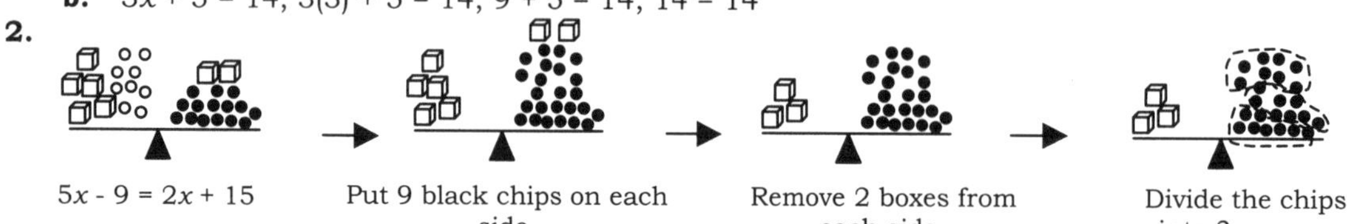

$5x - 9 = 2x + 15$

Put 9 black chips on each side.
$5x - 9 + 9 = 2x + 15 + 9$
$5x = 2x + 24$

Remove 2 boxes from each side.
$5x - 2x = 2x + 24 - 2x$
$3x = 24$

Divide the chips into 3 groups.
$(3x) \div 3 = 24 \div 3$ $x = 8$

3. **a.** The scale won't be balanced. **b.**

$2x + 4 < x + 6$

Put 4 red chips on each side.
$2x + 4 + (-4) < x + 6 + (-4)$
$2x < x + 2$

Remove 1 box from each side.
$2x - x < x + 2 - x$
$x < 2$

4. a.

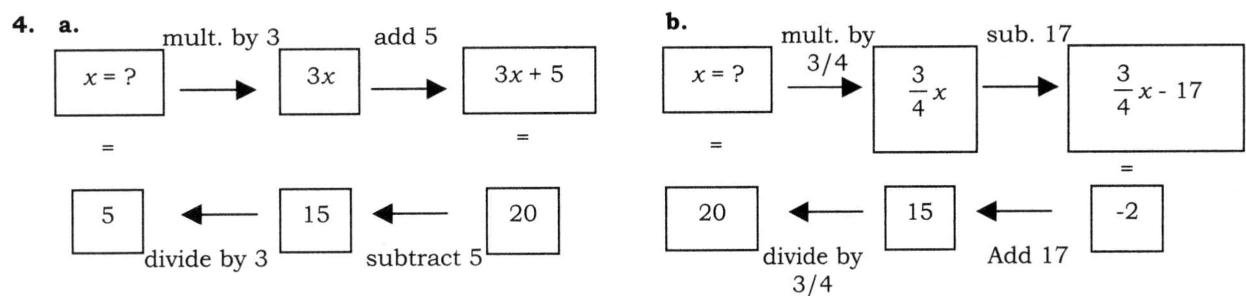

Activity 9.3.1

1. To each element in the domain there corresponds exactly one element in the codomain.
2. **a.** For each number of guests, there corresponds exactly one number of total cupcakes.
   b. 8, 16; 9, 18 e.g. **c.** $f(x) = 2x$; all whole numbers
   d. x-axis: "number of guests", y-axis: "number of cupcakes"; not appropriate to connect the points since you can have only a whole number of guests.
3. **a.** \$12.00, \$12.00, \$24.00, \$36.00, \$132.00 **b.** You always pay a multiple of \$12.00.
   c. Yes, every time corresponds to exactly one price.
   d. Open circles are needed to show which price corresponds to the times at the endpoints of each step of the graph. Without the open circles, this would not be a function (e.g. 15 minutes would correspond to both \$12.00 and \$24.00).
4. **a.** A function; every vertical line crosses the graph in at most one point; not a function, a vertical line may cross the graph more than once.
   b. Every vertical line crosses the graph in at most one point.
   c. Every multiple of 15 minutes would correspond to two different prices.
   d. The vertical line test tells you whether or not every element of the domain is assigned to exactly one element of the codomain.
5. **a.** Plan A is cheaper for few long-distance calls and plan B is cheaper for a lot of long-distance calls.
   b.

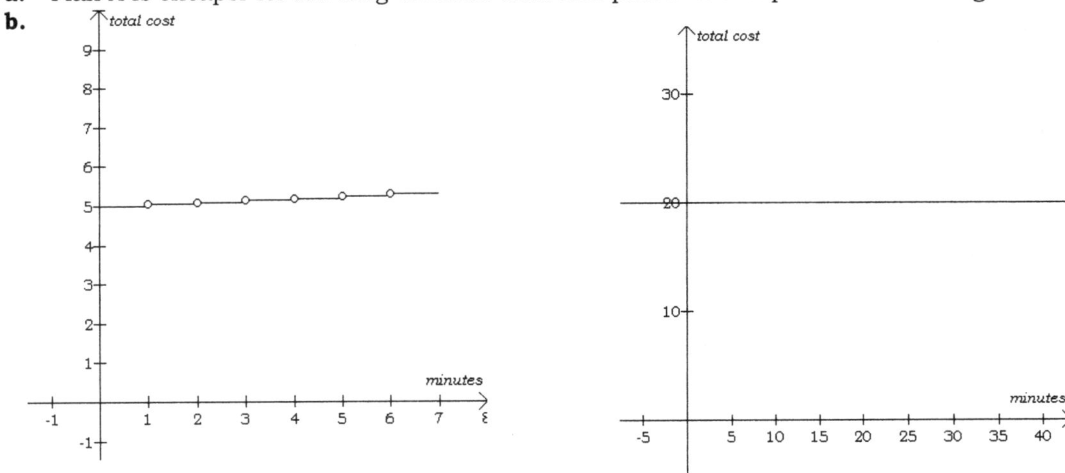

   c. Solve the equation: $0.05x + 4.95 = 20$, $x = 301$ minutes. This would be shown by the point of intersection of the two graphs, (301, 20), if both are plotted on the same set of axes.
   d. Both plans are functions; their graphs pass the vertical line test.

Exercise

THEY GET TOAD AWAY!

Connections to the Classroom

1. No, although there is a pattern, the same block of digits does not repeat over and over.
2. **a.**

   b. No, $3x$ was added to the right side but subtracted from the left side.

3. Maria's is true – if two points on a graph have the same x value, there is a vertical line that crosses the graph at both of these points. Josiah's is false; for example, consider the graph of a function in problem #4(a) in Activity 9.3.1.

Mental Math

Yes. (3(age + 12) + 9)/3 – 15 = age

CHAPTER 10

Hands-on Activities

Activity 10.1.1

1. Answers will vary
2. **a.** Answers will vary; 100%; due to rounding **b.** 360, 360, answers will vary
   c. Answers will vary **d.** We lose actual amounts - we see only the relative amounts.

Activity 10.1.2

1. **a.** In 60% snow cover there were 15 elk in the group **b.** Approximately 25
   c. Elk groups are larger when there is more snow on the ground.
   d. (20, 31), (50, 2), (70, 7) and (90, 17); they are far from the regression line.
2. Answers will vary

Activity 10.2.1

1. Answers will vary. You could put the two stacks together to make one stack of 8 squares, and then split it to make two stacks of 4 squares each.
2. **a.** 7 per stack **b.** 6 per stack **c.** 6 per stack
3. The usual method is to add the numbers (put the stacks together to form one stack) and divide the sum by the number of numbers (rearrange the squares so that you have an equal number of squares in each stack, while leaving the number of stacks the same).
4. **a.** 12 **b.** Answers will vary – need five numbers with sum of 35.

Activity 10.2.2

1. **a.** 5.6 **b.** 4 **c.** 4.5 **d.** Answers will vary **e.** 3; 13; 10
2. **a.** 8, 12 **b.** 10, 3 **c.** 10, 10 **d.** median
   e. A few people with either short or long names
   f. A group of people who share the same length of short or long names.
   g. Yes, if the bar graph were symmetrical (there is a vertical line that would act as a "mirror" line of symmetry) with the center name being the longest.
3. **a.** The mean will change to 5.4; there are fewer total letters.
   b. The mean will change to 5.15, and the median will change to 4.
   c. Mean will always change since the sum of the name lengths will change; mode changes if the most common value is changed; median changes if you exchange a value larger than the median for one smaller than the median, or vice versa; range changes if the highest or lowest value is changed.

Activity 10.2.3

1. **a.** 58, 94 **b.** 80 = (78 + 82) ÷ 2
   c. 70, 86; Q_1, the median, and Q_2 divide the data set approximately into quarters.
3. **a.** 7 **b.** median to lower quartile; $80 - 70 > 86 - 80$ **c.** left whisker; $70 - 58 > 94 - 86$
4. **a.** L = 38, M = 85, H = 94 **b.** $Q_1 = 70$, $Q_2 = 90$
   c. 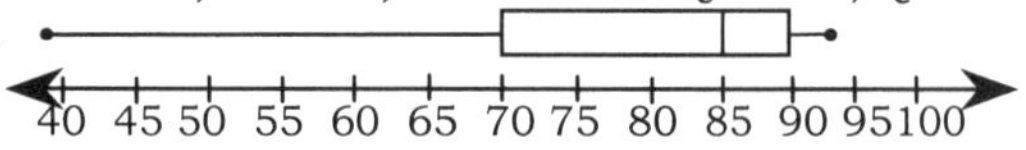
   40 45 50 55 60 65 70 75 80 85 90 95 100
   d. Because 38 is so low in relation to the rest of the scores.
   e. 38; 20 **f.** 1.5 × 20 = 30 and 70 – 30 = 40. $38 < 40$, so 38 is an outlier. Also, 90 + 30 = 120. Since there are no scores greater than 120, 38 is the only outlier.
5. **a.** Test 2: lowest score is higher on test 2, median is higher on test 2, and, if the outlier on test 2 is not considered, there is less spread in the scores on test 2. **b.** Answers will vary

Activity 10.3.1

1. a. The graph compares the starting salaries for new teachers vs. new college graduates for the years 1991 to 2001.
b. Yes, although you cannot determine the exact data values, it does appear that they are approximately the same for both graphs.
c. In the first graph, the vertical axis starts at $20,000 and a scale of $2000 is used. In the second graph, the vertical axis starts at $0 and a scale of $10,000 is used.
d. In the second graph, the salaries of beginning teachers look higher and the differences between the salaries for each year do not appear to be as large.

2. a. Yes, the titles match the steepness of each graph.
b. Yes, by approximating each data point, you can see these do represent the same data.
c. In the second graph, the horizontal axis has been stretched and all years are labeled.
d. Yes **e.** To convince the reader that there has been an increase in recycling.
f. To convince the reader that the recycling rate is not increasing fast enough.

3. a. Answers will vary: start the vertical axis at zero; use a scale of 2 on the vertical axis.
b. Answers will vary: stretch the horizontal axis to make changes appear less dramatic.
c. Answers will vary.
d. A political challenger who wants to convince the reader that U.S. unemployment rates rose dramatically and are still high.

Exercise

THEY ARE BOTH ILL-EAGLES!

Connections to the Classroom

1. No, the data is not written in order.

2. a. No, the median is 3 and the greatest number is three, so at least half the class has three siblings, not 2.
b. Ask the student to make up a small data set with his statistics.

3. a. The scale on the vertical axis is not consistent.
b. The percentage of hamburgers and pizza would look much larger when compared to the other four categories.
c. Same as (b).
d. Answers will vary: perhaps she realized, after starting with a scale of 2% for the vertical axis, that a scale of 2% for the entire vertical axis would make a very tall graph

4. a. 150, 300, twice
b. 8; the large building is 2 times as wide, 2 times as deep, and 2 times as tall as the small building.
c. Answers will vary: could make a large building that is the same width and depth, but twice as tall as the small building.

Mental Math

1. c **2.** b **3.** a **4.** c

CHAPTER 11

Hands-on Activities

Activity 11.1.1

1. Answers will vary
2. 1/3; 1/3; 1/3
3. a. 4/8; 3/8; 1/8 **b.** 12
4. Answers will vary
5. a. 0; 0/10 **b.** 10; 10/10 **c.** 0; 1
6. a. 0/3 **b.** 3/3
7. a. 2/12 **b.** 4/12 **c.** 6/12 **d.** 12/12 **e.** 6/12; 6/12; yes
f. 10/12; 8/12; 6/12, 1 – P(event) = P(complement of the event)
8. a. 0; 1 **b.** 0 **c.** 1 **d.** P(A) + P(B) **e.** 1 – P(A)

Activity 11.1.2

1. **a.** {R, Y, G, B} **b.** 1/4; 1/4; 2/4; 3/4

2. **a.** No, B sector is larger than G sector. **b.** 2/4; 3/4; you see that blue is 2 of 4.

c. P(B) = 3/6; P(Y) = 1/6; P(Y or B) = 4/6; P(not G) = 4/6

3. **a.** Answers will vary; 1/4; yes, all plots are the same size.

b. Answers will vary; answers will vary; 30 units^2; 66 units^2; 96 units^2; 30/96; 66/96

4. **c.** In the 1/4-inch strip down the middle between lines; 1/4 of the total area; 1/4 **d.** 1/16

5. Answers will vary

Activity 11.1.3

1. **a.** Answers will vary

b. There are more numbers on the dice which can be added to obtain those sums.

2.

+	1	2	3	4	5	6
1	2	3	4	5	6	7
2	3	4	5	6	7	8
3	4	5	6	7	8	9
4	5	6	7	8	9	10
5	6	7	8	9	10	11
6	7	8	9	10	11	12

; 1, 2, 3, 4, 5, 6, 5, 4, 3, 2, 1; 36

3. 6/36; 2/36; 1/36; 6/36; 21/36

Activity 11.2.1

1. **a.** {HT, HH, TT, TH}

b. First, all possible outcomes for the first coin were listed. Then, for each of those possibilities, all the possibilities for the second coin were listed; all the possible outcomes for flipping two coins; it has branches

c. 1/4, 3/4

d.

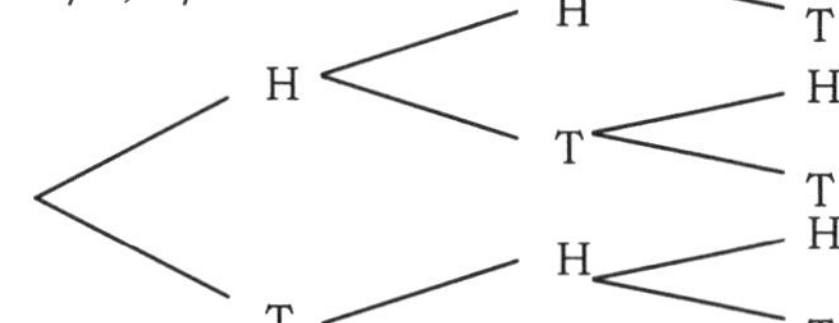

e. 3/8, 4/8, 2/8

2. **a.** Add E E F Y to the end of each branch given in the tree.

b. {BB,BB,BR,BY,BB,BB,BR,BY,RB,RB,RR,RY,YB,YB,YR,YY} **c.** 12/16; 6/16; 4/16

3. **a.** B, R, and Y are possibilities for the second draw, if a blue block is drawn on the first draw. B, B, and Y are possibilities for the second draw if a red block is drawn on the first draw.

b. 10/12; 2/12; 4/12

4. **a.** 2 × 2 × 2 × 2 = 16 **b.** 4 × 3 = 12 **c.** 6 × 6 = 36; 6 × 2 = 12; 4 × 6 = 24; 5 × 4 = 20

Activity 11.2.2

1. **a.** You can determine the probability of each outcome from both diagrams. The diagram on the left shows all the possible outcomes, while the diagram on the right does not list all the outcomes (only the probabilities of each).

b. Extend the tree by adding a set of three branches, identical to the one given, to the end of each of the three branches.

c. 1/16; P(R on the first draw) × P(R on the second draw) = 1/4 × 1/4 = 1/16.

d. The process for each outcome is similar to part (c).

e. The sum is 1; the probability of the sample space is 1.

2. **a.** 1/16; 4/16 **b.** (1/2 × 1/4) + (1/2 × 1/4) = 4/16 **c.** 3/4; 3/8

3. 6/30; 6/30; 18/30

Activity 11.2.3

1. a.

Outcomes	H	T
# of ways	1	1

Outcomes	HHH	HHT, HTH, THH	TTH, THT, HTT	TTT
# of ways	1	3	3	1

Outcomes	HHHH	HHHT, HHTH, HTHH, THHH	HHTT, HTHT, HTTH, THTH, TTHH, THHT	HTTT, THTT, TTHT, TTTH	TTTT
# of ways	1	4	6	4	1

b. Add the two numbers directly above the entry.

c.
1 5 10 10 5 1
1 6 15 20 15 6 1
1 7 21 35 35 21 7 1
1 8 28 56 70 56 28 8 1

Each entry in this row, from left to right, may be interpreted as the number of outcomes with 8 heads, 7heads, 6 heads, 5heads,...,1 head, and 0 heads, respectively.

d. 2^n; the total number of possible outcomes when you flip a coin n times.

2. a. $15/2^6$ **b.** $93/2^8$

3. a. 1/32 **b.** 31/32 **c.** 176/1024

Activity 11.3.1

1. a. {RBY, BRY, BYR, YBR, YRB} **b.** 3; 2; 1; 3 × 2 × 1

c. {RYBG, RYGB, RBYG, RBGY, RGYB, RGBY, YRBG, YRGB, YBGR, YBRG, YGBR, YGRB, BYRG, BYGR, BGYR, BGRY, BRGY, BRYG, GRYB, GRBY, GYBR, GYRB, GBYR, GBRY}; 24

d. 4! **e.** 5!; n!

2. a. BA means B is in 1st place and A is in 2nd place; a runner can't finish in both 1st and 2nd place; 20

b. 5, 20 **c.** 5 **d.** $\frac{5!}{(5-3)!}$

3. a. It's the same as {A,B}; order doesn't matter here; 10 **b.** ${}_5C_2 = {}_5P_2/2!$

c. 10; ABC, ABD, ABE, ACD, ACE, ADE, BCD, BCE, BDE, CDE **d.** ${}_nC_r = {}_nP_r/r!$

4. a. Combinations. A 5-card hand is the same no matter the order of the cards; 2,598,960

b. Permutations. Changing the order of letters changes the name of the radio station; 360

Activity 11.4.1

1. Answers will vary

2. a. There are 12 months; there are 5 people in the group; each month must have an equally-likely chance of being drawn.

b. Answers will vary: ex. Assign head and 1 through 6 to January – June, tails and 1 through 6 to July – December. Flip the coin and roll the die 5 times and note whether any outcome is repeated.

3. a. Answers will vary **b.** # of trials with a repeated month ÷ 20.

4. a. All 5 people have their birthday in a different month.

b. Two people can't have the same birth month.

c. 95,040 **d.** $1 - 95,040/12^5$, approx. 0.62 **e.** Answers will vary

Activity 11.4.2

1. a. Answers will vary **b.** 1/4, 1/4 **c.** No.

d. Give an equal number of points for match and no match.

2. a. $1 and $4 **b.** 50; 25; 25 **c.** A loss of $25 **d.** $0.25

e. No, in the long run you can expect to lose money.

f. "-3" indicates a loss of $3; on average, you will lose $0.25 per game.

3. c. approx. $0.11; no.

Exercise

A BIRD THAT KEEPS DROPPING PEARLS OF WISDOM!

Connections to the Classroom

1. Suppose you have a red die and a green die – rolling a 3 on the red die and 4 on the green die is a different outcome than rolling a 4 on the red die and a 3 on the green die.
2. Since the sample space S for an experiment contains all possible outcomes, $P(S) = \frac{\text{\# of outcomes in S}}{\text{\# of outcomes in S}} = 1$. Thus, the probability of an event must be less than or equal to 1.
3. No, 2 and 6 are even numbers *and* divisors of six; P(even or divisor of 6) = 3/6 + 4/6 – 2/6 = 5/6.
4. **a.** No, there are 36 possible outcomes.
   b. Ex: if the first die is even, then the number on the second die will tell you the month in Jan – June. If the first die is odd, then the number on the second die will tell you the month in July – Dec.
5. No, the probability of each flip of a coin is 1/2.

Mental Math

Hot Dog = 1/4; Hamburger = 1/2; Fishwich = 1/10; Not smiling on Pizza Day = 17/20

CHAPTER 12

Hands-on Activities

Activity 12.1.1

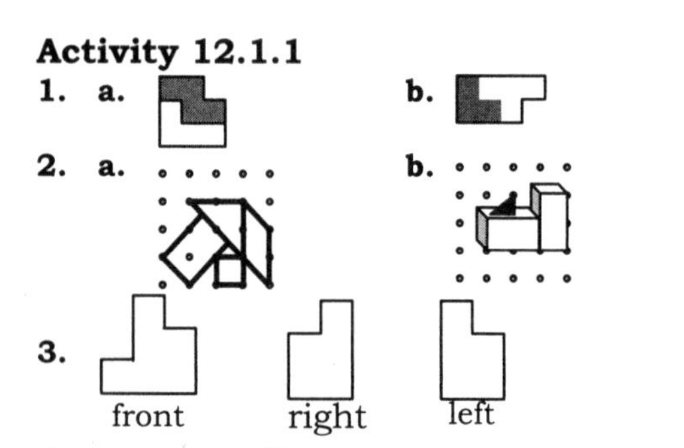

1. **a.** **b.** **c.**
2. **a.** **b.**
3.

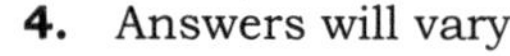

4. Answers will vary

Activity 12.1.2

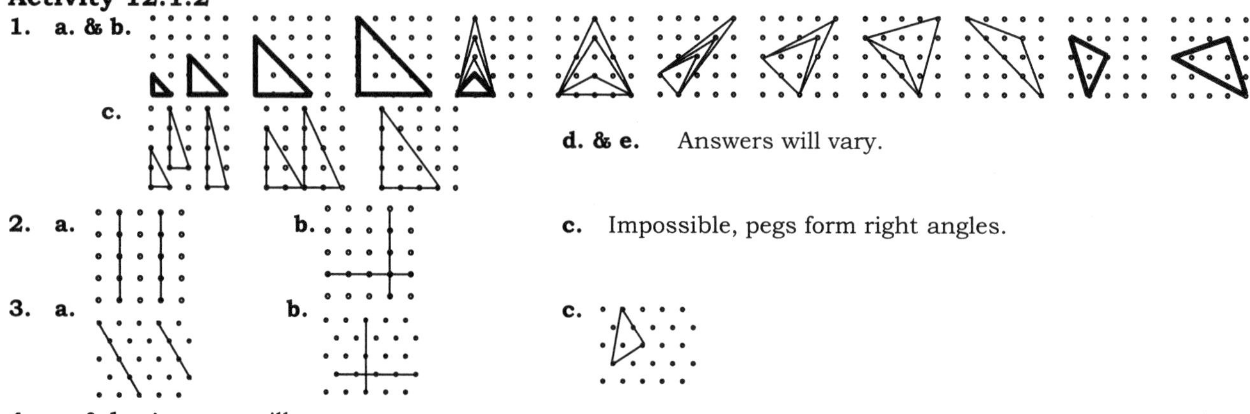

1. **a. & b.**
   c.
   d. & e. Answers will vary.
2. **a.** **b.** **c.** Impossible, pegs form right angles.
3. **a.** **b.** **c.**
4. **a. & b.** Answers will vary.
   c. Squares have four right angles. **d.** No, not all rectangles have four sides the same length.
   e. 4 right angles implies adjacent sides are perpendicular and opposite sides are parallel.
   f. No, it has only one pair of parallel sides.

Activity 12.2.1

1. **a.** C, circle, infinitely many; E, equilateral triangle, 3; F, isosceles triangle, 1; H, rhombus, 2; I, isosceles trapezoid, 1; J, pentagon, 1; K, kite, 1; R, rectangle, 2; S, square, 4.
   b. P, parallelogram; Q, quadrilateral; T, trapezoid.
2. 5
3. **a.** 1/4, 360, 90° **b.** 1/2, 3/4 **c.** 360

4. **a.** C, any fraction of a turn or any number of degrees; E, 1/3 turn = 120°, 2/3 turn = 240°; H, 1/2 turn = 180°; P, 1/2 turn = 180°; R, 1/2 turn = 180°; S, see example.
b. F, I, J, K, Q, T
5. **a. – c.**

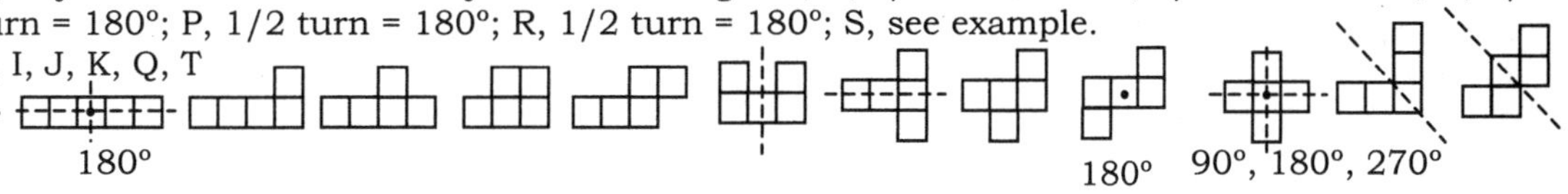

180° 180° 90°, 180°, 270°

Activity 12.2.2

1. **a.** 5 **b.** 2; 0
2. Answers may vary; the following are examples:
- **a.** Parallelogram, justification same as example, except you see the diagonals do not match.
- **b.** Square, a tracing of one central angle exactly matches the other 3 central angles. This implies the 4 central angles are all 90°; rectangle, a tracing of one central angle does not match 2 of the other central angles.
- **c.** Parallelogram, trace each diagonal and rotate 180° about the point of intersection of the diagonals – opposite halves of the diagonals match; kite, opposite halves of the diagonals do not match when traced and rotated.
- **d.** Parallelogram, the 2 triangles match when the parallelogram is folded over the diagonal; kite, one pair of triangles match when the kite is folded, but the other pair does not.

3. **a.** Yes, parallel. **b.** No, not perpendicular. **c.** b & c, d & f **d.** b & d, c & d, b & f, c & f

Activity 12.2.3

1. Answers may vary.

Activity 12.4.1

1. The 3 vertex angles fit together to form a straight line.
2. The 4 vertex angles fit together to form a 360° angle.
3. **a.** 3 **b.** 540°
4. **a.** 2; 360° **b.** 3; 540° **c.** 4; 720° **d.** $(n - 2) \times 180°$
5. **a.** 60° **b.** 90° **c.** 108° **d.** 120° **e.** $((n - 2) \times 180°) \div n$

Activity 12.4.2

1. **a.** Equilateral triangle **b.** 6
- **c.** Square, hexagon; (4, 4, 4, 4), (6, 6, 6); pentagon, octagon, decagon, dodecagon.
- **d.** With equilateral triangles, squares, and regular hexagons you can put a whole number of vertex angles together at a point to make exactly 360°, but this is not possible with the other regular polygons.
- **e.** The vertex angle measure must be a divisor of 360°, but the next largest divisor after 120 is 180. No polygon has an vertex angle of 180°.

2. **a.** Both use the same polygons but they are arranged in a different order.
b. & c. Answers will vary
3. **a.** Answers will vary **b.** Each vertex has the arrangement x,y,z,x,y,z
c. The 3 vertex angles of any triangle add to 180°, and 180° is a divisor of 360°.
4. **a.** 360°; draw one copy of each vertex angle of the quadrilateral is at each vertex of the tessellation.
b. Answers will vary. **c.** Yes

Activity 12.5.1

1. **a.** 6 **b.** 2 triangles and 3 rectangles
- **c.** Rectangles, rectangles; isosceles trapezoids, rectangles; pentagons, parallelograms.
- **d.** (a)Right square prism; (c) right rectangular prism, right trapezoidal prism, oblique pentagonal prism.

2. **b.** Right triangular pyramid.
3. **a.** Put 2 square pyramids together.

5.

	F	V	E
Cube	6	8	12
Triangular prism	5	6	9
Square pyramid	5	5	8
Tetrahedron	4	4	6
Octahedron	8	6	12
Dodecahedron	12	20	30
Cube Octahedron	14	12	24

(F + V) – 2 = E

Exercise

VERTEX

Connections to the Classroom

1. **a.** The student does not understand the meaning of the term "base angles."
b. Answers will vary, ex: "If you rotate the triangle on the left, where is the base then?"
c. The triangles are congruent.
2. Yes; use the Pythagorean theorem: $(\sqrt{18})^2 + (\sqrt{2})^2 = (\sqrt{20})^2$
3. **a.** No; if you fold on the line, the 2 sides don't match. **b.** Tracing and paper-folding.
4. The 12 angles are not congruent - 6 are 90° and 6 are 270°.
5. Trace lines and rotate about the point of intersection - vertical angles match up.

Mental Math

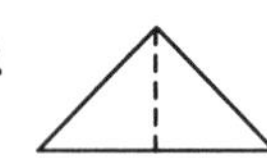 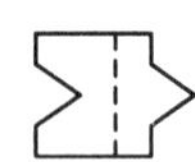 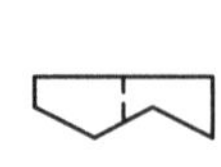 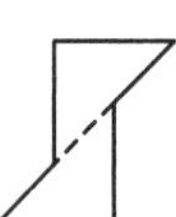

CHAPTER 13

Hands-on Activities

Activity 13.1.1

1. **a. & b.** Answers will vary **c.** Thumbs; easier to approximate a fraction of a unit than hand spans.
d. Answers will vary
2. Answers will vary
3. **c.** 10
4. Answers will vary

Activity 13.1.2

1. $28.00
2. **a.** The units of feet will "cancel out." **b.** This is the number of yards that 13 feet is equal to.
3. **a.** 30 rolls equals 1box
b. 90 rolls × (25 nickels/1 roll) = 2250 nickels; 2250 nickels × (1 dollar/20 nickels) = 112.50 dollars
c. $17,500
4. **a.** 11 feet/second **b.** 17.5 weeks/year

Activity 13.2.1

1. $\sqrt{2}$
2. **a.** 12 **b.** $6 + \sqrt{8}$ **c.** $\sqrt{2} + \sqrt{8} + \sqrt{10} = 3\sqrt{2} + \sqrt{10}$
3. It has the shape of a square.
4. **a.** 5 **b.** 4
5. **a.** 2; The area of the triangle is half the area of the rectangle, 1.
b. triangle + triangle + square = 1 + 1 + 4 = 6.
c. 8; 2; 6
d.

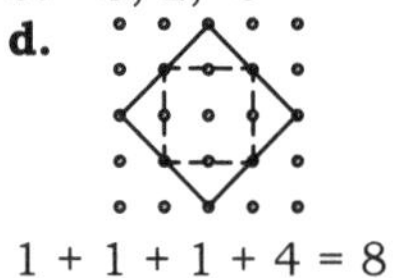  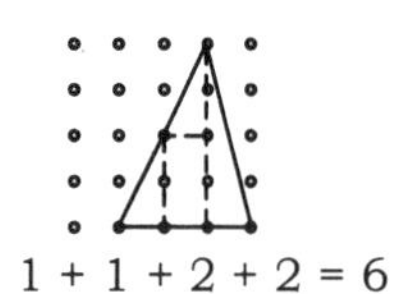 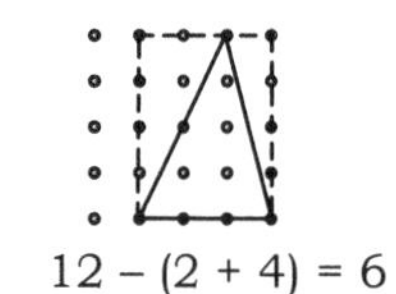  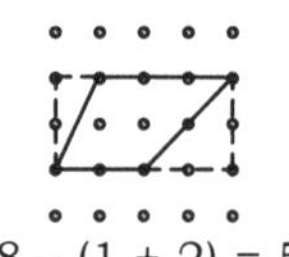

1 + 1 + 1 + 4 = 8 16 – 4(2) = 8 1 + 1 + 2 + 2 = 6 12 – (2 + 4) = 6 1 + 2 + 2 = 5 8 – (1 + 2) = 5

6. **a.** 1.5; $3 + \sqrt{2} + \sqrt{17}$ **b.** 6.5; $7 + \sqrt{5} + \sqrt{2} + \sqrt{8} = 7 + \sqrt{5} + 3\sqrt{2}$

Activity 13.2.2

1. **a.** 6 **b.** Area of rectangle = area of parallelogram; 6 **c.** 3; 2 **d.** area **e.** $b \times h$
2. **a.** The area of triangle is 1/2 the area of the parallelogram.
b. $1/2(b \times h)$ **c.** 6; 6; 6
d. Each base is 3 units and each altitude is 4 units; areas are equal; answers will vary
3. **a.** Yes; rearranging the parts of a shape does not change its area.
b. The base of the parallelogram is the sum of the trapezoid's two bases;
c. It is 1/2 the height of the trapezoid. **d.** 5; 5 **e.** base × height = $(a + b) \times (1/2h)$
4. Find the sum of the areas of the two triangles: $1/2(a \times h) + 1/2(b \times h) = (1/2a + 1/2b) \times h = 1/2(a + b) \times h$

Activity 13.2.3

1. **a.** Answers will vary.
b. One method is to trace the circle onto a piece of paper. Then fold the circle in half and crease. The fold line is a diameter. **c.** π **d.** $\pi \times d$

2. **a.** Parallelogram
b. $1/2C = 1/2(\pi \times d) = \pi \times r$; one-half the circle's sectors are on the bottom of the parallelogram.
c. r **d.** πr^2

Activity 13.3.1

1. **a.** A – cube; B – right triangular prism; C – right trapezoidal prism.
b. 9.375 in.^{2}; 5.25in.^{2}; 9.75in.^{2}
c. Add the area of the 2 bases and the areas of all the lateral faces.

2. **a. & b.**

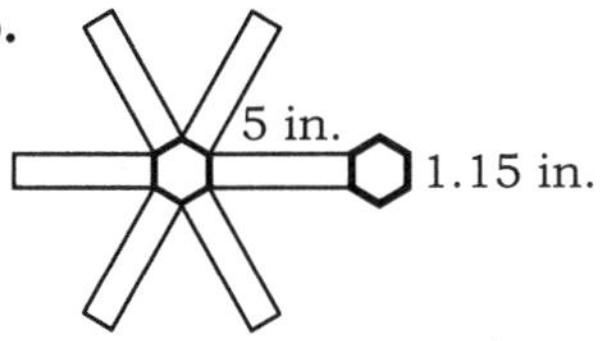

c. 37.95units^2

3. **a** D: right square pyramid; E: oblique square pyramid **b.** 5 in.^{2}; 8 in.^{2}
c. Add the area of the base and the areas of all the triangular faces.

4. **a.** 2 circles and a rectangle.
b. The radius and circumference of the circular base & height of rectangle.
c. $2\pi r^2 + C \times h = 2\pi r^2 + 2\pi r \times h$

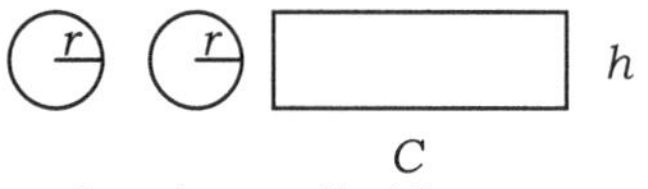

Activity 13.4.1

1. **a.** A cubic unit is a cube that measures 1 unit on all sides.
b.

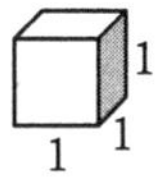

c. A cubic unit is 3-dimensional.

2. **a.** 6; 2; 12 **b.** 2, 4, 8; 5, 2, 10; 8, 3, 24
c. Area of the base is the same as the number of cubes in the bottom layer and you multiply this by the number of layers (the height).

3. **a.** They both have a square base with side length 2 cm. and their heights are both 3 cm.
b. It is 1/3 the volume of the prism. **c.** 1/3(Area of the base) × (height)

Activity 13.4.2

1. **a.** 9π cm^2 **b.** 9π cm^3 **c.** 10; 90π cm^3
d. Area of base = the number of cubes in one layer and height = the number of layers.

2. **a. b. & c.** Answers will vary
d. Changing the radius of a cylinder affects the volume more than changing the height because the radius is squared in the formula.

3. **a.** Same size base and height. **b.** Answers will vary **c.** 3; it is 1/3 the volume of the prism.
d. $1/3(\pi r^2 \times h)$

Exercise

Answers will vary

Connections to the Classroom

1. **a.** 14 **b.** No, he counted each corner twice.

2. **a.** Incorrect, examples will vary.
b. Incorrect, examples will vary; the area will be 4 times as big.

3. **a.** Different formula from activity.
b. Same formula: $1/2(ch) + ah + 1/2(dh) = 1/2(c + 2a + d)h = 1/2(c + a + d + a)h = 1/2(a + b)h$

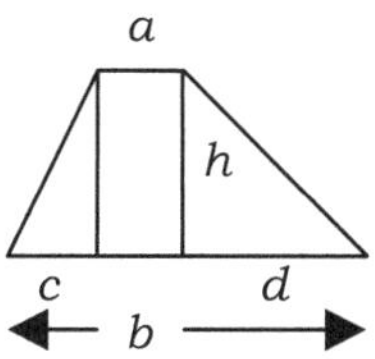

4. **a.** No, it will not be a right circular cone.
b. Cut out a sector of a circle.

5. **a.** Yes, each triangular face has a base of 5 and height of 6, so one triangle has an area of $1/2(5 \times 6)$. Multiply by 4 for the four identical lateral faces.

b. No, it only works if the pyramid is a right regular pyramid.
c. Yes, but the perimeter of the base is the circumference of a circle.

Mental Math

A→3, B→1, C→2

CHAPTER 14

Hands-on Activities

Activity 14.1.1

1. **a.** Yes **b.** Vertex L; vertex A; vertex S; $\overline{LA}$, $\overline{AS}$, $\overline{SL}$ **c.** 1st and 3rd
2. **a.** $\Delta TOP \cong \Delta YJP$ **b.** $\Delta TOY \cong \Delta JYO$ **c.** $\Delta TOY \cong \Delta YJO$

Activity 14.1.2

1. **b.** m **e.** Yes; yes
f. If two triangles have a correspondence of vertices such that all three pairs of corresponding sides are congruent, then the triangles are congruent.
2. **d.** Yes; yes
e. If two triangles have a correspondence of vertices such that two pairs of corresponding sides and the pair of included angles are congruent, then the triangles are congruent.
3. **a.** No; for example:

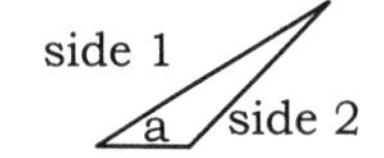

b. Yes, the two sides with unknown length can only meet in one point.
c. Yes, if the measures of two angles are known, then the third is also known, so this is ASA.
d. No; for example:

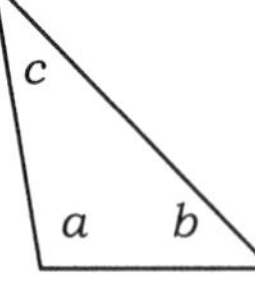

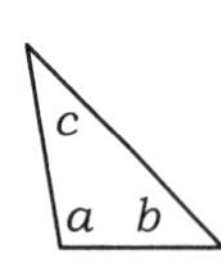

e. SSS, SAS, and AAS are congruence properties, but SSA and AAA are not.

Activity 14.2.1

1. **a.** 55° **b.** No, the corresponding sides are not congruent.
c. All approximately 0.75. **d.** Yes.
2. **a.** Whether two pairs of corresponding sides are proportional in length and the corresponding included angles are congruent **b.** The two triangles are similar.
c. No, the third side is unique because you must connect the endpoints of the other two sides.
3. **a.** No; for example: **b.** Yes.

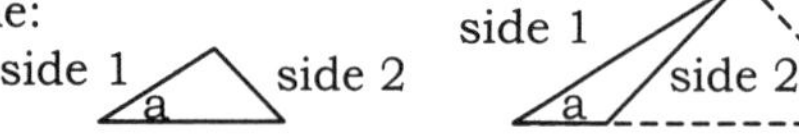

Activity 14.2.2

1. **a.** Answers will vary **b.**

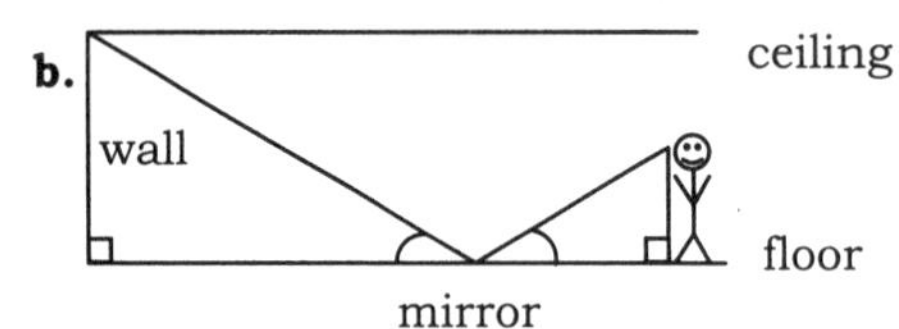

c. AA Similarity
d. & e. Answers will vary
2. Answers will vary; ratios of corresponding sides are equal; shadow length will change, but ratios of corresponding sides will remain equal.
3. **a. & b.** Answers will vary **c.** AA Similarity **d.** Answers will vary

Activity 14.3.1

1. See constructions in Section 14.3 of your textbook.
2. Construct a pair of perpendicular lines to form a right angle of a triangle; on one leg measure off two units with your compass, on the other leg measure off one unit; connect these two points; this third side of the triangle has length square root of 5 since $1^2 + 2^2 = 5$.
3. **a.** Construction #6: construct a line perpendicular to a line (the base of the triangle) through a point not on the line (the opposite vertex).

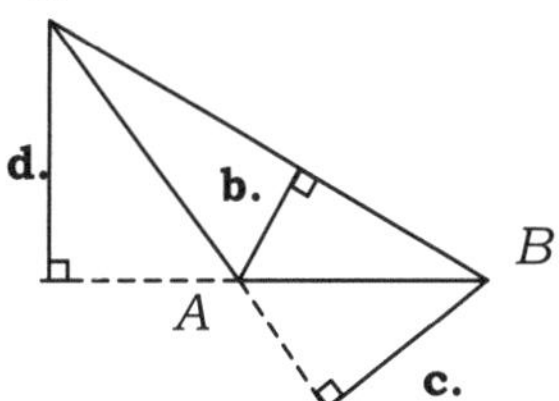

4. **a.** Construction #3: construct the perpendicular bisector of a line segment to find the midpoint of a side. Then connect the midpoint to the opposite vertex.
   b. Answers will vary
   c. When you have an equilateral triangle.
   d. Two of the altitudes coincide with the legs of the triangle.

Activity 14.3.2

1. **d.** The perpendicular bisector of $\overline{PQ}$; it is the reverse of the Perpendicular Line Segments Test.
2. **a.** Fold the paper at the vertex of the angle so that the two sides of the angle are superimposed.
   b. The angle on one side of fold exactly matches the angle on the other side of the fold.
3. **c.** A perpendicular to line through a point on the line.
4. **c.** The circumcenter of an acute triangle is inside the triangle. For an obtuse triangle, it is outside. The circumcenter of a right triangle is the midpoint of the hypotenuse.
5. **c.** The incenter is located inside the triangle for all types of triangles.
6. **c.** The orthocenter falls inside an acute triangle and outside an obtuse triangle. The orthocenter of a right triangle is the vertex of the right angle.

Activity 14.4.1

1. **e.** The circumcenter, the centroid, and the orthocenter.

Activity 14.4.2

1. **a.** By construction, all three sides are the same length.
   c. Bisecting the 3 angles created 3 more points on the circle that are halfway between the 3 vertices of the triangle – this makes the 6 sides the same length.
   d. For a regular 12-gon, bisect the 6 central angles of the regular hexagon and connect the points on the circle. For a regular 24-gon, bisect the 12 central angles of the regular 12-gon.
2. **b.** Answers will vary; the fewest number of right angles that must be constructed is 1 – construct the angle first and mark off lengths on two sides of the angle. With the same compass setting, construct arcs from these endpoints that meet to form the fourth vertex.
   c. The circumscribed circle; because each central angle of a square is 90°; angles were bisected forming eight equal length sides.
   d. Bisect the 8 central angles of the regular octagon and connect the points on the circle.
3. **c.** $\frac{\sqrt{5}-1}{2}$ **e.** Decagon **f.** Connect every other point on the circle.
   g. Bisect each of the ten central angles to form 10 more points on the circle and connect all 20 points.

Exercise

Answers will vary

Connections to the Classroom

1. No; the two triangles are congruent, but the student has not written the congruence relationship correctly – corresponding vertices should be in the same order: ΔABC is congruent to ΔEFD.
2. **a.** No, the measure of angle X is 30°. **b.** 1/2 is the ratio of corresponding sides.
   c. Trace angle X and compare the tracing to the corresponding 30° angle in the other triangle.

3. Two pairs of corresponding angles congruent is enough to know that two triangles are similar.

4. a. She constructed a perpendicular to a line through a point not on the line; because an altitude is perpendicular to the base of a triangle through the opposite vertex.

b. She should have used side $\overline{BC}$ as the base (*A* as the vertex); instead she used side $\overline{AC}$ as the base (*B* is the opposite vertex).

c. Answers will vary.

Mental Math 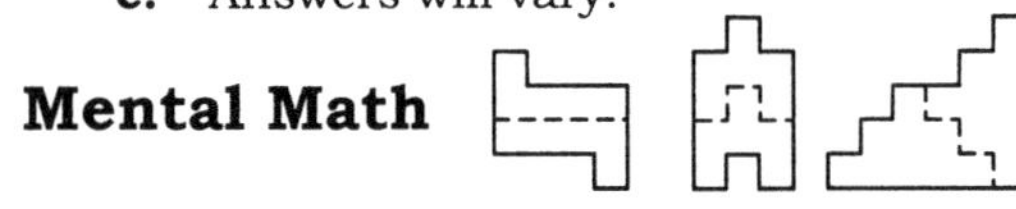

CHAPTER 15

Hands-on Activities

Activity 15.1.1

1. a. (4, 1) **b.** (1,2) **c.** (0, 0)

2. b. parallelogram

3. b. A: I; B: I; C: II; D: IV; E: III; F: IV

c. Negative *x*-coordinate and positive y-coordinate; both coordinates are negative; positive *x*-coordinate and negative y-coordinate.

4. b. (1, 1) or (6, 5) **c.** 4 or 5, 5 or 4 **d.** $\sqrt{41}$ units **e.** $\sqrt{41}$ units

5. a. (c, b) **b.** $c - a$, $d - b$ **c.** $\sqrt{(c-a)^2 + (d-b)^2}$ **d.** ≈ 11.4

6. a. 14; answers will vary **b.** Yes; for example $4^2 + 3^2 = 25$, $\sqrt{25} = 5$

c. Rational → 1, 2, 3, 4, 5; irrational → $\sqrt{2}$, $\sqrt{5}$, $\sqrt{8}$, $\sqrt{10}$, $\sqrt{13}$, $\sqrt{17}$, $\sqrt{18}$, $\sqrt{20}$, $\sqrt{32}$

Activity 15.1.2

1. a. 1 **b.** 1; the slope is always 1.

c. $\overline{AB}$ is steeper; answers will vary; answers will vary; a steeper line has a greater slope.

2. a. -3/4 **b.** The horizontal change is negative.

c. A line with positive slope goes up from left to right, and a line with negative slope goes down from left to right.

3. a. 0/3; 0 **b.** 3; 0

4. b. 1/3 **c.** $y_1 - y_2$; $x_1 - x_2$; $(y_1 - y_2)/(x_1 - x_2)$

d. $y_2 - y_1$; $x_2 - x_1$; $(y_2 - y_1)/(x_2 - x_1)$ **e.** The first and second points are reversed.

f. $(y_2 - y_1)/(x_2 - x_1) = (y_1 - y_2)/(x_1 - x_2)$

Activity 15.1.3

1. a. 6/5

b. Choose any point not on the given line and go up 6 and right 5 to get another point on the new line. **c.** They are parallel.

2. a. 2/1; -1/2 **b.** Their product is -1. **c.** 4/3; yes **d.** 4/3

3. a. Answers will vary; the product of the slopes of one pair of adjacent sides must equal 1.

b. Answers will vary; use the distance formula to find *AB* and *BC* and show that $AB^2 + BC^2 = AC^2$.

4. a. Check whether opposite sides have the same slope.

b. Check whether the product of the slopes of adjacent sides equals 1.

5. a. The origin. **c.** Yes, there are two: (7, -9) and (13, -1).

Activity 15.2.1

1. a. -4, 0, 4, -2, 2, -6 **b.** A line.

c. 2; Yes, it is the coefficient of *x*; (0, -4); the *y*-intercept is the -4 in the equation.

d. Points on the line may vary, ex: (-1, 14/3), (0, 4), (1, 10/3); slope = -2/3, y-intercept = 4

e. *m* is the slope, *b* is the *y*-intercept.

2. a. 2 **b.** 3/4; from (0, 2), go up 3 and right 4 OR down 3 and left 4.

3. b. The *x*-coordinate = 3, for every point on the line. **c.** $x = a$

4. b. The *y*-coordinate = 2, for every point on the line. $y = 2$ **c.** $y = b$

d. Yes, there is no *mx* term, because $m = 0$.

e. No, the slope of this line is undefined.

5. a. -4 **b.** 7; yes. **c.** $y = -4x + 7$

Exercise

A. Reflection symmetry in the y-axis. B. Reflection symmetry in the x-axis.
C. Reflection symmetry in the x-axis and y-axis; rotation symmetry in the origin.

Connections to the Classroom

1. Hayes: denominator should be -2 – (-5) = -2 + 5; Farzaneh: has $(y_1 - y_2)/(x_2 - x_1)$.
2. **a.** No, it does have a slope, slope = 0.
b. Answers will vary, ex: Ask the student to write the equation of a line in slope-intercept form.
3. Ask the student to graph a line with slope 1 and a line with slope -1. How does the steepness of these lines compare? What does a negative slope mean?
4. **a.** No, on a horizontal line, all the y-coordinates are the same so its equation has the form $y = a$; on a vertical line, all the x-coordinates are the same so its equation has the form $x = \text{b}$.
b. Ask the student to write the coordinates of some points on each of the given lines.

Mental Math

1. p **2.** o **3.** n **4.** l **5.** m **6.** q

CHAPTER 16

Hands-on Activities

Activity 16.1.1

2. **b.** It is congruent and parallel. **c.** Yes **d.** D
3. **a. & b.** They are congruent. **c.** They are parallel and point in the same direction.
4. **d.** It is equal **e.** They are congruent.
5. **a. & b.** They are congruent **c.** Same shape and size.
6. **d.** They are the same point **e.** They are perpendicular.
7. **a.**

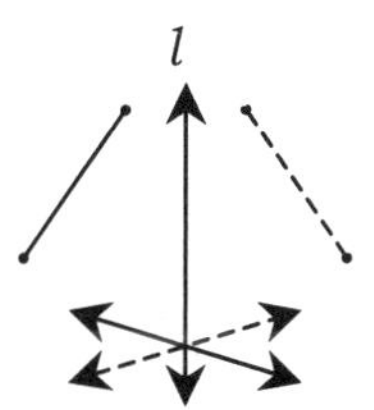

b.

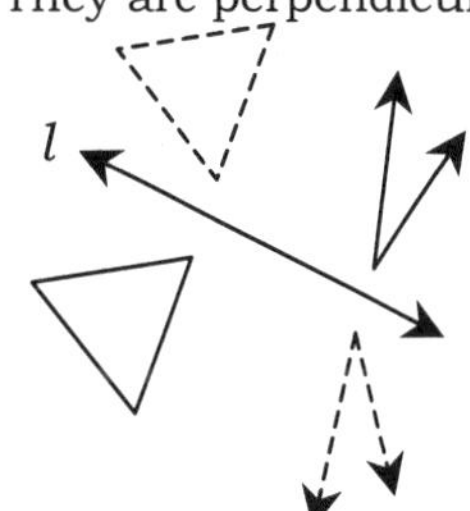

8. **a.** No, every point in the plane moves the same distance and the same direction.
b. Yes, the vertex of the directed angle.
c. Yes, any point on the reflection line.

Activity 16.1.2

1. **b.** P" is the same point, no matter the order of the translation and reflection.
2. **c.** Yes
3. **c.** They are different.
4. **e.** (d); when combining two reflections, order doesn't matter if reflection lines are perpendicular.

Activity 16.1.3

2. **a.** Yes **b.** $\overline{P'S'}$ is 3 times as long as $\overline{PS}$. **c.** They are parallel.
3. **a.** Yes **b.** They are equal. **c.** They are 1/2 the size.
d. They have the same shape and the image triangle is 1/2 the size of the original.

Activity 16.2.1

1. **b.** Fold so that the triangles coincide.
2. **a.** Translation **b.** They are parallel and congruent. **c.** Yes **d.** Same
3. **a.** Rotation **c.** Yes **d.** Same
4. **a.** Translation **b.** Rotation **c.** Reflection

Activity 16.2.2

1. **a.** Length of a side in image triangle/length of corresponding side in original triangle; 0.75
b. *A*

2. **a.** Translation **b.** Translate according to the directed line segment $\overrightarrow{AA'}$.

3. **a.** $k = 2$ or $1/2$; reflection **b.** $k = 4$ or $1/4$; rotation **c.** $k = 1.5$ or $2/3$; translation

4. **a.** No, the line segments formed by connecting corresponding vertices are not parallel; also, vertices have opposite orientation. **b.** No, vertices have opposite orientation.
c. (B) No, the line segments formed by connecting corresponding vertices are not parallel; (E) translation; (F) size transformation with $k = 3/4$, followed by a glide-reflection (reflection and a translation).

Exercise

A. Translation only **B.** Rotation, horizontal and vertical reflections **C.** Rotation
D. Vertical Reflection **E.** Horizontal reflection **F.** Vertical reflection, glide reflection
G. Glide reflection

Connections to the Classroom

1. **a.** Yes, $\overline{OA}$ and $\overline{OA'}$ are perpendicular; also *A'* seems to be in the right place for a 90° roation.
b. Slopes of perpendicular lines have a product of -1.
d. For example, to find *A'*, she would go up 1 and left 2.

2. **a.** Ask the student to try doing his size transformation using one of the vertices of the triangle. Then ask him to choose another point as the center. What does he notice about the two image triangles?
b. Construct 3 lines that connect the center point and extend through each vertex of the triangle; measure the distance from the center to each vertex and mark off 3 more of each of these distances on each line to find the image of the three vertices; connect the three new vertices.

3. **a.** No, reflecting twice would produce an image with the same orientation as the original figure, but these two shapes have opposite orientations.
b. Yes, reflect over a horizontal line and then translate to the right.

Mental Math

1. D **2.** C **3.** A **4.** B

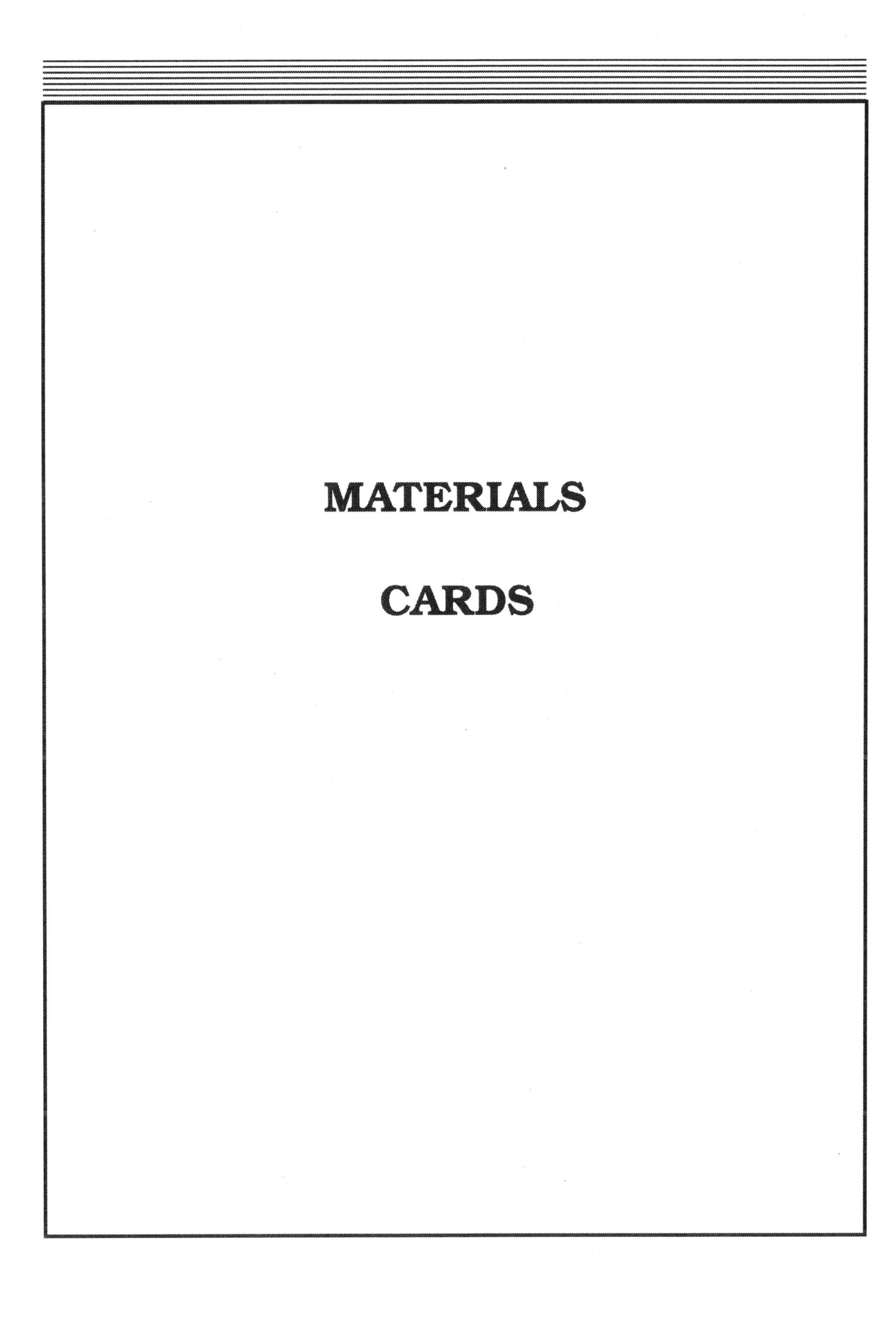
MATERIALS
CARDS

MATERIALS CARD 1.1.3

Remove each of these pieces.

1.1.3 1

1.1.3 2

1.1.3 3

1.1.3 4

1.1.3 5

1.1.3 6

1.1.3 7

1.1.3 8

1.1.3 9

1.1.3 10

1.1.3 11

1.1.3 12

MATERIALS CARD 1.2.1

You may do this activity by placing the circles over the designated letter or you may construct the towers by following the directions given below.

YOU WILL NEED: 3 paper clips and tape. Straighten one end of the paper clip and insert it through the base at point A. Repeat for points B and C. Bend curved portion of clip flat against the bottom of the base and tape in place. Your finished base should look like this:

Remove each circle and punch a small hole in the center of each.

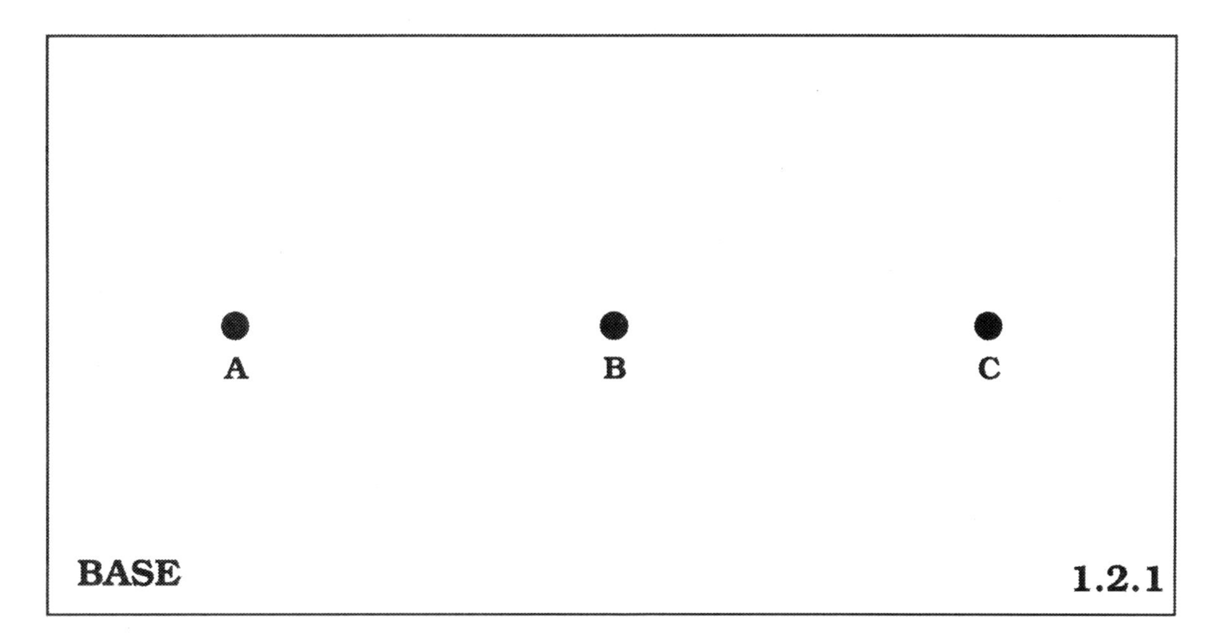

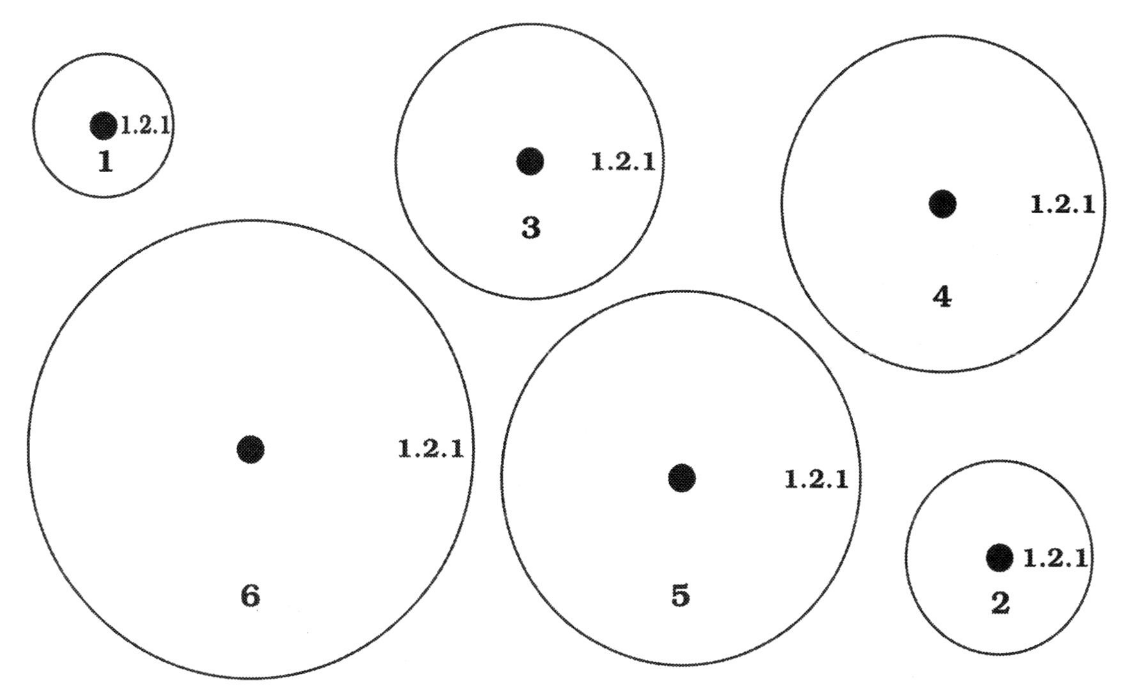

MATERIALS CARD 2.1.1

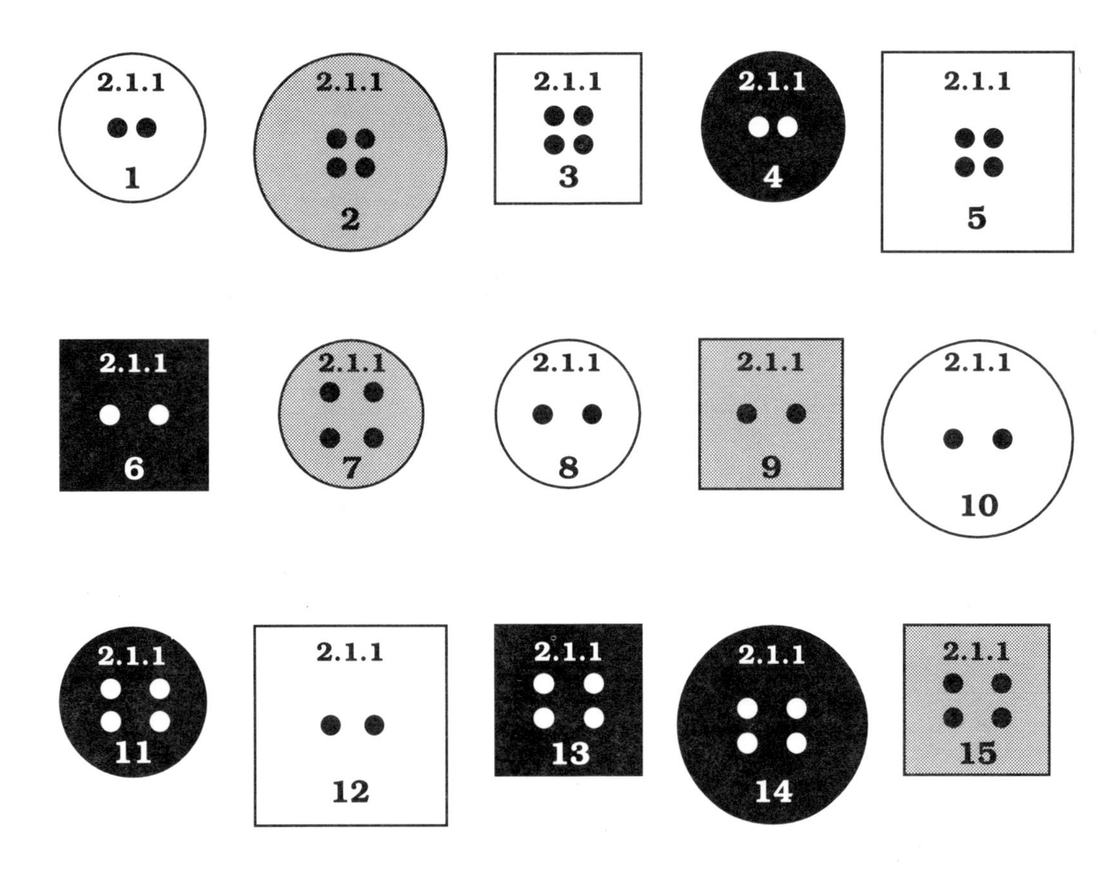

Venn diagram #1

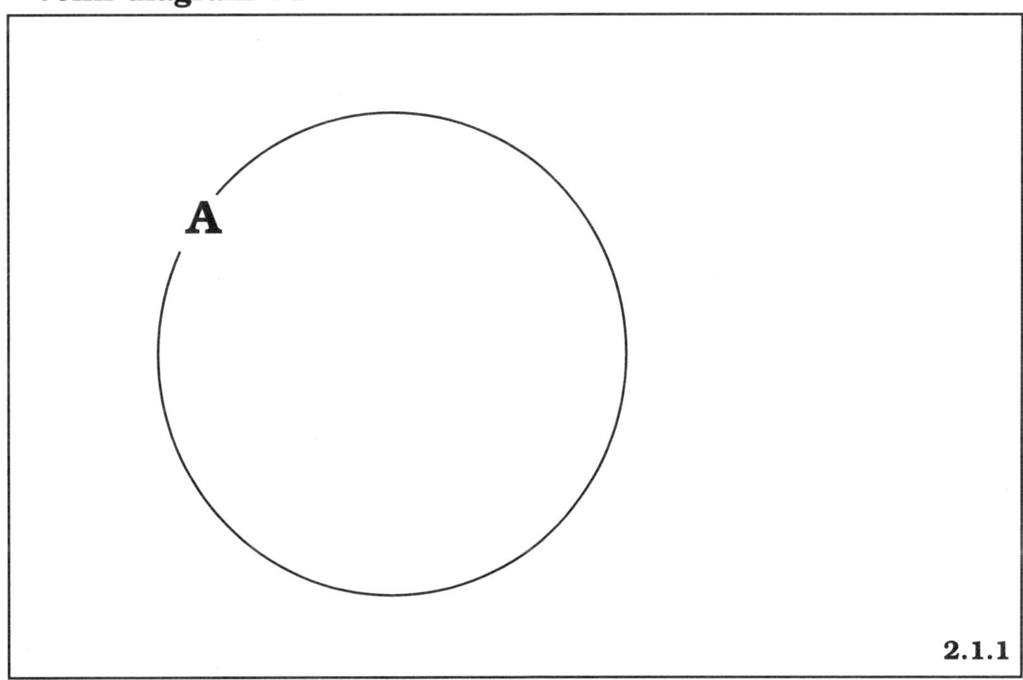

MATERIALS CARD 2.1.1, cont.

Venn diagram #2

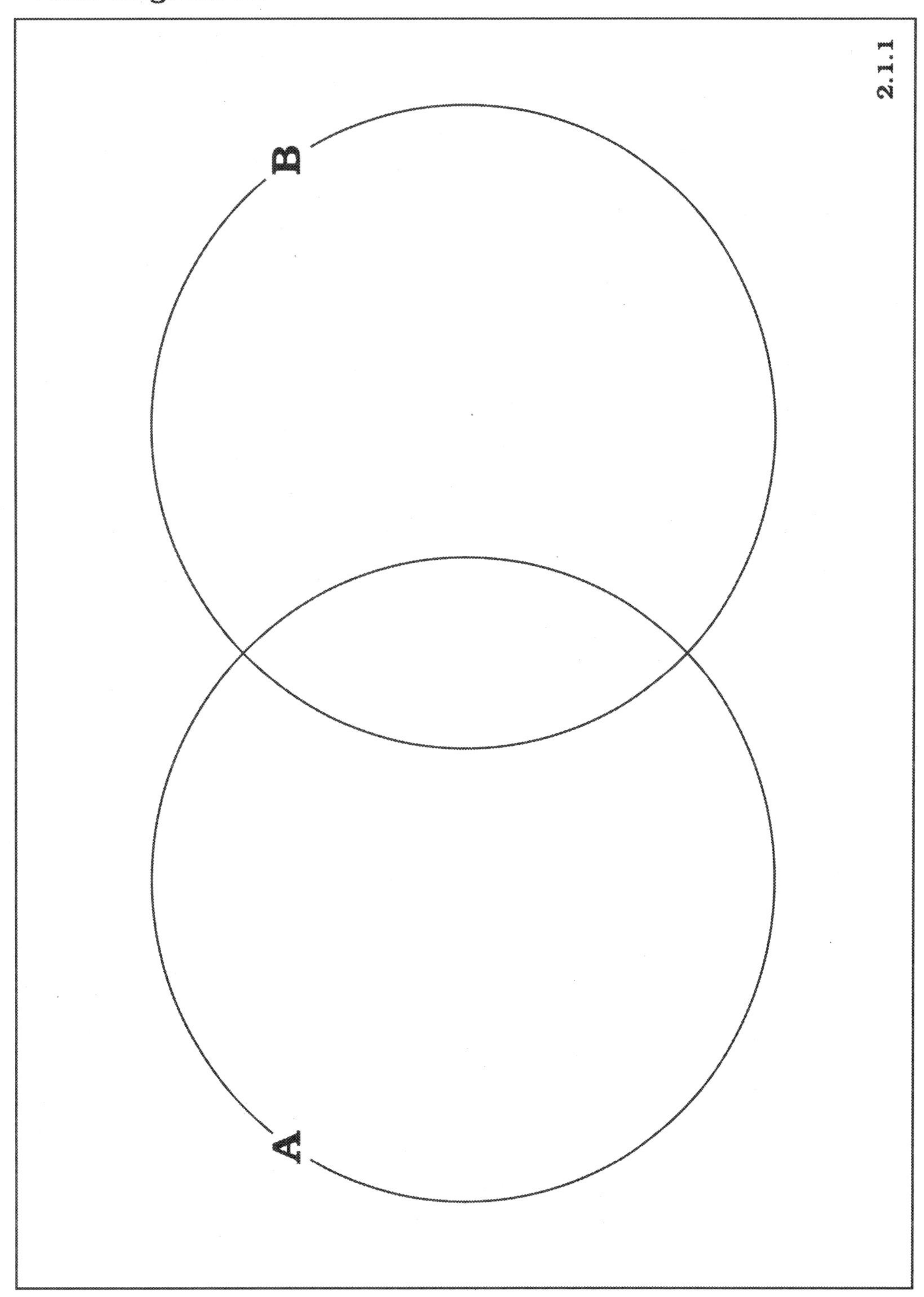

MATERIALS CARD 2.3.1

Remove the following pieces from the materials card.

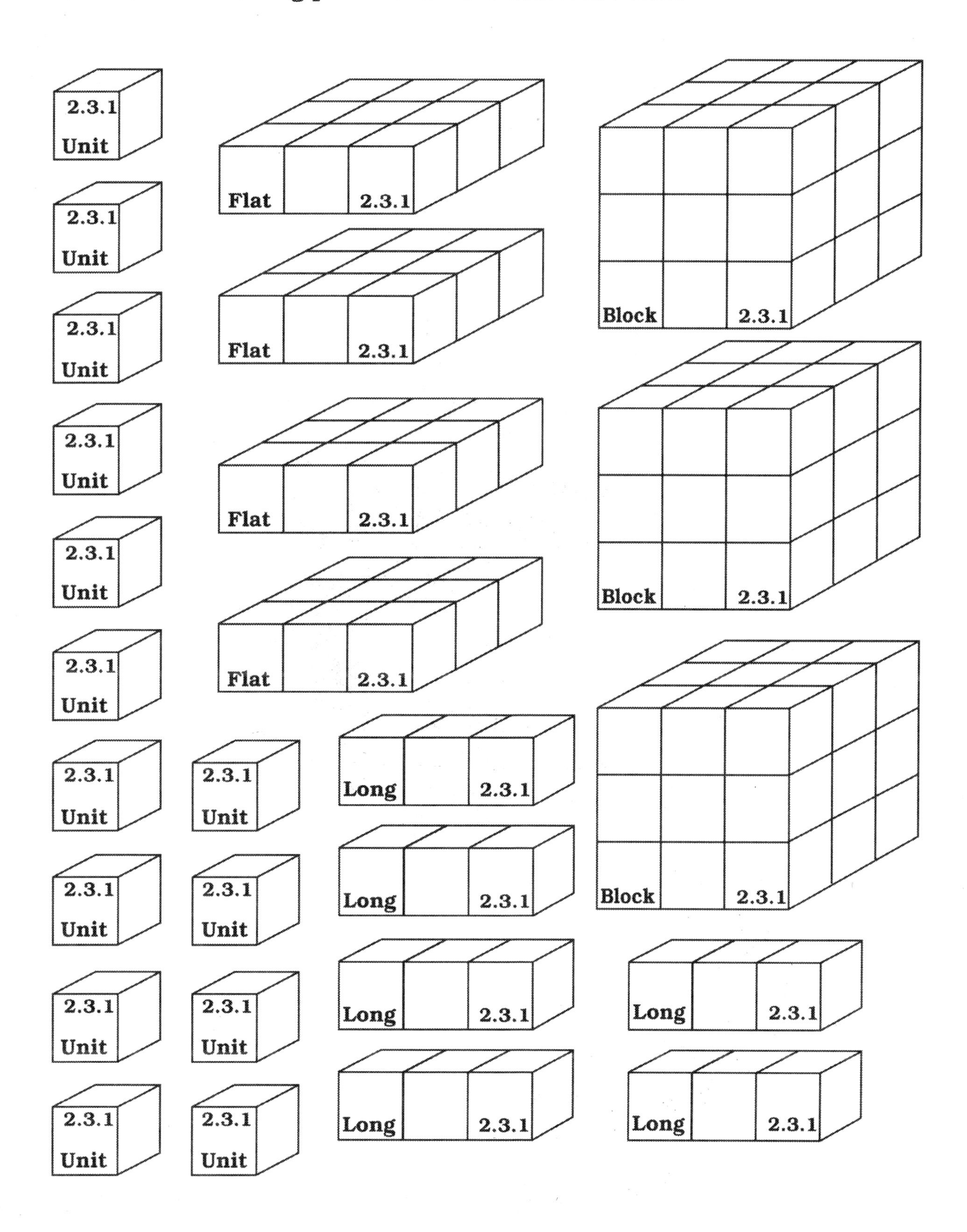

MATERIALS CARD 3.1.1

Remove these strips and color them as labeled.

- **W = White**
- **R = Red**
- **G = Light Green**
- **P = Purple**
- **Y = Yellow**
- **D = Dark Green**
- **K = Black**
- **N = Brown**
- **E = Blue**
- **O = Orange**

Save these pieces for use in Activities 3.1.3, 5.2.1, 6.1.2, and 6.3.3.

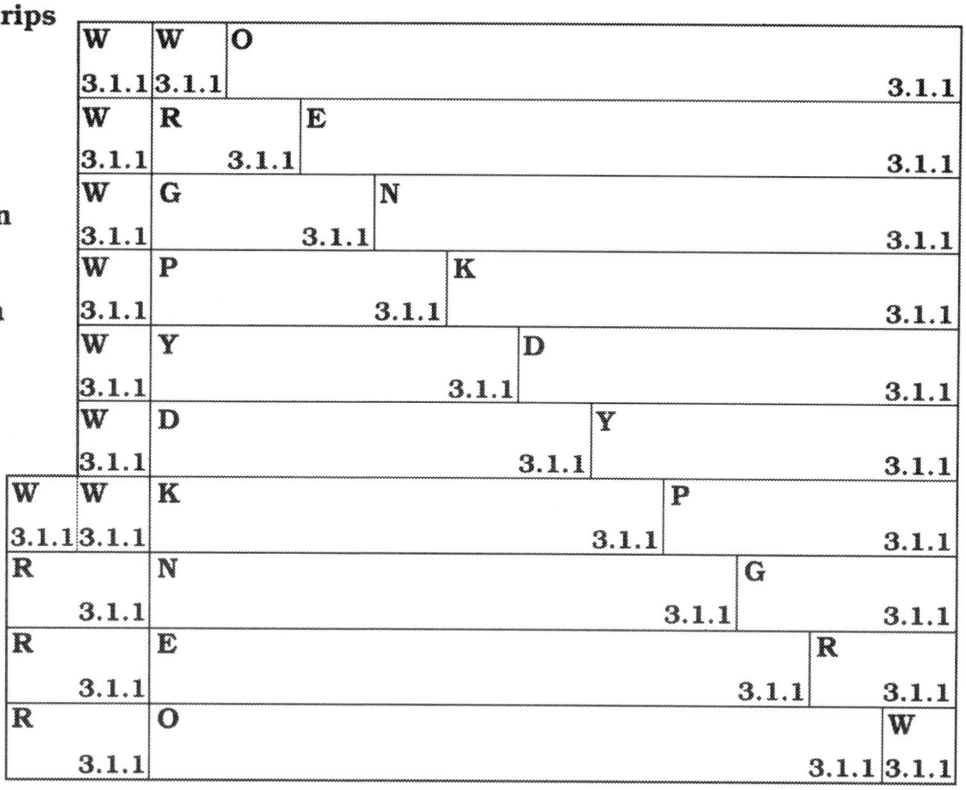

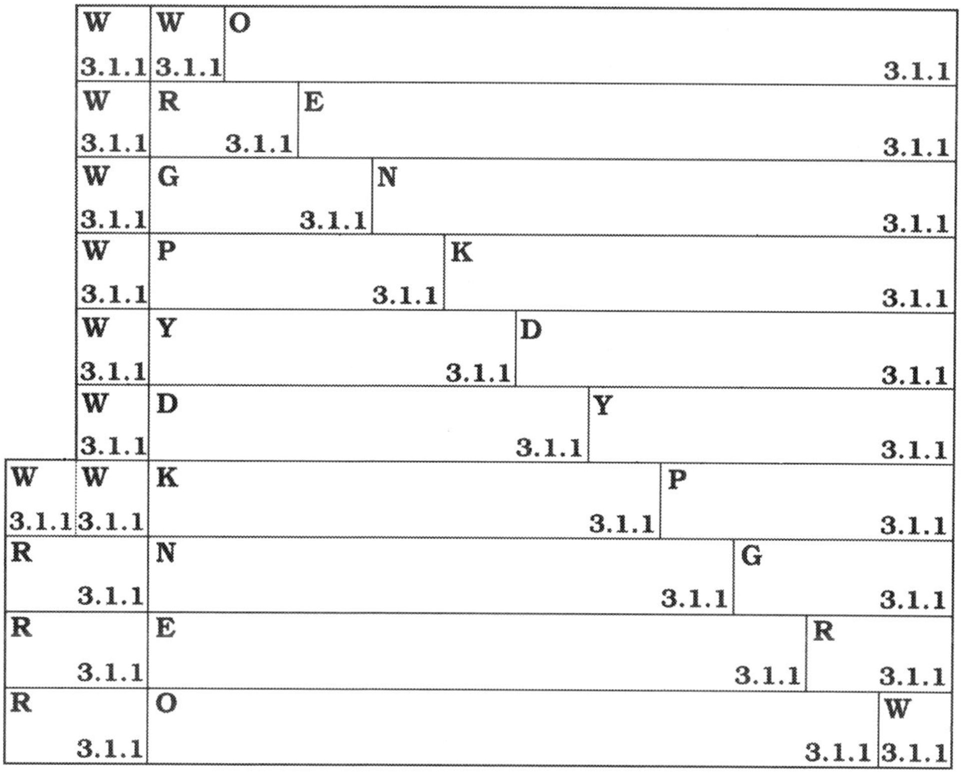

W 3.1.1	W 3.1.1	O 3.1.1	
W 3.1.1	R 3.1.1	E 3.1.1	
W 3.1.1	G 3.1.1	N 3.1.1	
W 3.1.1	P 3.1.1	K 3.1.1	
W 3.1.1	Y 3.1.1	D 3.1.1	
W 3.1.1	D 3.1.1	Y 3.1.1	
W 3.1.1	W 3.1.1	K 3.1.1	P 3.1.1
R 3.1.1	N 3.1.1	G 3.1.1	
R 3.1.1	E 3.1.1	R 3.1.1	
R 3.1.1	O 3.1.1	W 3.1.1	

W 3.1.1	W 3.1.1	O 3.1.1	
W 3.1.1	R 3.1.1	E 3.1.1	
W 3.1.1	G 3.1.1	N 3.1.1	
W 3.1.1	P 3.1.1	K 3.1.1	
W 3.1.1	Y 3.1.1	D 3.1.1	
W 3.1.1	D 3.1.1	Y 3.1.1	
W 3.1.1	W 3.1.1	K 3.1.1	P 3.1.1
R 3.1.1	N 3.1.1	G 3.1.1	
R 3.1.1	E 3.1.1	R 3.1.1	
R 3.1.1	O 3.1.1	W 3.1.1	

MATERIALS CARD 3.1.1 cont.

	W 3.1.1	W 3.1.1	O 3.1.1
	W 3.1.1	R 3.1.1	E 3.1.1
	W 3.1.1	G 3.1.1	N 3.1.1
	W 3.1.1	P 3.1.1	K 3.1.1
	W 3.1.1	Y 3.1.1	D 3.1.1
	W 3.1.1	D 3.1.1	Y 3.1.1
W 3.1.1	W 3.1.1	K 3.1.1	P 3.1.1
R 3.1.1		N 3.1.1	G 3.1.1
R 3.1.1		E 3.1.1	R 3.1.1
R 3.1.1		O 3.1.1	W 3.1.1

MATERIALS CARD 3.1.2

Separate into squares. Save these pieces for use later in Activities 3.2.1, 3.2.2, 5.1.1, 10.2.1, and 12.2.1.

3.1.2	3.1.2	3.1.2	3.1.2	3.1.2	3.1.2	3.1.2	3.1.2
3.1.2	3.1.2	3.1.2	3.1.2	3.1.2	3.1.2	3.1.2	3.1.2
3.1.2	3.1.2	3.1.2	3.1.2	3.1.2	3.1.2	3.1.2	3.1.2
3.1.2	3.1.2	3.1.2	3.1.2	3.1.2	3.1.2	3.1.2	3.1.2
3.1.2	3.1.2	3.1.2	3.1.2	3.1.2	3.1.2	3.1.2	3.1.2
3.1.2	3.1.2	3.1.2	3.1.2	3.1.2	3.1.2	3.1.2	3.1.2
3.1.2	3.1.2	3.1.2	3.1.2	3.1.2	3.1.2	3.1.2	3.1.2
3.1.2	3.1.2	3.1.2	3.1.2	3.1.2	3.1.2	3.1.2	3.1.2
3.1.2	3.1.2	3.1.2	3.1.2	3.1.2	3.1.2	3.1.2	3.1.2

MATERIALS CARD 4.2.1

Remove the pieces on the next three pages. Separate the flats, longs, and units on the heavy dark lines.

Save these pieces for use in Activities 4.2.2 and 7.2.2.

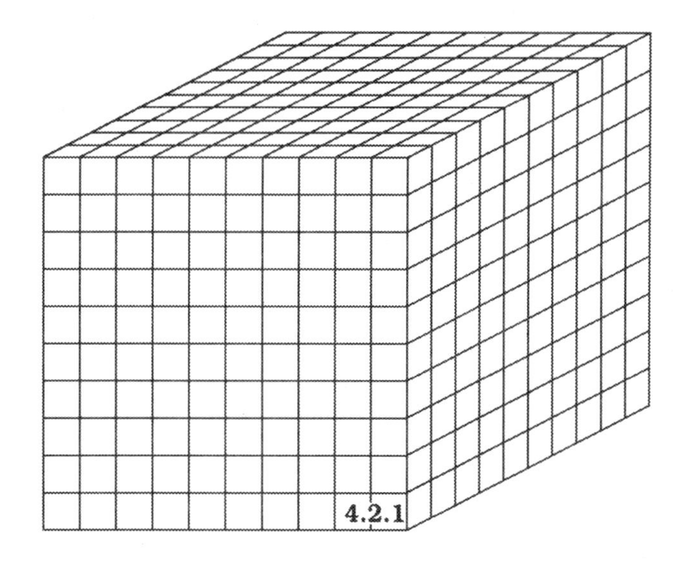

									4.2.1
									4.2.1
									4.2.1
									4.2.1
									4.2.1
									4.2.1
									4.2.1
4.2.1	4.2.1	4.2.1	4.2.1	4.2.1	4.2.1	4.2.1	4.2.1	4.2.1	4.2.1
4.2.1	4.2.1	4.2.1	4.2.1	4.2.1	4.2.1	4.2.1	4.2.1	4.2.1	4.2.1
4.2.1	4.2.1	4.2.1	4.2.1	4.2.1	4.2.1	4.2.1	4.2.1	4.2.1	4.2.1
4.2.1	4.2.1	4.2.1	4.2.1	4.2.1	4.2.1	4.2.1	4.2.1	4.2.1	4.2.1
4.2.1	4.2.1	4.2.1	4.2.1	4.2.1	4.2.1	4.2.1	4.2.1	4.2.1	4.2.1
4.2.1	4.2.1	4.2.1	4.2.1	4.2.1	4.2.1	4.2.1	4.2.1	4.2.1	4.2.1
4.2.1	4.2.1	4.2.1	4.2.1	4.2.1	4.2.1	4.2.1	4.2.1	4.2.1	4.2.1
4.2.1	4.2.1	4.2.1	4.2.1	4.2.1	4.2.1	4.2.1	4.2.1	4.2.1	4.2.1

MATERIALS CARD 4.2.1, cont.

4.2.1

4.2.1

4.2.1

4.2.1

4.2.1

4.2.1

MATERIALS CARD 4.2.1, cont.

4.2.1 4.2.1

4.2.1 4.2.1

4.2.1 4.2.1 4.2.1 4.2.1 4.2.1 4.2.1 4.2.1 4.2.1 4.2.1 4.2.1 4.2.1 4.2.1 4.2.1 4.2.1 4.2.1 4.2.1 4.2.1 4.2.1 4.2.1 4.2.1

MATERIALS CARD 4.3.1

Use coins or paper clips as chips. Save this for use later in Activity 4.3.2.

			4.3.1

MATERIALS CARD 6.1.1A

Remove the circles and then cut on all solid lines. Save these circle pieces for use later in Activities 6.1.3, 6.2.1 and 6.3.1.

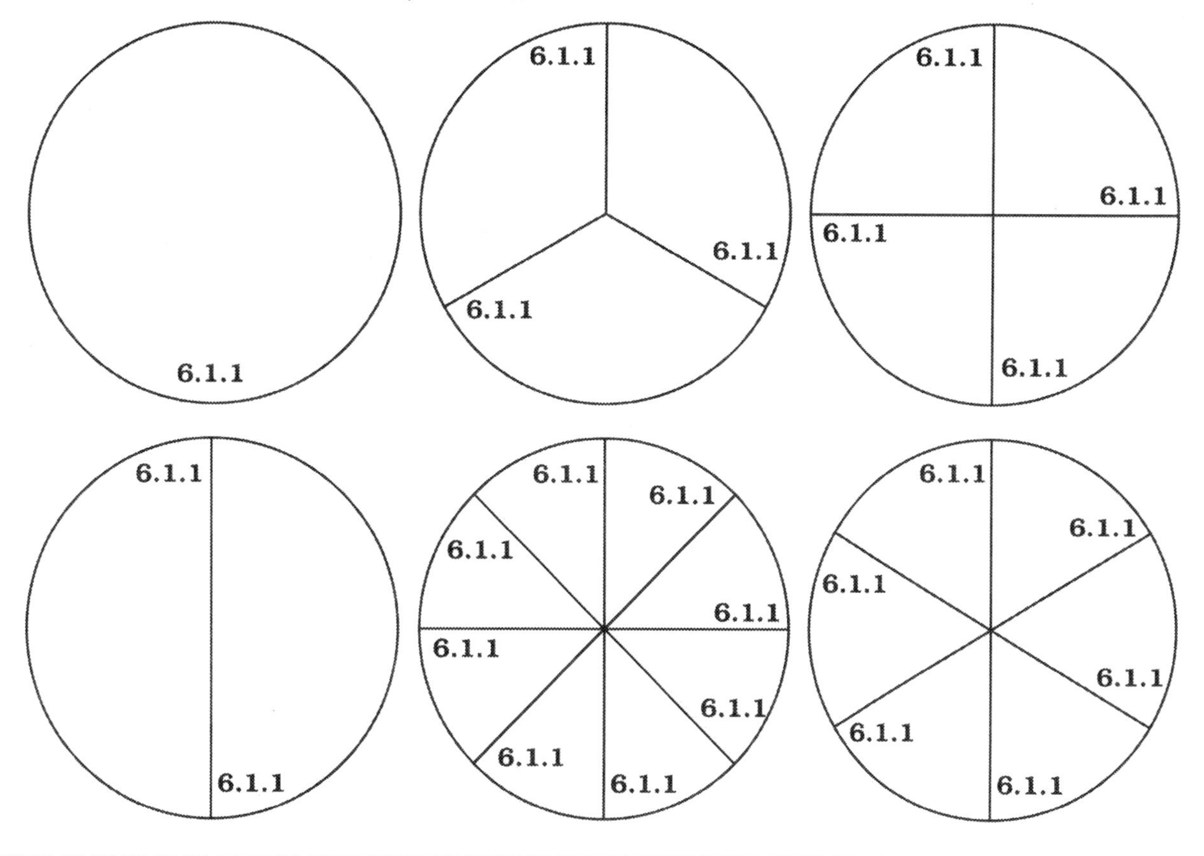

MATERIALS CARD 6.1.1B

Remove the fraction strips on this and the next card. Separate them at the heavy dark lines. Save these strips for use later in the Activities 6.1.3 and 6.2.2.

					6.1.1
					6.1.1
					6.1.1
					6.1.1
					6.1.1
					6.1.1
					6.1.1

MATERIALS CARD 6.1.1B, cont.

6.1.1

6.1.1

6.1.1

6.1.1

6.1.1

6.1.1

6.1.1

6.1.1

6.1.1

6.1.1

6.1.1

6.1.1

6.1.1

6.1.1

6.1.1

6.1.1

6.1.1

6.1.1

6.1.1

6.1.1

6.1.1

6.1.1

6.1.1

6.1.1

MATERIALS CARD 7.1.1

Use the circle as a decimal point marker.
Save these for use later in Activities 7.1.3 and 7.2.1.

				7.1.1

MATERIALS CARD 7.3.1

Separate into single triangles, rhombuses, trapezoids, and hexagons.

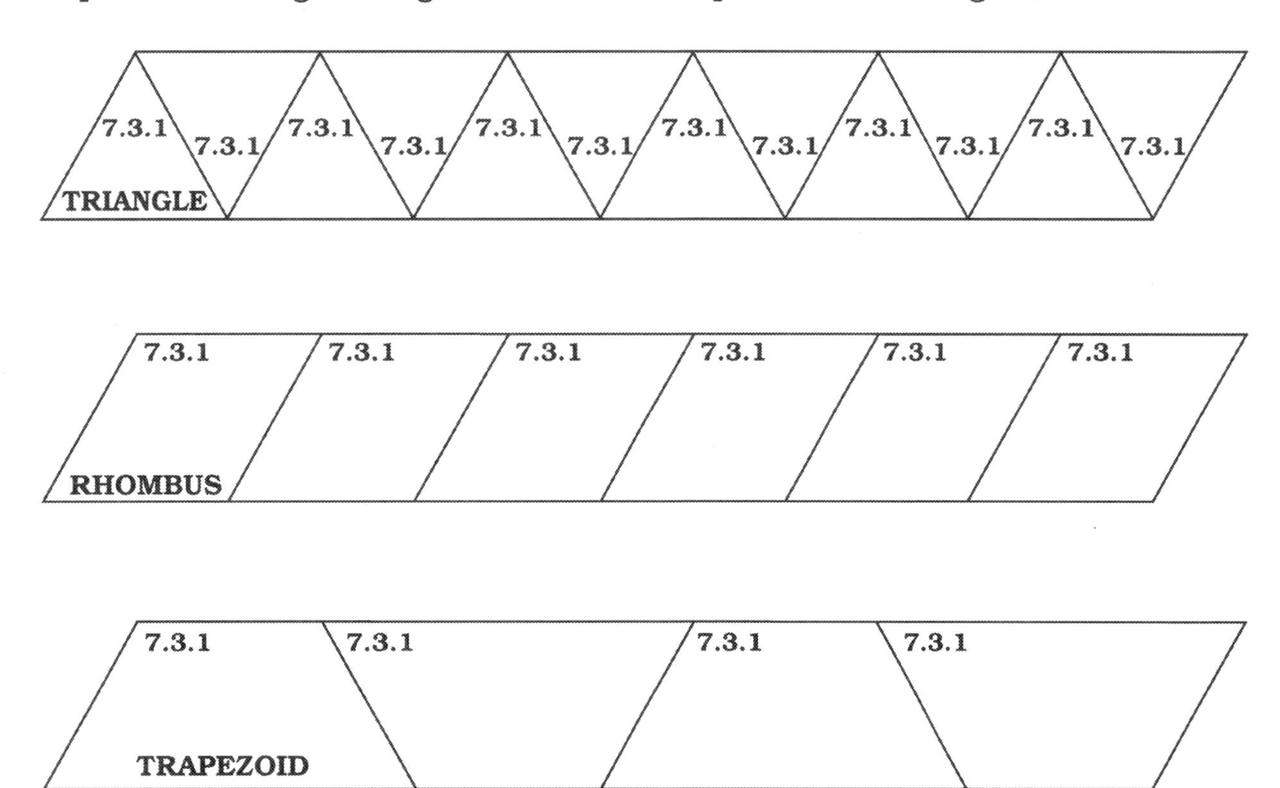

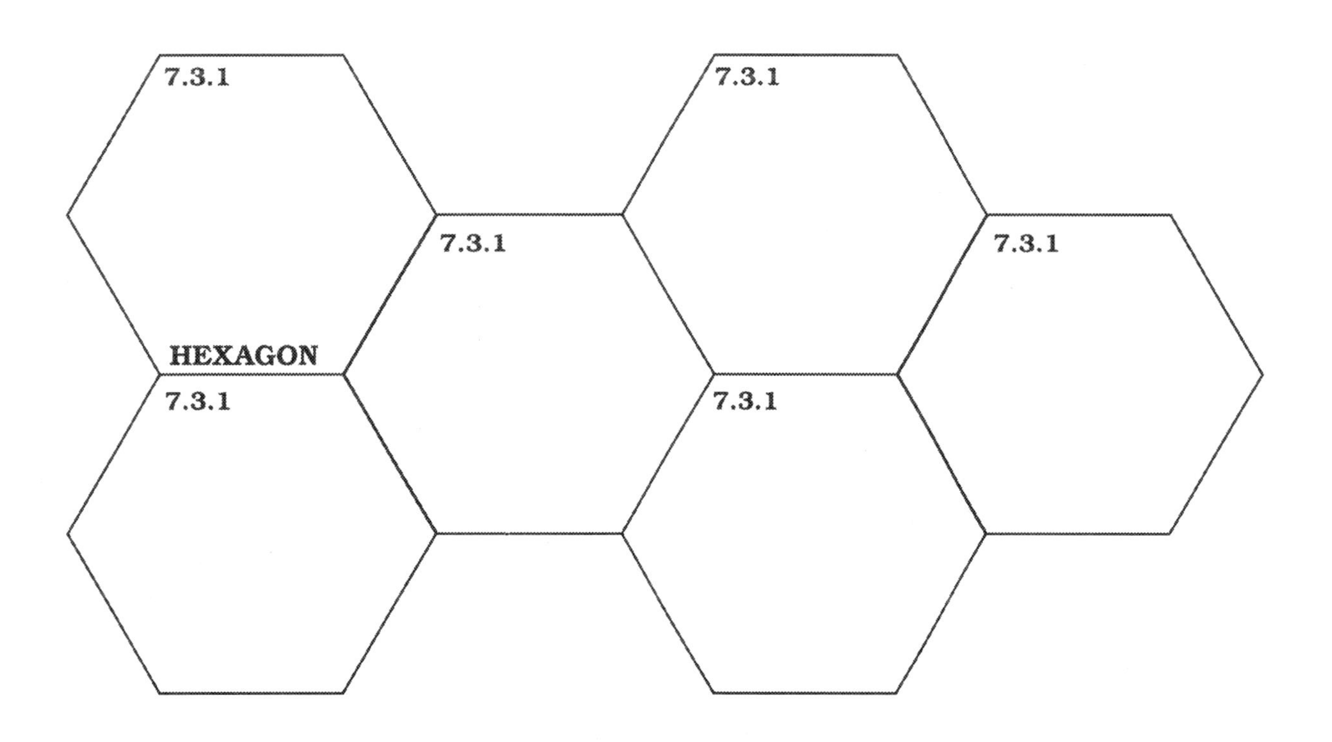

MATERIALS CARD 8.1.1 Remove these chips. Save for use later in Activities 8.1.2, 8.1.3, 8.2.1 and 8.2.2.

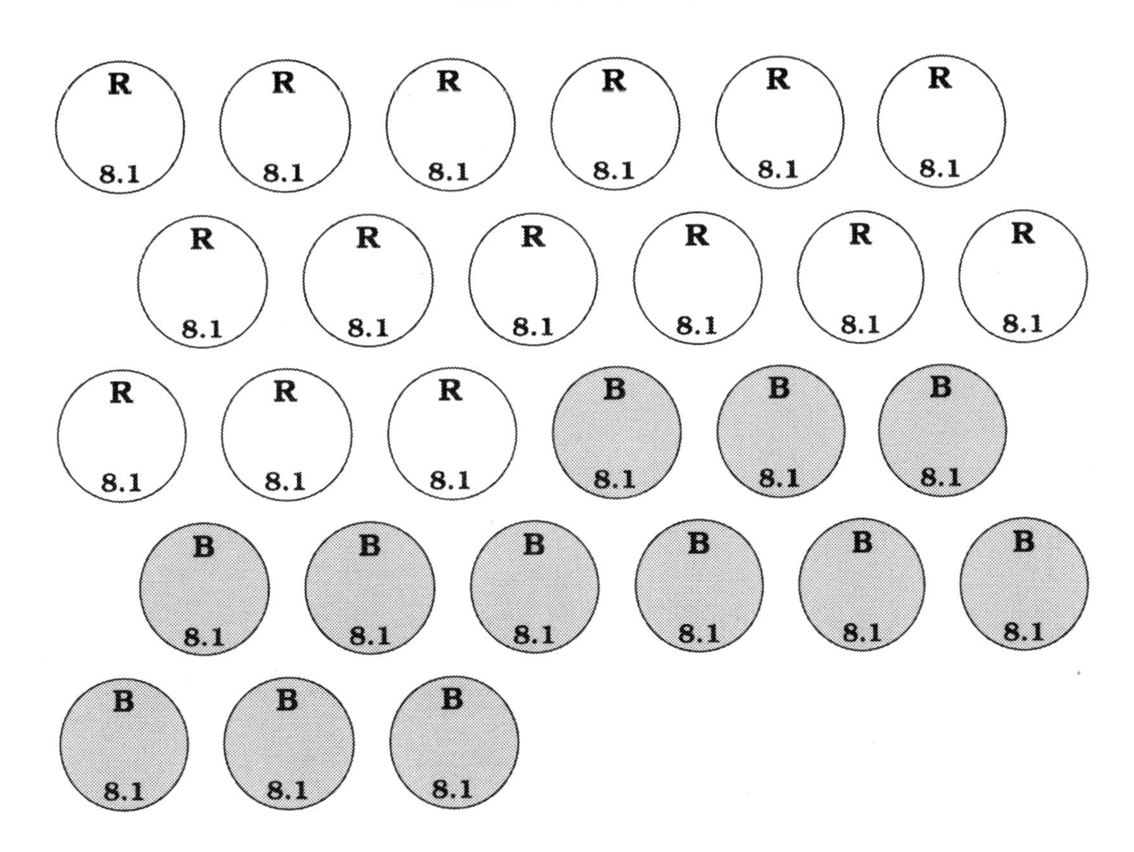

MATERIALS CARD 9.2.1

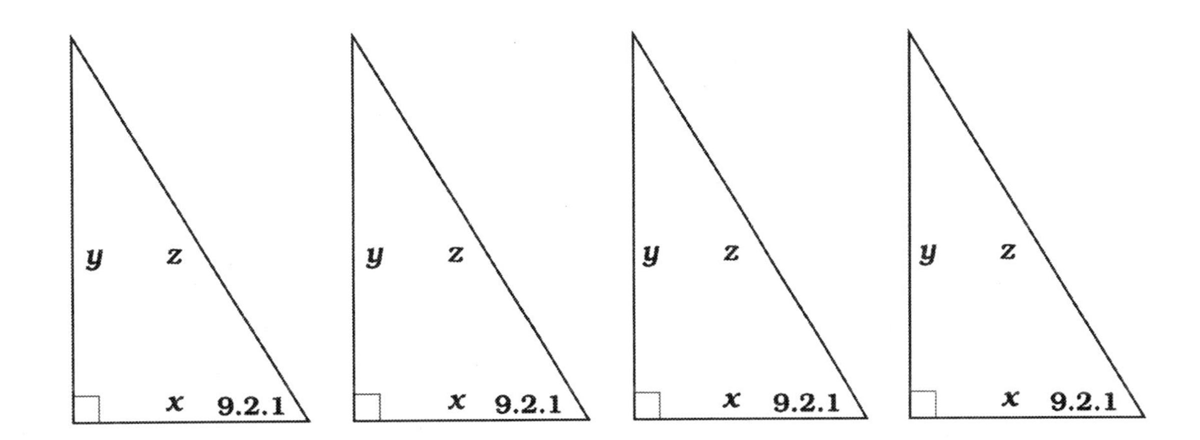

MATERIALS CARD 10.1.1

Save this protractor for use in Activity 14.1.2.

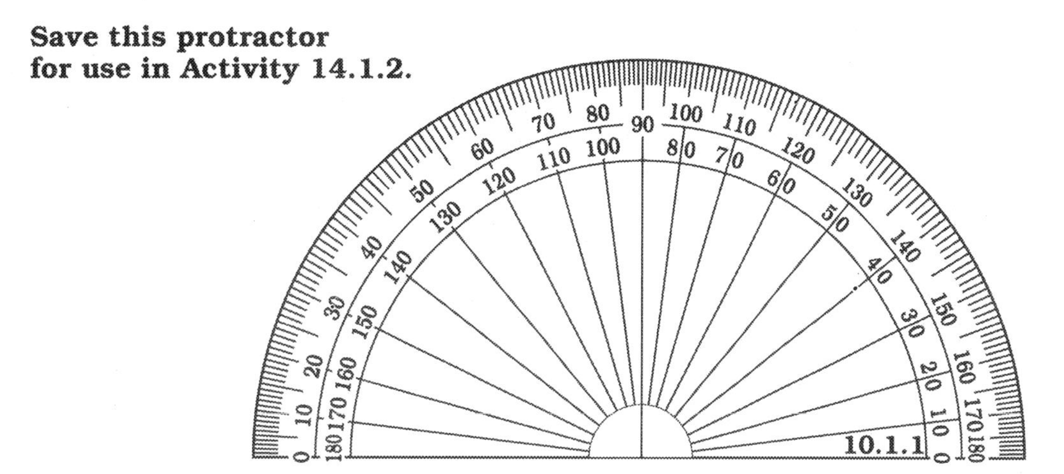

MATERIALS CARD 11.1.1

Remove these cubes. Save these for use later in Activities 11.2.1 and 11.4.1.

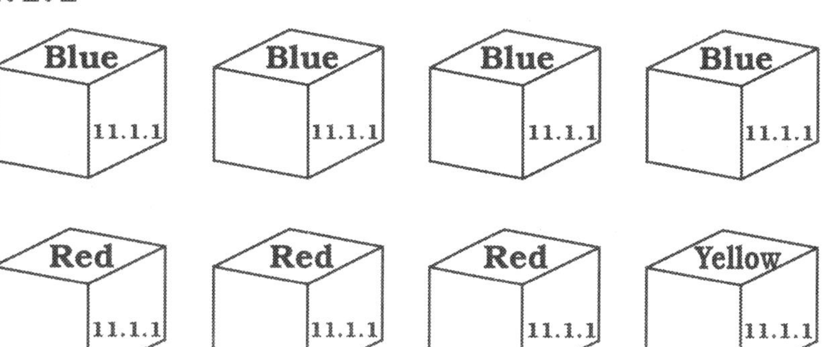

MATERIALS CARD 11.1.2

MATERIALS CARD 11.3.2

Place your pencil point at the center of the disk through the end of a paper clip or safety pin as illustrated.

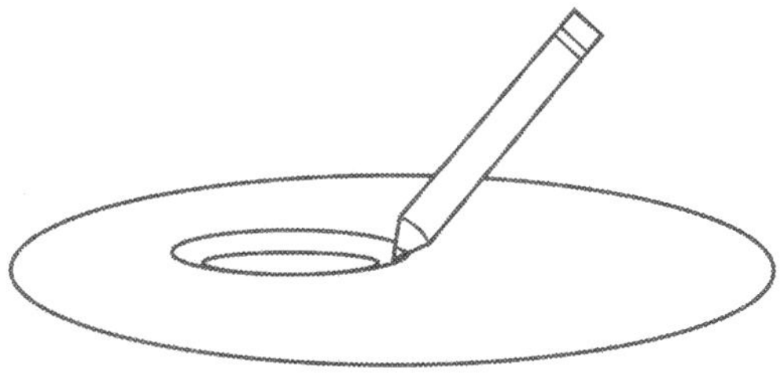

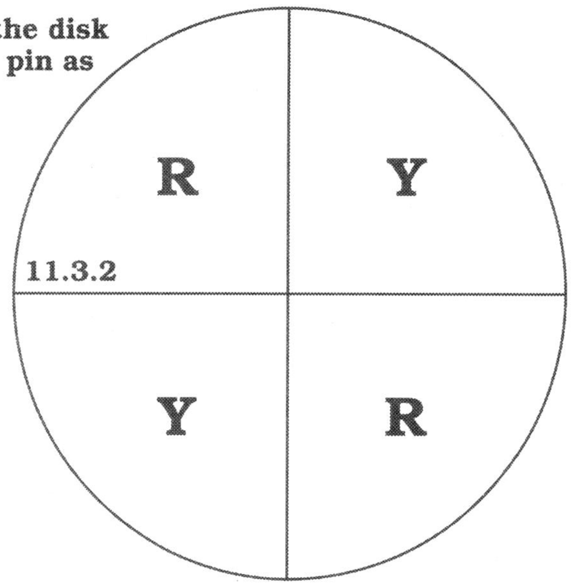

R Y

11.3.2

Y R

MATERIALS CARD 12.2.1

Cut out these shapes and save for use later Activity 12.2.2.

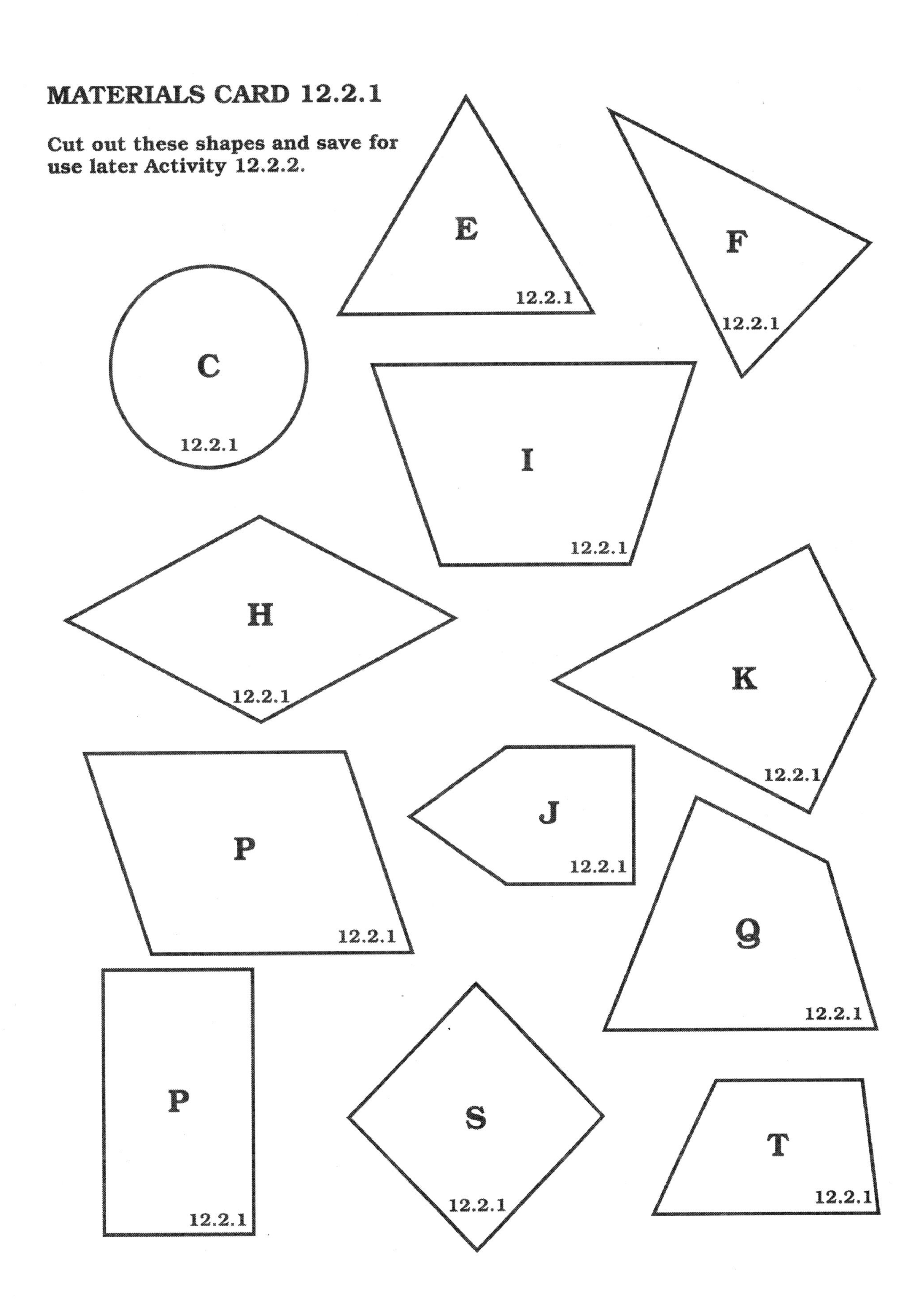

MATERIALS
CARD 12.2.3

Remove the Tangram Puzzle and separate on all of the solid lines. You will have 7 pieces.

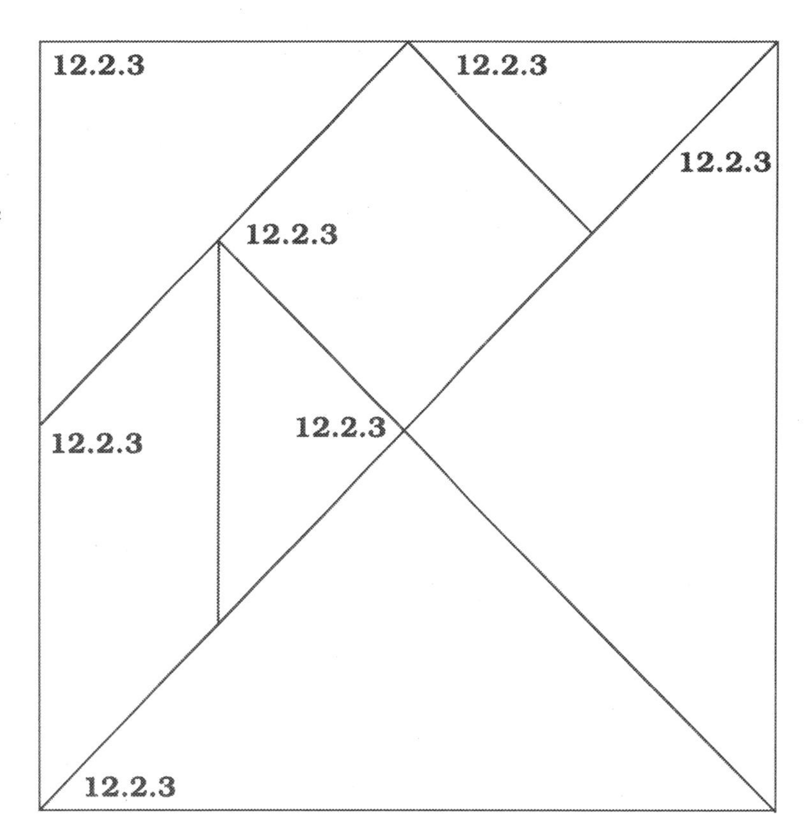

MATERIALS CARD 12.4.2

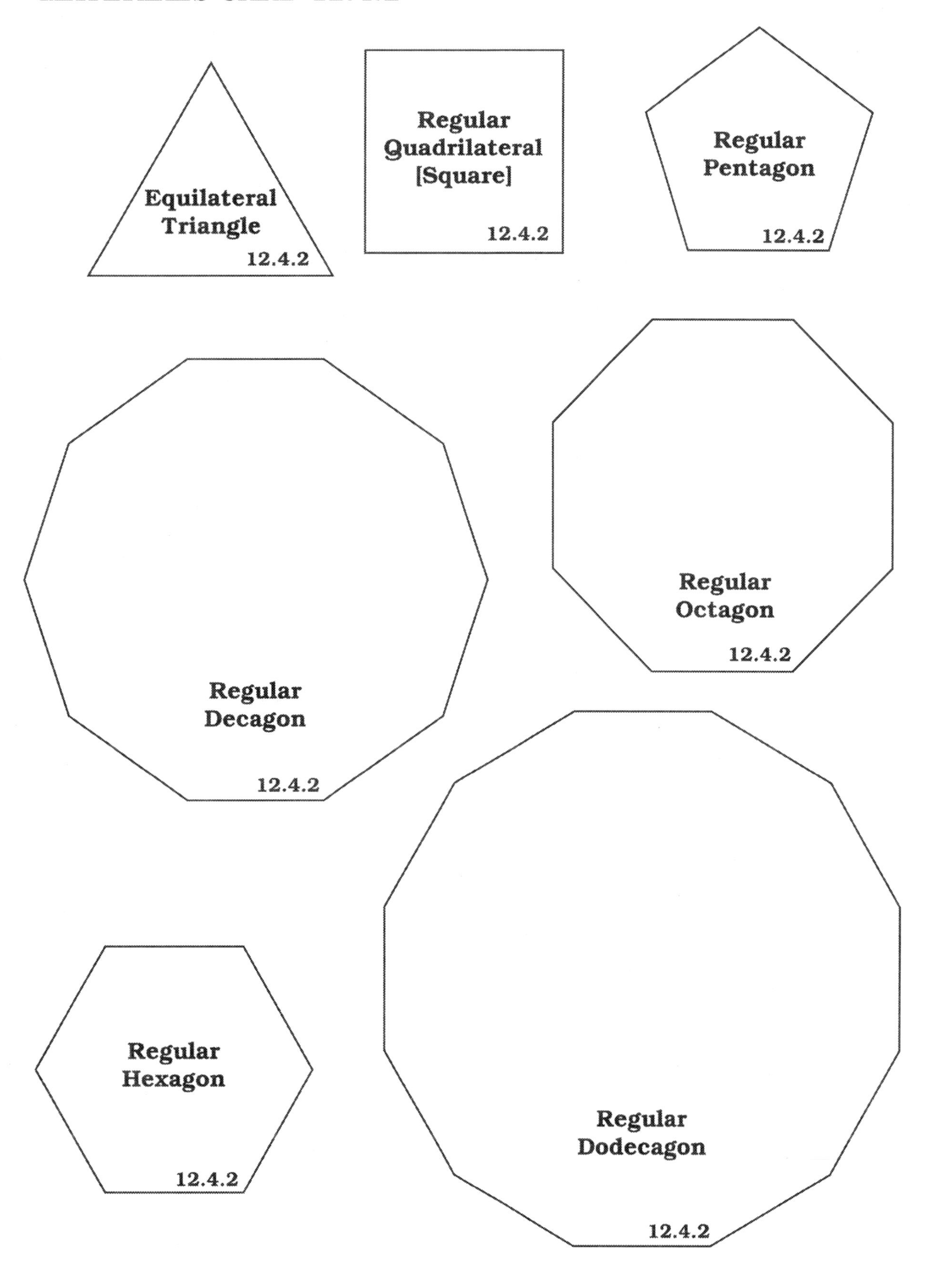

MATERIALS CARD 12.5.1

Use these circles as patterns. Make the number of copies indicated. Fold along the dotted lines and put together with paper clips or tape--flaps on the outside!

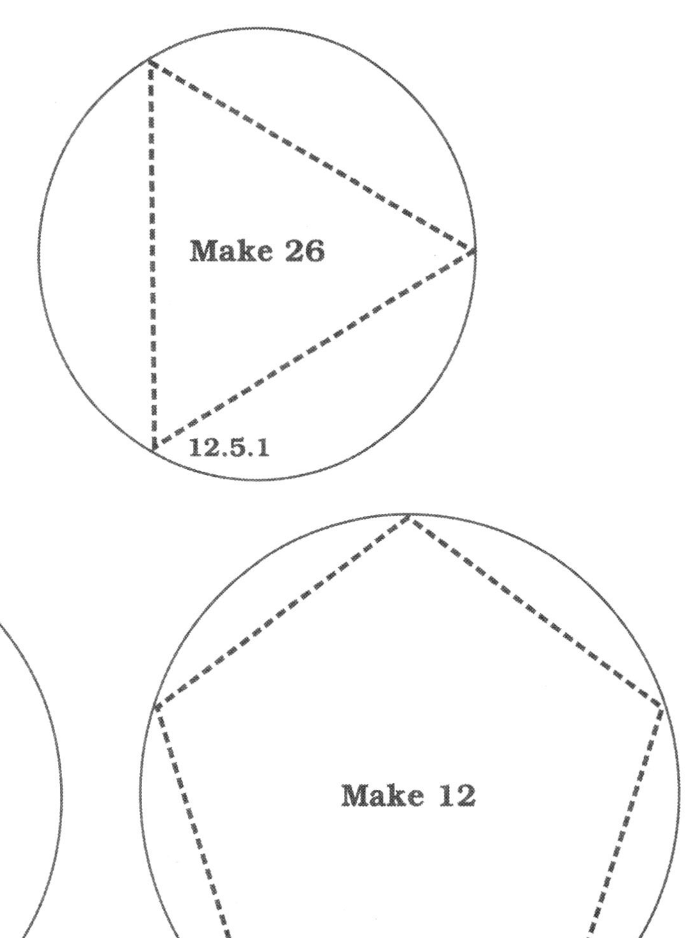

Make 16

12.5.1

Make 12

12.5.1

MATERIALS CARD 13.1.1

Cut into strips and tape the flap marked X under the next strip each time to construct a meter strip.

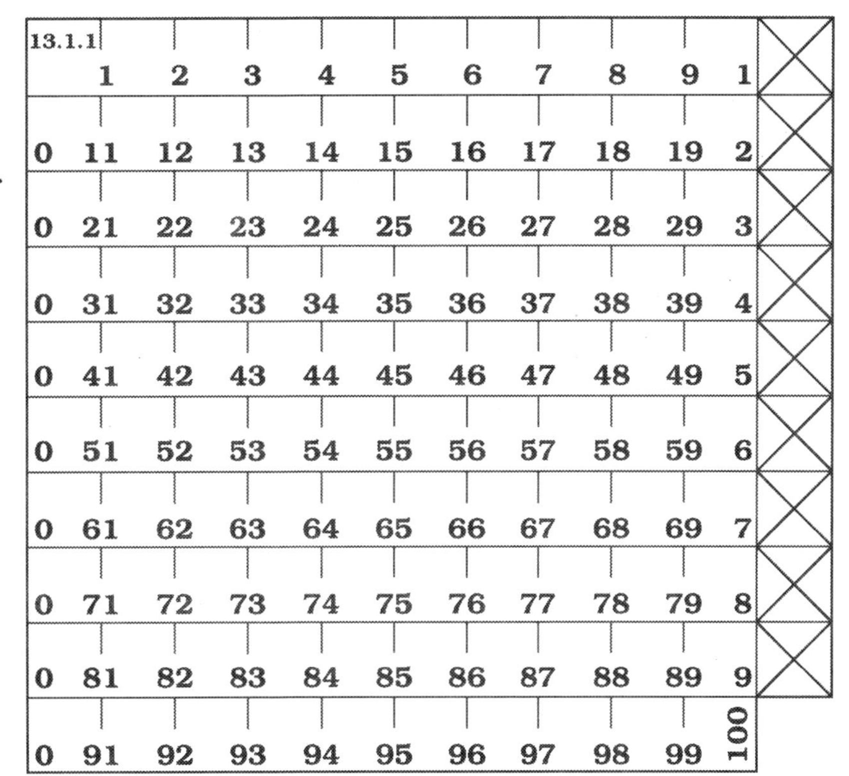

MATERIALS CARD 13.2.3

Cut out these strips.

MATERIALS CARD 13.2.3 cont.

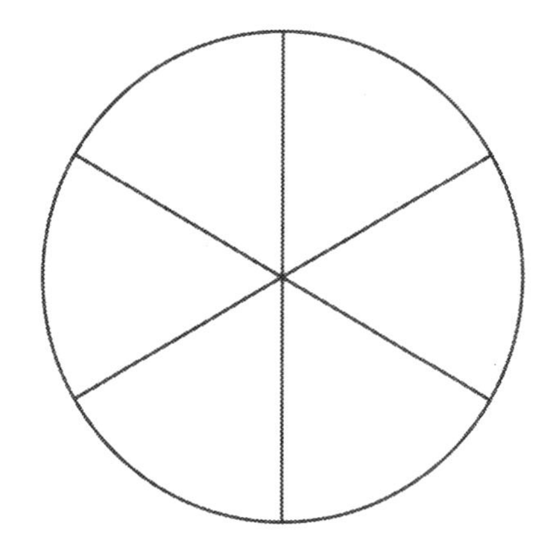

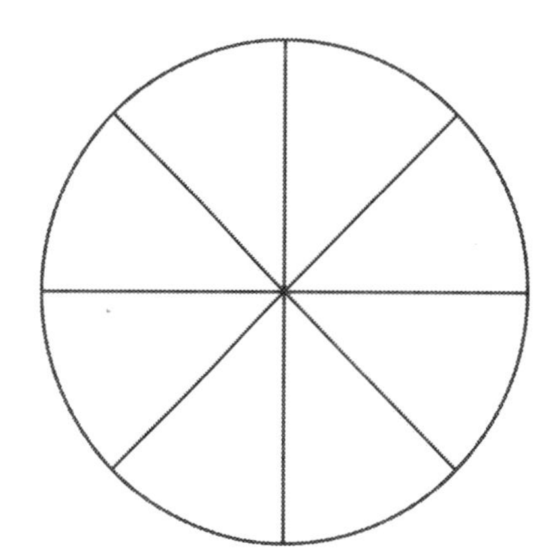

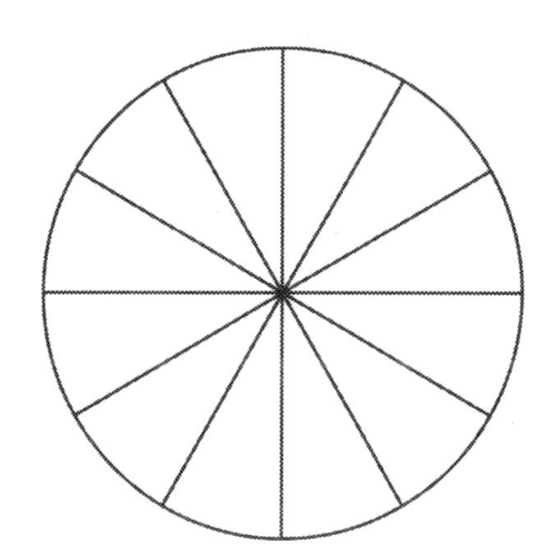

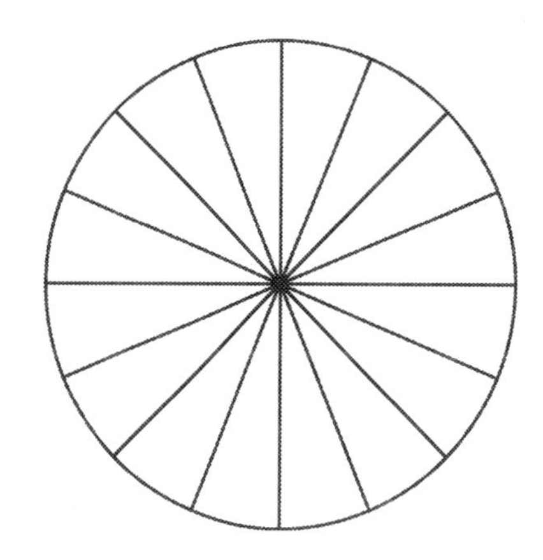

MATERIALS CARD 13.3.1

Cut out these nets along the outer lines and fold them along the inside lines.

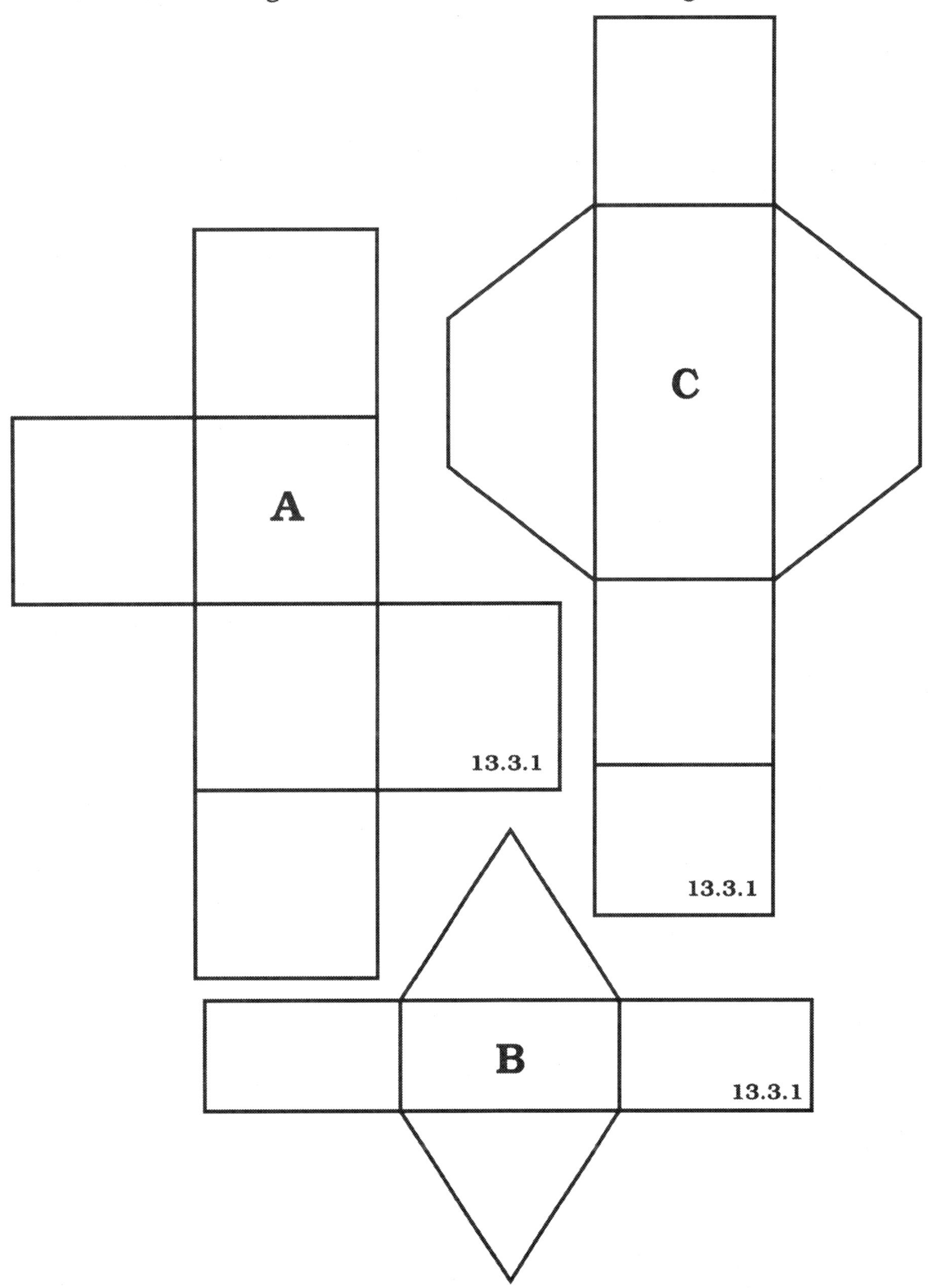

MATERIALS CARD 13.3.1, continued

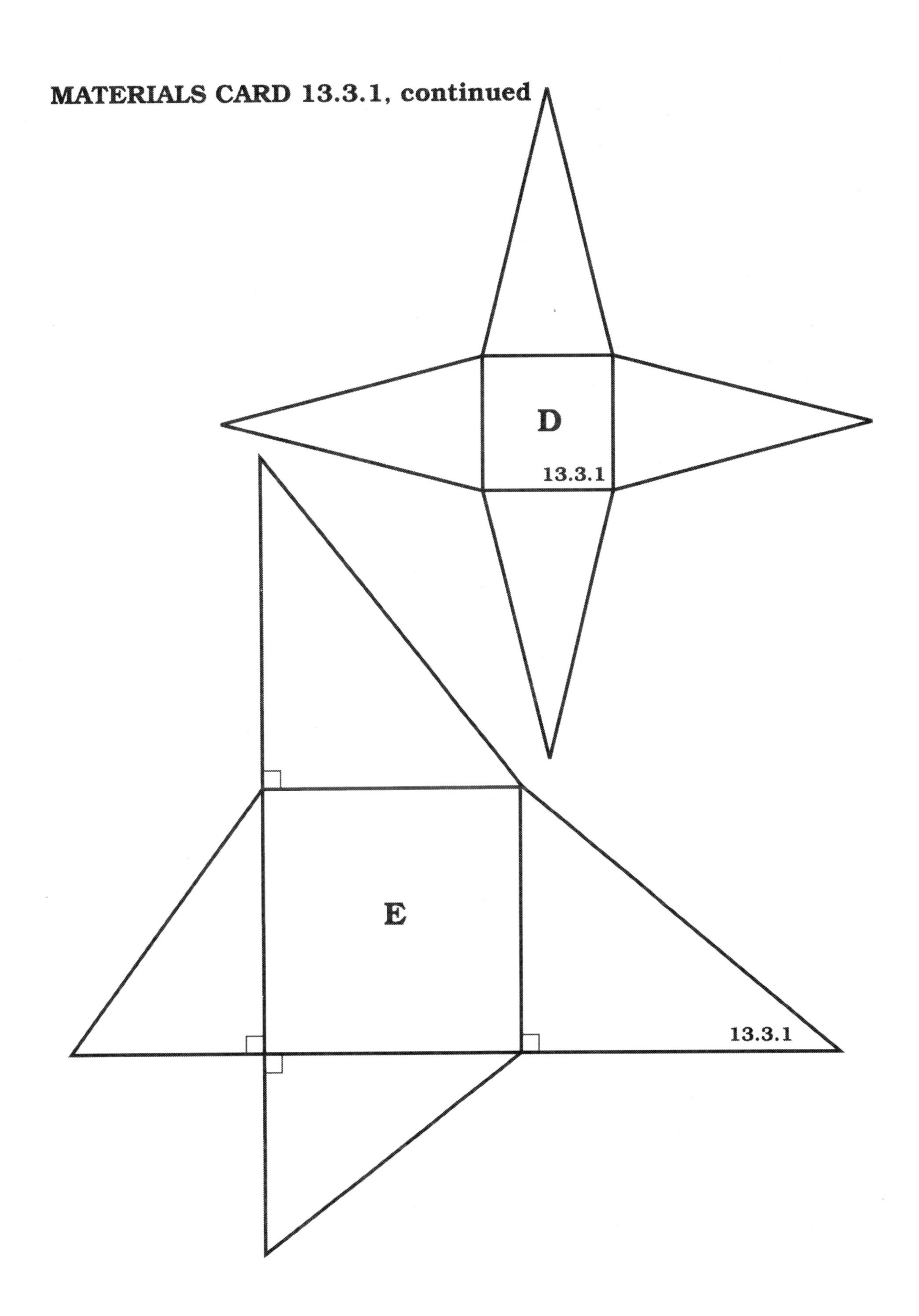

MATERIALS CARD 13.4.1

Cut out these nets along the outer lines and fold along the inside solid lines.

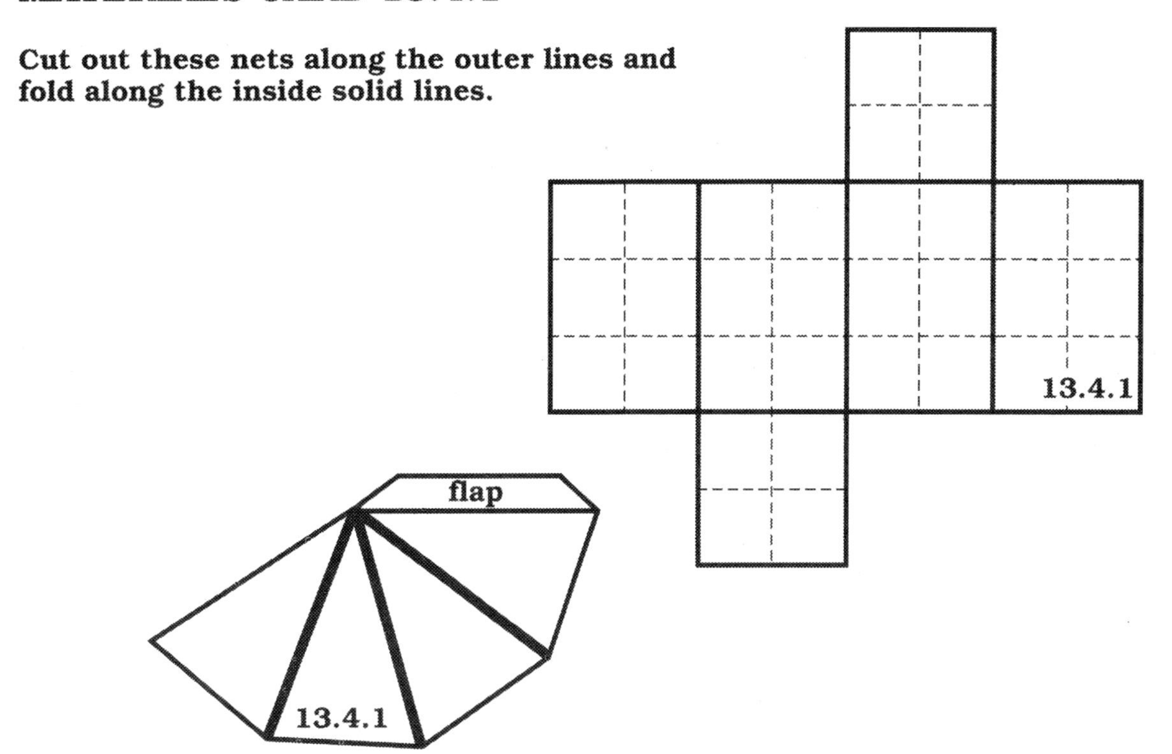

MATERIALS CARD 13.4.2

13.4.2

Overlap and tape securely

MATERIALS CARD 13.4.2, continued

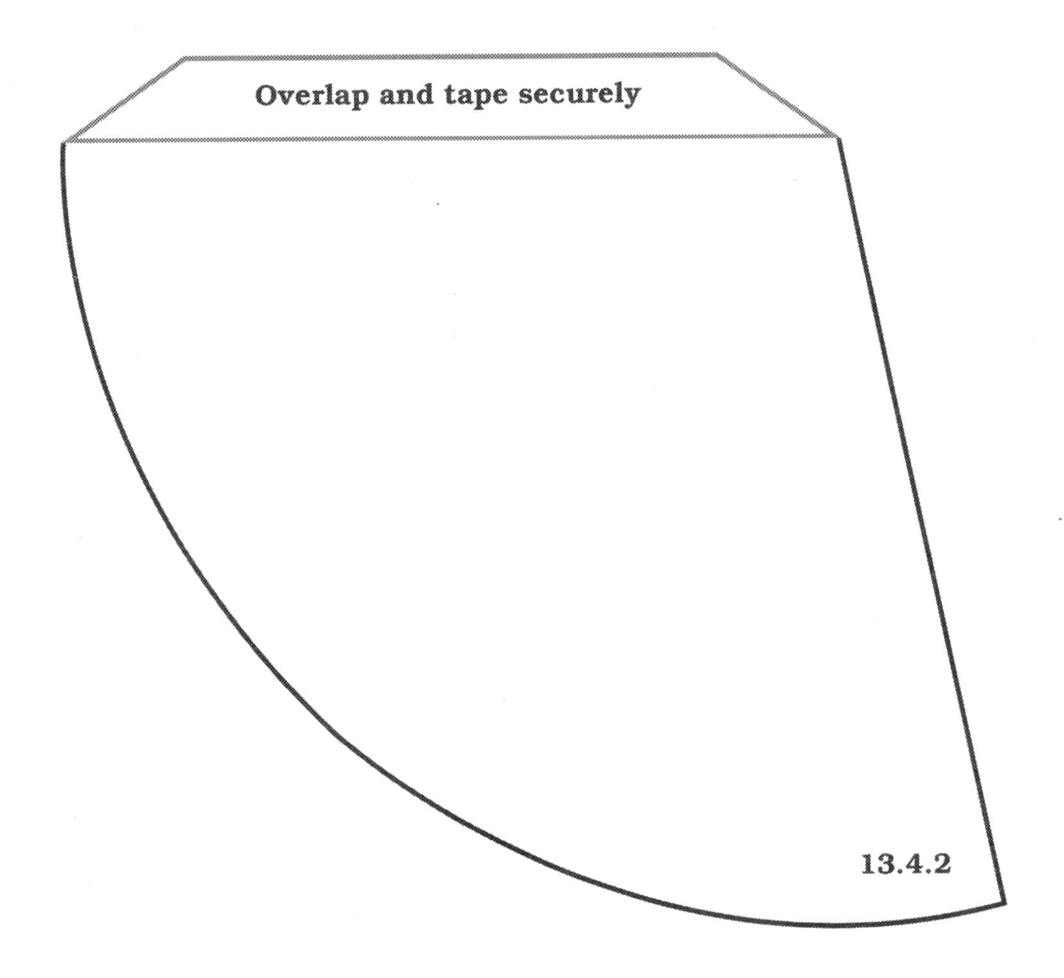

MATERIALS CARD 14.2.1

Cut out this millimeter ruler.

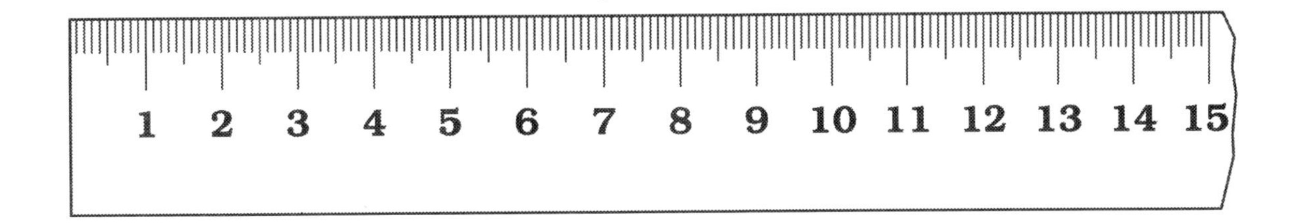

MATERIALS CARD 15.1.3